Good Pharmaceutical Manufacturing Practice

Rationale and Compliance

Good Pharmaceutical Manufacturing Practice

Rationale and Compliance

John Sharp

CRC Press
Taylor & Francis Group
Boca Raton London New York

CRC Press is an imprint of the
Taylor & Francis Group, an **informa** business

CRC Press
Taylor & Francis Group
6000 Broken Sound Parkway NW, Suite 300
Boca Raton, FL 33487-2742

First issued in paperback 2019

© 2010 by Taylor & Francis Group, LLC
CRC Press is an imprint of Taylor & Francis Group, an Informa business

No claim to original U.S. Government works

ISBN-13: 978-0-8493-1994-5 (hbk)
ISBN-13: 978-0-367-39377-9 (pbk)

A CIP record for this book is available from the British Library.

Library of Congress Cataloging-in-Publication Data available on application

Visit the Taylor & Francis Web site at
http://www.taylorandfrancis.com

and the CRC Press Web site at
http://www.crcpress.com

PREFACE

Worldwide, there are more than 20 different official regulatory statements, national and international, on Good Manufacturing Practice (GMP) for pharmaceutical (or "drug" or "medicinal") products. These regulatory statements may be termed "regulations," "guides," or "guidelines," and of the 20 or so of them, 2 stand out as being the most influential and most frequently referenced: The United States Current Good Manufacturing Practice Regulations (cGMPs) and the European Commission's "Good Manufacturing Practices for Medicinal Products for Human and Veterinary Use" (the **EC GMP guide**). The full titles of these documents are as follows:

1. Code of Federal Regulations 21 CFR, Part 210: Current Good Manufacturing Practice in Manufacturing, Processing, Packing or Holding of Drugs; General. CFR, and Part 211: Current Good Manufacturing Practice for Finished Pharmaceuticals. Revised April 1, 2002 (may be viewed on FDA website, www.fda.gov).
   For the purposes of this book, and for convenience, 21 CFR Part 210 and Part 211 will be referred to as the **US cGMPs**.
2. "The rules governing medicinal products in the European Union, Volume 4, Good Manufacturing Practices, Medicinal Products for Human and Veterinary Use" (European Commission, Directorate General — Industry, Pharmaceuticals and Cosmetics), 1998 Edition. Updated September 27, 2002. Published in Luxembourg by the Office for Official Publications of the European Communities (may be viewed at http://pharmacos.eudra.org/F2/eudralex/vol-4/pdfs-en/). The English version of this text is also published in full as part of the U.K. Medicines Control Agency's (MCA) "Rules and Guidance for Pharmaceutical Manufacturer's and Distributors 2002," London, The Stationery Office, 2002, ISBN 011 322559 8.

The object of this book is to present the major substance of these two publications in a way that will permit a comparative consideration of the reasoning behind each of the main requirements and to offer suggestions for their practical implementation.

There is another set of related guidelines, which are given in full in the above MCA publication. These are the Guidelines on Good Distribution Practice (GDP) of Medicinal Products for Human Use. These guidelines were issued in support of the European Directive 2001/83/EC, which introduced the requirement for wholesale dealers to obtain a formal authorization to engage in such activities. The relevant paragraph (77/1) in that directive reads:

> Member States shall take all appropriate measures to ensure that the wholesale distribution of medicinal products is subject to the possession of an authorisation to engage in activity as a wholesaler in medicinal products, stating the place for which it is valid.

Although these regulatory texts are freely available in the public domain and no formal permission is required to quote from them, I would like to express my personal thanks to both the US FDA and the European Commission for making them so readily available. While every attempt has been made in this book to offer advice and guidance that will assist manufacturers in regulatory compliance, it would be foolhardy to claim that it will ensure survival of all levels of regulatory scrutiny. Further, while substantial portions of both the US cGMPs and the EC GMP Guide are quoted here, the full texts are not presented. To ensure compliance, reference must be made to the full original texts.

Note: Although the correct current title of the interstate partnership that exists in Europe is the European Union (EU), readers of this book will notice references to the titles of earlier manifestations of this assembly, notably the European Economic Community (EEC) and the European Community (EC, also used to signify the European Commission). This is because these earlier titles are still in use, both in common European parlance, and in the titles of various European directives and guidelines. The policy adopted for the purposes of this book is to use for the title of a document the one by which it is most commonly known.

John Sharp
September 2004

THE AUTHOR

John Sharp, BSc, CChem, FRSC, Hon. MRPharmS, CBiol, MiBiol, FIQA, worked nearly 20 years in the pharmaceutical industry — in the practice and management of the chemical analysis, QA/QC, production, and distribution of most major categories of medicinal products and of the bulk production of a number of active substances. He joined the U.K. Medicines Inspectorate in 1971 and was promoted to principal inspector in 1972. In addition to his inspection duties, he compiled and edited the 1977 and 1983 editions of the *U.K. Guide to Good Pharmaceutical Manufacturing Practice* or "The Orange Guide," which later formed the basis for the EC/EU GMP Guidelines. Since leaving the Inspectorate in 1985, he has held appointments with the Association of the British Pharmaceutical Industry (as project manager) and has returned briefly to the Industry (as technical director of Waverley Pharmaceutical). Since 1987, Sharp has run his own consultancy business. He is the author of more than 70 published papers, books, booklets, and manuals on many aspects of pharmaceutical technology, QA/QC, etc. (including three entire modules in the Manchester University Advanced Pharmaceutical Training series and four modules for the ABPI NVQ Operator Training series). He has presented papers at technical, professional, and academic seminars and symposia in Britain, Continental Europe, Canada, the U.S., Africa, and Asia. He is a member of the Editorial Board of the European Journal of Parenteral and Pharmaceutical Sciences, and a referee for the U.S. Parenteral Drug Association's *Journal of Pharmaceutical Science and Technology.*

Sharp is an honors science graduate of London University, a fellow of the Royal Society of Chemistry (and for many years a member of the RSC's panel of assessors for qualified persons), an honorary member of the Royal Pharmaceutical Society (for services to Pharmacy in the field of GMP), a member of the Institute of Biology, a fellow of the Institute of Quality Assurance, an honorary life member of the Parenteral Society, and an honorary life member of the Pharmaceutical BFS Operators Association. In 1996 he received the U.S. PDA's Korczynski Award for international contributions to the pharmaceutical sciences. In 2003 he received the U.K. Parenteral Society's George Sykes Memorial Award for his paper on sterile products manufacture. He is a member of the Middlesex County Cricket Club and the Harlequin Football Club.

TABLE OF CONTENTS

1

INTRODUCTION: STATUS AND APPLICABILITY OF US REGULATIONS/ EU GUIDELINES — GENERAL QUALITY ISSUES

The main objective of this book is to consider and compare the principle requirements of Good Manufacturing Practice (GMP), most notably in:

- The US Current Good Manufacturing Practice for Finished Pharmaceuticals regulations (the "US cGMPs")
- The Guide to Good Manufacturing Practice for Medicinal Products of the European Union (the "EC GMP Guide")

This book will also discuss the rationale behind these requirements and will propose ways and means of complying with them.

The approach generally adopted for this book, after the following consideration of the status and applicability of these two major regulatory statements on GMP, is to base each chapter on one of the subparts of the US cGMPs (Organization and Personnel, Buildings, Equipment, etc.), which are quoted (often in a somewhat abridged version) and then followed by the (also often abridged) corresponding sections of the EC GMP Guide. These quotes are then followed by a comparative discussion, which enlarges on the regulatory requirements and proposes steps toward implementation and then compliance. The reader should note that, although the quotes are verbatim, they may not represent the full original regulatory texts, which should be consulted whenever necessary.

STATUS AND APPLICABILITY

US cGMPs

Part 210 — Current Good Manufacturing Practice in Manufacturing, Processing, Packaging, and Holding of Drugs:

Sec. 210.1 Status of current good manufacturing practice regulations.

(a) The regulations set forth in this part and in parts 211 through 226 of this chapter contain the minimum current good manufacturing practice for methods to be used in, and the facilities or controls to be used for, the manufacture, processing, packing, or holding of a drug to assure that such drug meets the requirements of the act as to safety, and has the identity and strength and meets the quality and purity characteristics that it purports or is represented to possess.

(b) The failure to comply with any regulation set forth in this part and in parts 211 through 226 of this chapter in the manufacture, processing, packing, or holding of a drug shall render such drug to be adulterated under section 501(a)(2)(B) of the act and such drug, as well as the person who is responsible for the failure to comply, shall be subject to regulatory action.

Sec. 210.2 Applicability of current good manufacturing practice regulations.

(a) The regulations in this part and in parts 211 through 226 of this chapter as they may pertain to a drug and in parts 600 through 680 of this chapter as they may pertain to a biological product for human use, shall be considered to supplement, not supersede, each other, unless the regulations explicitly provide otherwise. In the event that it is impossible to comply with all applicable regulations in these parts, the regulations specifically applicable to the drug in question shall supersede the more general.

(b) If a person engages in only some operations subject to the regulations in this part and in parts 211 through 226 and parts 600 through 680 of this chapter, and not in others, that person need only comply with those regulations applicable to the operations in which he or she is engaged.

EC GMP Guide — Foreword

The Pharmaceutical Industry of the European Community maintains high standards of Quality Assurance in the development, manufacture and control of medicinal products. A system of Marketing Authorisations ensures that all medicinal products are assessed by a Competent Authority to ensure compliance with contemporary requirements of safety, quality and efficacy. A system of Manufacturing Authorisations ensures that all products authorised on the European market are manufactured only by authorised manufacturers, whose activities are regularly inspected by the Competent Authorities. Manufacturing Authorisations are required by all pharmaceutical manufacturers in the European Community whether the products are sold within or outside of the Community.

Two directives laying down principles and guidelines of good manufacturing practice (GMP) for medicinal products were adopted by the Commission in

1991, the first for medicinal products for human use (Directive 91/356/EEC), the second one for veterinary use (Directive 91/412/EEC). Detailed guidelines in accordance with those principles are published in the Guide to Good Manufacturing Practice which will be used in assessing applications for manufacturing authorisations and as a basis for inspection of manufacturers of medicinal products.

The principles of GMP and the detailed guidelines are applicable to all operations which require the authorisation referred to in Article 16 of Directive 75/319/EEC and in Article 24 of Directive 81/851/EEC as modified. They are also relevant for all other large-scale pharmaceutical manufacturing processes, such as that (*sic*) undertaken in hospitals, and for the preparation of products for use in clinical trials.

All Member States and the Industry itself are agreed that the GMP requirements applicable to the manufacture of veterinary medicinal products are the same as those applicable to the manufacture of medicinal products for human use. Certain detailed adjustments to the GMP guidelines are set out in two annexes specific to veterinary medicinal products and to immunological veterinary medicinal products.

The Guide is presented in chapters, each headed by a principle. Chapter 1 on Quality Management outlines the fundamental concept of Quality Assurance as applied to the manufacture of medicinal products. Thereafter each chapter has a principle outlining the Quality Assurance objectives of that chapter and a text which provides sufficient detail for manufacturers to be made aware of the essential matters to be considered when implementing the principle.

In addition to the general matters of Good Manufacturing Practice outlined in the 9 chapters of this guide, a series of annexes providing detail about specific areas of activity is included. For some manufacturing processes, different annexes will apply simultaneously (e.g., annex on sterile preparations and on radio pharmaceuticals and/or on biological medicinal products).

A glossary of some terms used in the Guide has been incorporated after the annexes.

The first edition of the Guide was published in 1989, including an annex on the manufacture of sterile medicinal products.

The second edition was published in January 1992 ... [it] ... also included 12 additional annexes.

The basic requirements in the main guide have not been modified. 14 annexes on the manufacture of medicinal products have been included in this third edition....

The Guide is not intended to cover security aspects for the personnel engaged in manufacture. This may be particularly important in the manufacture of certain medicinal products such as highly active, biological and radioactive medicinal products, but they are governed by other provisions of Community or national law.

Throughout the Guide it is assumed that the requirements of the Marketing Authorisation relating to the safety, quality and efficacy of the products, are systematically incorporated into all the manufacturing, control and release for sale arrangements of the holder of the Manufacturing Authorisation.

The manufacture of medicinal products has for many years taken place in accordance with guidelines for Good Manufacturing Practice and the manufacture of medicinal products is not governed by CEN/ISO standards. Harmonised standards as adopted by the European standardisation organisations CEN/ISO may be used at industry's discretion as a tool for implementing a quality system in the pharmaceutical sector. The CEN/ISO standards have been considered but the terminology of these standards has not been implemented in this third edition of the Guide.

It is recognised that there are acceptable methods, other than those described in the Guide, which are capable of achieving the principles of Quality Assurance. The Guide is not intended to place any restraint upon the development of any new concepts or new technologies which have been validated and which provide a level of Quality Assurance at least equivalent to those set out in this Guide.

Comment

Preliminary notes for non-European readers: The European Marketing Authorisation referred to above is similar to the US NDA procedure. In addition, each manufacturing site is required to hold a Manufacturing Authorization. This European system of Manufacturing Authorization and Inspection may be considered comparable to the US system of registration and inspection of manufacturers under Section 510(b) and (c) of the Federal Food Drug, and Cosmetic Act. In Britain, a Marketing Authorization is often termed a "Product License" and a Manufacturing Authorization a "Manufacturer's License."

Some readers may also be puzzled by the term "competent authority." Competent, in this context, is not, it seems, used in the usual sense of meaning capable, appropriately qualified, or effective. Here it means responsible. That is competent authority is Eurospeak for Government Health Authority.)

The legal status of the US cGMPs is clear and unequivocal. They are regulations, enforceable in US law. As they clearly state, "failure to comply ... in the manufacture, processing, packing, or holding of a drug shall render such drug to be adulterated ... such drug, as well as the person responsible for the failure to comply, shall be subject to regulatory action."

On the surface, the EC GMP Guide (note 'Guide') does not appear to have such legal weight and force. However, there is another European regulatory document that does have the full force of European law — the European Commission Directive of 13 June 1991 laying down the principles and guidelines of good manufacturing practice for medicinal products for human use (91/356/EEC). (Another similar directive, 91/412/EEC, applies to veterinary medicinal products.)

There is a certain ambiguity in the title of this directive, for these are hardly guidelines. The essence of this directive is a set of brief GMP principles with which manufacturers in the European Union are legally required to comply. There is another, possibly confusing, oddity. Contrary to what might be expected, the "principles" given in this directive are not the same as the principles that appear at the head of each chapter of the EC GMP Guide (see the fifth paragraph of the foreword to the EC GMP Guide above). This disarray is no doubt a consequence of the Byzantine, often ad hoc, committee structure through which the European Union conducts this type of business. The principles of GMP are set out in this European Directive in nine articles, Article 6 to Article 14. (Articles 1 through 5 are concerned with general and administrative matters.) In outline form, these nine articles, slightly abridged, read:

EC Principles of Good Manufacturing Practice (Digest of EC Directive 91/356/EEC)

Article 6

Quality Management

The manufacturer shall establish and implement an effective pharmaceutical quality assurance system, involving the active participation of the management and personnel … involved.

Article 7

Personnel

1. At each manufacturing site, the manufacturer shall have competent and appropriately qualified personnel at his disposal in sufficient number…
2. The duties of managerial and supervisory staff responsible for implementing and operating good manufacturing practice shall be defined in job descriptions. Their hierarchical relationships shall be defined in an organization chart.
3. Staff referred to in paragraph 2 shall be given sufficient authority to discharge their responsibilities correctly.
4. Personnel shall receive initial and continuing training including the theory and application of the concept of quality assurance and good manufacturing practice.
5. Hygiene programmes adapted to the activities to be carried out shall be established and observed. These programmes include procedures relating to health, hygiene and clothing of personnel.

Article 8

Premises and Equipment

1. Premises and manufacturing equipment shall be located, designed, constructed, adapted, and maintained to suit the intended operations.
2. Lay out, design, and operation must aim to minimize the risk of errors and permit effective clearing and maintenance …

3. Premises and equipment intended to be used for manufacturing operations which are critical for the quality of the products shall be subjected to appropriate qualification.

Article 9

Documentation

1. The manufacturer shall have a system of documentation based upon specifications, manufacturing formulae and processing and packaging instructions, procedures and records covering the various manufacturing operations that they perform... Pre-established procedures for general manufacturing operations and conditions shall be available, together with specific documents for the manufacture of each batch. This set of documents shall make it possible to trace the history of the manufacture of each batch. The batch documentation shall be retained for at least one year after the expiry date of the batches to which it relates or at least five years after the (release of the product) whichever is the longer.
2. When electronic, photographic or other data processing systems are used instead of written documents, the manufacturer shall have validated the systems...The electronically stored data shall be protected against loss or damage of data (e.g., by duplication or back-up and transfer onto another storage system).

Article 10

Production

The different production operations shall be carried out according to pre-established instructions and procedures and in accordance with good manufacturing practice. Adequate and sufficient resources shall be made available for the in-process controls.

Appropriate technical and/or organizational measures shall be taken to avoid cross contamination and mix-ups.

Any new manufacture or important modification of a manufacturing process shall be validated. Critical phases of manufacturing processes shall be regularly revalidated.

Article 11

Quality Control

1. The manufacturer shall establish and maintain a quality control department. This department shall be placed under the authority of a person having the required qualifications and shall be independent of the other departments.
2. The quality control department shall have at its disposal one or more quality control laboratories appropriately staffed and equipped to carry out the necessary examination and testing of starting materials, packaging

materials and intermediate and finished products testing. Resorting to outside laboratories may be authorized...

3. During the final control of finished products before their release for sale or distribution, in addition to analytical results, the quality control department shall take into account essential information such as the production conditions, the results of in-process controls, the examination of the manufacturing documents and the conformity of the products to their specifications (including the final finished pack).

4. Samples of each batch of finished products shall be retained for at least one year after the expiry date. Unless in the Member State... a longer period is required, samples of starting materials (other than solvents, gases and water) used shall be retained for at least two years after the release of the product. This period may be shortened if their stability, as mentioned in the relevant specification, is shorter. All these samples shall be maintained at the disposal of the competent authorities.

Article 12

Work Contracted Out

1. Any manufacturing operation ... which is carried out under contract, shall be the subject of a written contract between the contract giver and the contract acceptor.

2. The contract shall clearly define the responsibilities of each party and in particular the observance of good manufacturing practice by the contract acceptor ...

3. The contract acceptor shall not subcontract any of the work entrusted to him by the contract giver without the written authorization of the contract giver.

4. The contract acceptor shall respect the principles and guidelines of good manufacturing practice and shall submit to inspections carried out by the competent authorities ...

Article 13

Complaints and Product Recall

The manufacturer shall implement a system for recording and reviewing complaints together with an effective system for recalling promptly and at any time medicinal products in the distribution network. Any complaint concerning a defect shall be recorded and investigated by the manufacturer. The competent authority shall be informed by the manufacturer of any defect that could result in a recall or abnormal restriction on the supply...

Article 14

Self-Inspection

The manufacturer shall conduct repeated self-inspections as part of the quality assurance system in order to monitor the implementation and respect of good manufacturing practice and to propose any necessary corrective measures.

> Records of such self-inspections and any subsequent corrective action shall be maintained.

At first sight, it might appear that it is the European Directives 91/356/EEC and 91/412/EEC (relating, respectively, to human and veterinary medicines) that have the full force of regulatory law, with the considerably more detailed EC GMP Guide providing more "friendly" guidance to manufacturers on the implementation of those directives. However, careful attention needs to be paid to the statement that appears in the second paragraph of the foreword to the EC GMP Guide (see above), that is:

> … (this) Guide to Good Manufacturing Practice which will be used in assessing applications for manufacturing authorisations and as a basis for inspection of manufacturers of medicinal products.

Since the result of an adverse assessment of an application, or of an inspection, can be the refusal by the regulatory (or competent) authority to grant a manufacturing authorization, or to suspend or revoke such an authorization, the EC GMP Guide can be said to have significantly powerful teeth. As such, and in potential practice, it hardly has less regulatory force in Europe than do the US cGMPs in the U.S.

There are, nevertheless, differences in the applicability of these two sets of GMPs. The US cGMPS (21 CFR) declare that they apply only to human medicines. The EC GMP Guide applies to both human and veterinary medicines. The US cGMPs (21 CFR) do not cover the manufacture of active pharmaceutical ingredients (APIs). However, in August 2001 the US Food and Drug Administration (FDA) issued what are, in effect, cGMPs for APIs. This document is entitled "Guidance for Industry — Q7A Good Manufacturing Practice for Active Pharmaceutical Ingredients."[1] In the introductory material to this Q7A document it is stated that "this guidance represents the FDA's current thinking on this topic," and that "this document is intended to provide guidance regarding GMP for the manufacturing of APIs under an appropriate system for managing quality."

Although in the introductory material to EC GMP Guide it is not explicitly stated that it applies to APIs, in July 2001 the European Commission issued a detailed Annex 18 to the Guide, called "Good Manufacturing Practice for Active Pharmaceutical Ingredients." This annex is very closely similar to the Q7A Guidance as issued by the FDA. On July 18, 2001, the European Commission adopted a new legislative proposal that introduced a requirement for pharmaceutical manufacturers to use only active substances that have been manufactured according to GMP. At the time of writing, this new legislation has yet to be implemented. When it is, the application of Annex 18 is expected to become mandatory in the EU.

NOTE ON THE GENESIS OF THE EC GMP GUIDE

Although it has been a requirement of US law since 1962 that all drug products (or "finished pharmaceuticals") marketed in the US must be manufactured in accordance with Current Good Manufacturing Practice, in Europe, even into the 1980s, only a minority of nations had published official guidelines on GMP.

They were Britain, Denmark, France, Italy, and of those, only the British and French publications were documents of any great substance or detail.

The first edition of the British Guide to Good Pharmaceutical Manufacturing Practice, prepared by the UK Medicines Inspectorate, was published in 1971. It was a relatively slight volume of 20 pages, and was reissued as a third impression in 1972, with the addition of a 2-page appendix on sterile medicinal products. Because of the color of its cover, it became known as the Orange Guide. A second, more substantial edition (52 pages, including five appendices) was published in 1977. A third edition (110 pages, five appendices) was published in 1983 and came to be regarded as the definitive British Orange Guide.

The first move toward harmonization of GMP requirements in the EC was made in the early 1980s, when the working group of the European Commission's Pharmaceutical Committee was first set up to look at the feasibility of producing a community guide to GMP. Armed with copies of various then-existent guidelines and regulations, this working party set about preparing a draft guide for consultation. This was eventually published in summer 1987 and was fully reviewed by industry both at a national and European level. The draft drew most heavily on the Orange Guide of the U.K., and to a lesser extent on the French guidelines on GMP, the Bonnes Pratiques de Fabrication (BPF). Consultation with the European industry continued through the early part of 1988 and the final document was published in January 1989 (as volume IV of The Rules Governing Medicinal Products in the European Community — Guide to Good Manufacturing Practice for Medicinal Products), with an annex on the manufacture of sterile medicinal products. In January 1992, the document was reprinted, with a few textual amendments and was issued with a number of additional annexes on more specific topics (e.g., the manufacture of biological medicinal products, radiopharmaceuticals, and medicinal gases). At the time of writing, the latest edition, currently available on the Internet (go to http://pharmacos.eudra.org/F2 and click on Eudral.ex collection, and then on Volume 4), is dated 1998 (and was updated September 27, 2002).

In 1993, the UK Medicines Control Agency (MCA) issued, as a forth edition of the Orange Guide — the full text of the EC Guide, plus other material — under the title Rules and Guidance for Pharmaceutical Manufacturers 1993. In 1997, the MCA issued a fifth edition, under the title Rules and Guidance for Pharmaceutical Manufacturers and Distributors 1997. This edition, in addition to the full text of the EC GMP Guide, plus a total of 14 more specific annexes, contains the texts of the two GMP Directives (91/456/EEC and 91/412/EEC — on products for human use and for animal use, respectively), the UK standard provisions for manufacturer's licenses, and the Code of Practice for Qualified Persons. It also contains UK Guidance on Certificates of Analysis, the UK Standard Provisions for Wholesale Dealer's Licenses, Guidance on the Appointment and Duties of the Responsible Person, the Directive 92/25/EEC on Wholesale Distribution, with the EC Guidelines on Good Distribution Practice, and a note on the UK Defective Medicines Report Centre. In 2002, an updated sixth UK edition was issued, now with 18 annexes.

Perhaps the most noticeable feature of the EC GMP Guide (in both its original manifestations and as incorporated in the later editions of the UK MCA's Orange Rules and Guidance) is its obvious debt to the 1983 British edition. It is, however, as if the latter is being seen in a distorting mirror. The language, at times, seems to have been shifted to a strange, grammatically dubious, and unidiomatic Pidgin English, and a number of ambiguities, imprecisions, and equivocations have intruded.

This doubtless reflects the long deliberations of a large international committee: an example of the horse/camel committee design effect. It has been rumored that the original English text was translated into French, for the benefit of the French-speaking members of the working group. The final draft, it is said, was prepared in French and then translated into English by a native French speaker who was not exactly an expert in English. There is no documentary evidence to support this version of events, but it certainly looks as if this is what happened.

GENERAL QUALITY ISSUES/QUALITY MANAGEMENT

Regulatory Statements

US cGMPs

These contain no initial discussion on the nature of "Quality" or on "Quality Management," nor are the terms "Quality Assurance," "GMP," and "Quality Control" defined. However, a definition of "Quality Control" is implicit in Section 211.22 ('There shall be a quality control unit…") (see Chapter 2).

EC GMP Guide

Principle

> The holder of a Manufacturing Authorisation must manufacture medicinal products so as to ensure that they are fit for their intended use, comply with the requirements of the Marketing Authorisation and do not place patients at risk due to inadequate safety, quality or efficacy. The attainment of this quality objective is the responsibility of senior management and requires the participation and commitment by staff in many different departments and at all levels within the company, by the company's suppliers and by the distributors. To achieve the quality objective reliably there must be a comprehensively designed and correctly implemented system of Quality Assurance incorporating Good Manufacturing Practice and thus Quality Control. It should be fully documented and its effectiveness monitored. All parts of the Quality Assurance system should be adequately resourced with competent personnel, and suitable and sufficient premises, equipment and facilities. There are additional legal responsibilities for the holder of the Manufacturing Authorisation and for the Qualified Person(s).

> 1.1 The basic concepts of Quality Assurance, Good Manufacturing Practice and Quality Control are interrelated. They are described here in order to emphasise their relationships and their fundamental importance to the production and control of medicinal products.

Quality Assurance

1.2   Quality Assurance is a wide ranging concept which covers all matters which individually or collectively influence the quality of a product. It is the sum total of the organised arrangements made with the object of ensuring that medicinal products are of the quality required for their intended use. Quality Assurance therefore incorporates Good Manufacturing Practice plus other factors outside the scope of this Guide.

The system of Quality Assurance appropriate for the manufacture of medicinal products should ensure that:

i.   medicinal products are designed and developed in a way that takes account of the requirements of Good Manufacturing Practice and Good Laboratory Practice;
ii.   production and control operations are clearly specified and Good Manufacturing Practice adopted;
iii.   managerial responsibilities are clearly specified;
iv.   arrangements are made for the manufacture, supply and use of the correct starting and packaging materials;
v.   all necessary controls on intermediate products, and any other in-process controls and validations are carried out;
vi.   the finished product is correctly processed and checked, according to the defined procedures;
vii.   medicinal products are not sold or supplied before a Qualified Person has certified that each production batch has been produced and controlled in accordance with the requirements of the Marketing Authorisation and any other regulations relevant to the production, control and release of medicinal products;
viii.   satisfactory arrangements exist to ensure, as far as possible, that the medicinal products are stored, distributed and subsequently handled so that quality is maintained throughout their shelf life;
ix.   there is a procedure for Self-Inspection and/or quality audit which regularly appraises the effectiveness and applicability of the Quality Assurance system.

Good Manufacturing Practice for Medicinal Products (GMP)

1.3 Good Manufacturing Practice is that part of Quality Assurance which ensures that products are consistently produced and controlled to the quality standards appropriate to their intended use and as required by the Marketing Authorisation or product specification. Good Manufacturing Practice is concerned with both production and quality control. The basic requirements of GMP are that:

i.   all manufacturing processes are clearly defined, systematically reviewed in the light of experience and shown to be capable of consistently manufacturing medicinal products of the required quality and complying with their specifications;
ii.   critical steps of manufacturing processes and significant changes to the process are validated;
iii.   all necessary facilities for GMP are provided including:
       a. appropriately qualified and trained personnel;

    b. adequate premises and space;

    c. suitable equipment and services;

    d. correct materials, containers and labels;

    e. approved procedures and instructions;

    f. suitable storage and transport;

iv. instructions and procedures are written in an instructional form in clear and unambiguous language, specifically applicable to the facilities provided;

v. operators are trained to carry out procedures correctly;

vi. records are made, manually and/or by recording instruments, during manufacture which demonstrate that all the steps required by the defined procedures and instructions were in fact taken and that the quantity and quality of the product was as expected. Any significant deviations are fully recorded and investigated;

vii. records of manufacture including distribution which enable the complete history of a batch to be traced, are retained in a comprehensible and accessible form;

viii. the distribution (wholesaling) of the products minimises any risk to their quality;

ix. a system is available to recall any batch of product, from sale or supply;

x. complaints about marketed products are examined, the causes of quality defects investigated and appropriate measures taken in respect of the defective products and to prevent reoccurrence.

Quality Control

1.4 Quality Control is that part of Good Manufacturing Practice which is concerned with sampling, specifications and testing, and with the organisation, documentation and release procedures which ensure that the necessary and relevant tests are actually carried out and that materials are not released for use, nor products released for sale or supply, until their quality has been judged to be satisfactory.

The basic requirements of Quality Control are that:

i. adequate facilities, trained personnel and approved procedures are available for sampling, inspecting and testing starting materials, packaging materials, intermediate, bulk, and finished products, and where appropriate for monitoring environmental conditions for GMP purposes;

ii. samples of starting materials, packaging materials, intermediate products, bulk products and finished products are taken by personnel and by methods approved by Quality Control;

iii. test methods are validated;

iv. records are made, manually and/or by recording instruments, which demonstrate that all the required sampling, inspecting and testing procedures were actually carried out. Any deviations are fully recorded and investigated;

v. the finished products contain active ingredients complying with the qualitative and quantitative composition of the Marketing Authorisation, are of the purity required, and are enclosed within their proper containers and correctly labelled;

vi. records are made of the results of inspection and that testing of materials, intermediate, bulk, and finished products is formally assessed against specification. Product assessment includes a review and evaluation of relevant production documentation and an assessment of deviations from specified procedures:

vii. no batch of product is released for sale or supply prior to certification by a Qualified Person that it is in accordance with the requirements of the Marketing Authorisation;

viii. sufficient reference samples of starting materials and products are retained to permit future examination of the product if necessary and that the product is retained in its final pack unless exceptionally large packs are produced.

Comment

Chapter 1 of the EC GMP Guide, which is quoted in full above, is an attempt to encapsulate the essence of the concepts of Quality Assurance and Quality Systems as distinct from mere analytical control or quality control, and as relevant to the manufacture of medicinal products, or drug products. It is thus neither more, and certainly not less, than a reflection of the current view that to ensure the quality of these, and indeed of other manufactured products, much more is necessary than the simple testing of samples and comparison of the results against a specification. Few, if any, would argue that this is not the right, proper, effective and indeed rational approach.

"Quality" is not explicitly defined, but the meaning of the term is implicit in the opening sentence of the Principle — Quality is fitness for intended use. In the semantic scheme set out in the EC GMP Guide, Quality Assurance embraces GMP and includes such additional factors as original product design and development, and GMP in turn includes Quality Control (see Figure 1.1).

In a sense, the remainder of the EC GMP Guide is a series of expansions on, or enlargements of, the matters covered in this initial chapter.

Although the US cGMPs do not have an opening section like this, which lays down basic quality principles and issues, the concept and application of Quality Assurance is implicit in the document as a whole. For example, Part 211, Subpart B — Organization and Personnel, Section 211-22 Responsibilities of quality control unit states:

(a) There shall be a quality control unit that shall have the responsibility and authority to approve or reject all components, drug product containers, closures, in-process materials, packaging material, labeling, and drug products, and the authority to review production records to assure that no errors have occurred or, if errors have occurred, that they have been fully investigated. The quality control unit shall be responsible for approving or rejecting drug products manufactured, processed, packed, or held under contract by another company.

(b) Adequate laboratory facilities for the testing and approval (or rejection) of components, drug product containers, closures, packaging materials, in-process materials, and drug products shall be available to the quality control unit.

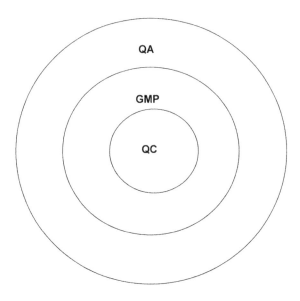

Figure 1.1   QA/GMP/QC Relationships

(c) The quality control unit shall have the responsibility for approving or rejecting all procedures or specifications impacting on the identity, strength, quality, and purity of the drug product.

(d) The responsibilities and procedures applicable to the quality control unit shall be in writing; such written procedures shall be followed.

Also noteworthy is the FDA's Compliance Program Guidance Manual for FDA Staff: Drug Manufacturing Inspections Program 7356.002 (implementation date February 1, 2002), which establishes a systems approach to inspection of manufacturing operations. It states:

Currently there are not enough FDA resources to audit every aspect of cGMP in every manufacturing facility during every inspection visit. Profile classes generalize inspection coverage from a small number of products to all the products in that class. This program establishes a systems approach to further generalize inspection coverage from a small number of profile classes to an overall evaluation of the firm.

This program identifies a number of systems (Quality System, Facilities and Equipment System, Materials System and so on). The section on the Quality System reads:

Quality System

Assessment of the Quality System is two phased. The first phase is to evaluate whether the Quality Control Unit has fulfilled the responsibility to review and approve all procedures related to production, quality control, and quality assurance and assure the procedures are adequate for their intended use. This also includes the associated record keeping systems. The second phase is to assess the data collected to identify quality problems and may link to other major systems for inspectional coverage.

For each of the following, the firm should have written and approved procedures and documentation resulting therefrom. The firm's adherence to procedures should be verified through observation whenever possible These areas are not limited to finished products, but may also incorporate components and in-process materials. These areas may indicate deficiencies, not only in this system but also in other major systems that would warrant expansion of coverage. All areas under this system should be covered; however the depth of coverage may vary depending upon inspectional findings.

- Product reviews: at least annually; should include information from areas listed below as appropriate; batches reviewed, for each product representative of all batches manufactured; trends are identified;
- Complaint reviews (quality and medical): documented; evaluated investigated in a timely manner; includes corrective action where appropriate;
- Discrepancy and failure investigations related to manufacturing and testing: documented; evaluated; investigated in a timely manner; includes corrective action where appropriate;
- Change Control: documented; evaluated; approved; need for revalidation assessed;
- Product Improvement Projects: for marketed products;
- Reprocess/Rework: evaluation, review and approval; impact on validation and stability;
- Returns/Salvages: assessment; investigation expanded where warranted; disposition;
- Rejects: investigation expanded where warranted; corrective action where appropriate;
- Stability Failures: investigation expanded where warranted; need for field alerts evaluated; disposition;
- Quarantine products;
- Validation: status of required validation/revalidation (e.g., computer, manufacturing process, laboratory methods);
- Training/qualification of employees in quality control unit function.

Discussion

Both the EC GMP Guide and the US cGMPs, either explicitly or by implication and extension, are founded upon the concept of Quality Assurance, as distinct from Quality Control alone. The EC GMP Guide is, perhaps the more explicit, with its definition of these terms, and its drawing of a distinction between them. As such, both reflect the salutary realization, over say the last 30 to 40 years, that to ensure the quality of pharmaceuticals, more is needed than just Quality Control, whatever that may be. And therein lies something of a problem, since ideas of what Quality Control *is* have varied considerably.

These various ideas may be resolved into three main types:

    a. Quality Control (QC) as the application of statistical methods and techniques to the selection and carrying-out of tests and measurements on samples (Engineering or Statistical QC)

    b. Quality Control as an overall "systems" approach, in which statistical methods are the tools, but not the totality, of the enterprise

    c. Quality Control as mere testing of samples in the laboratory (i.e., for "Quality Control Department" read "Analytical Laboratory")

Type (c) above is an extreme expression of a position that might be described as the old, traditional industrial pharmaceutical view. It was a view that was still to be encountered (and widely) well into the 1970s, but it was also one that had long-since been abandoned by the more enlightened pharmaceutical manufacturers. Nevertheless, in the early 1970s more than one UK pharmaceutical company (faced with the need to comply with new legislation) implemented Quality Control by painting over "Chief Analyst" or "Head of Analytical Laboratory" on the laboratory manager's door and replacing it with "Quality Control Manager." (Well, it was all the same thing really, wasn't it?)

There had, of course, been stirrings toward more acceptably modern views much earlier. A prime example is the paper entitled "Quality Control" by Taylor (1947).[2] Despite the title of his paper, Taylor writes about what he terms "control procedures." Side-stepping the semantics, what he was discussing was something similar to Quality Assurance, approximately 30 years before (in the expressed views of some current pharmaceutical quality "experts") Quality Assurance actually hit the pharmaceutical industry. Taylor states:

> Control procedures must encompass all things that may influence the quality of the completed medicinal preparations; they must permit inquiry into every phase of purchasing, manufacturing, packaging storage and labeling...

Taylor also refers to the quality implications of research, development, and distribution activities. Taylor's paper should be read by experts and beginners alike. It deserves to be regarded as a landmark paper.

There is no need to argue the case for the Quality Assurance (or Quality Management) approach to ensuring the quality of manufactured pharmaceuticals. It is one that is accepted by all enlightened manufacturers worldwide. As a sort of encapsulation of the essence of this approach, the steps toward implementation may be summarized as:

- First, design and develop the product.
- Scale-up the process, design the plant, install equipment, and develop relevant documentation (formulas, manufacturing methods, material and product specifications, maintenance schedules, etc.).
- Validate process(es) — as scaled-up.
- Make and package product (with trained staff, using confirmed quality materials, following validated processes, and in accordance with the preestablished formulas and methods).
- While doing so, carry-out in-process checks and controls.
- Record all process details.
- Record details of in-process checks and controls.

- Test end product against established specification.
- Make release/reject decision a basis of

    a. end product test results
    b. review of batch manufacturing and batch packaging records,
    c. review of in-process control records, and
    d. all other information relevant to the quality of the product.

REFERENCES

1. US Food and Drug Administration, *Guidance for Industry — Q7A Good Manufacturing Practice for Active Pharmaceutical Ingredients*, Rockville, MD. Also available at http://www.fda.gov/cder/guidance/index.htm, 2001.
2. Taylor, F.O., Quality Control, *JAPA,* 3, 3, 1947.

2

PERSONNEL, ORGANIZATION, AND TRAINING

The relevant part of the US cGMPs does not pause to deal with more general matters, but leads straight in with "Responsibilities of quality control unit." This passage has already been considered in Chapter 1 and is set out again here for ease of comparison.

SUBPART B — ORGANIZATION AND PERSONNEL

Sec. 211.22 Responsibilities of quality control unit

(a) There shall be a quality control unit that shall have the responsibility and authority to approve or reject all components, drug product containers, closures, in-process materials, packaging material, labeling, and drug products, and the authority to review production records to assure that no errors have occurred or, if errors have occurred, that they have been fully investigated. The quality control unit shall be responsible for approving or rejecting drug products manufactured, processed, packed, or held under contract by another company.

(b) Adequate laboratory facilities for the testing and approval (or rejection) of components, drug product containers, closures, packaging materials, in-process materials, and drug products shall be available to the quality control unit.

(c) The quality control unit shall have the responsibility for approving or rejecting all procedures or specifications impacting on the identity, strength, quality, and purity of the drug product.

(d) The responsibilities and procedures applicable to the quality control unit shall be in writing; such written procedures shall be followed.

Comparison of this extract from the US cGMPs with the EC GMP Guide's definition of "Quality Control" (QC) and "basic requirements of Quality Control," as set out in the last chapter, will reveal that the concepts of Quality Control embodied in both documents are similar.

Characteristically, the EC GMP Guide chapter on Personnel starts with a statement of general principle:

EC GMP GUIDE

Chapter 2 Personnel

Principle

The establishment and maintenance of a satisfactory system of quality assurance and the correct manufacture of medicinal products relies upon people. For this reason there must be sufficient qualified personnel to carry out all the tasks which are the responsibility of the manufacturer. Individual responsibilities should be clearly understood by the individuals and recorded. All personnel should be aware of the principles of Good Manufacturing Practice that affect them and receive initial and continuing training, including hygiene instructions, relevant to their needs.

General

2.1   The manufacturer should have an adequate number of personnel with the necessary qualifications and practical experience. The responsibilities placed on any one individual should not be so extensive as to present any risk to quality.

2.2   The manufacturer must have an organisation chart. People in responsible positions should have specific duties recorded in written job descriptions and adequate authority to carry out their responsibilities. Their duties may be delegated to designated deputies of a satisfactory qualification level. There should be no gaps or unexplained overlaps in the responsibilities of those personnel concerned with the application of Good Manufacturing Practice.

Thus the EC GMP Guide establishes its own rationale for its requirements regarding Personnel by stressing that a "satisfactory system of quality assurance" and "correct manufacture of medicinal products" is dependent upon people, that is "the quality of a product ultimately depends on the quality of those producing it" (Sir Derek Dunlop, 1971).

The EC GMP Guide continues with statements about Key Personnel (the Head of the Production Department, the Head of the Quality Control Department and the Qualified Person):

EC GMP GUIDE

Key Personnel

2.3   Key Personnel include the head of Production, the head of Quality Control, and if at least one of these persons is not responsible for the duties described in Article 22 of Directive 75/319/EEC, the Qualified Person(s) designated for the purpose. Normally key posts should be occupied by full-time personnel. The heads of Production and Quality Control must be independent from each other. In large organisations, it may be necessary to delegate some of the functions listed in 2.5, 2.6 and 2.7.

2.4   The duties of the Qualified Person(s) are fully described in Article 22 of Directive 75/319/EEC, and can be summarised as follows:

(a) for medicinal products manufactured within the European Community, a Qualified Person must ensure that each batch has been produced and tested/checked in accordance with the directives and the marketing authorisation;

(b) for medicinal products manufactured outside the European Community, a Qualified Person must ensure that each imported batch has undergone, in the importing country, the testing specified in paragraph 1 (b) of Article 22;

(c) a Qualified Person must certify in a register or equivalent document, as operations are carried out and before any release, that each production batch satisfies the provisions of Article 22.

The persons responsible for these duties must meet the qualification requirements laid down in Article 23 of the same Directive, they shall be permanently and continuously at the disposal of the holder of the Manufacturing Authorisation to carry out their responsibilities. Their responsibilities may be delegated, but only to other Qualified Person(s).

2.5   The head of the Production Department generally has the following responsibilities:
   i. to ensure that products are produced and stored according to the appropriate documentation in order to obtain the required quality;
   ii. to approve the instructions relating to production operations and to ensure their strict implementation;
   iii. to ensure that the production records are evaluated and signed by an authorised person before they are sent to the Quality Control Department;
   iv. to check the maintenance of his department, premises and equipment;
   v. to ensure that the appropriate validations are done;
   vi. to ensure that the required initial and continuing training of his department personnel is carried out and adapted according to need.

2.6   The head of the Quality Control Department generally has the following responsibilities:
   i. to approve or reject, as he sees fit, starting materials, packaging materials, and intermediate, bulk and finished products;
   ii. to evaluate batch records;
   iii. to ensure that all necessary testing is carried out;
   iv. to approve specifications, sampling instructions, test methods and other Quality Control procedures;
   v. to approve and monitor any contract analysts;
   vi. to check the maintenance of his department, premises and equipment;
   vii. to ensure that the appropriate validations are done;
   viii. to ensure that the required initial and continuing training of his department   personnel is carried out and adapted according to need.

2.7 The heads of Production and Quality Control generally have some shared, or jointly exercised, responsibilities relating to quality. These may include, subject to any national regulations:

■ the authorisation of written procedures and other documents, including amendments;
■ the monitoring and control of the manufacturing environment;
■ plant hygiene;
■ process validation;
■ training;
■ the approval and monitoring of suppliers of materials;
■ the approval and monitoring of contract manufacturers;
■ the designation and monitoring of storage conditions for materials and products;
■ the retention of records;
■ the monitoring of compliance with the requirements of Good Manufacturing Practice;
■ the inspection, investigation, and taking of samples, in order to monitor factors which may affect product quality.

Note for non-European readers: The term "Qualified Person," as used in the context of the EC GMP Guide and in related legislation, has a specialized meaning. It does not mean, in a general sense, a person who is qualified. The role and function of a Qualified Person (QP) in this specialized sense is outlined in the passage ("Key Personnel" 2.4) quoted above from the EC GMP Guide. More information is given in an Annex to this chapter.

Concerning qualification of personnel (in the standard general sense) the US cGMPs state:

Subpart B — Organization and Personnel

Sec. 211.25 Personnel qualifications

(a) Each person engaged in the manufacture, processing, packing, or holding of a drug product shall have education, training, and experience, or any combination thereof, to enable that person to perform the assigned functions. Training shall be in the particular operations that the employee performs and in current good manufacturing practice (including the current good manufacturing practice regulations in this chapter and written procedures required by these regulations) as they relate to the employee's functions. Training in current good manufacturing practice shall be conducted by qualified individuals on a continuing basis and with sufficient frequency to assure that employees remain familiar with CGMP requirements applicable to them.

(b) Each person responsible for supervising the manufacture, processing, packing, or holding of a drug product shall have the education, training, and experience, or any combination thereof, to perform assigned functions in such a manner as to provide assurance that the drug product has the safety, identity, strength, quality, and purity that it purports or is represented to possess.

(c)   There shall be an adequate number of qualified personnel to perform and supervise the manufacture, processing, packing, or holding of each drug product.

With regard to training, the EC GMP Guide has a separate subsection as follows:

EC GMP Guide

Training

2.8   The manufacturer should provide training for all the personnel whose duties take them into production areas or into control laboratories (including the technical, maintenance and cleaning personnel), and for other personnel whose activities could affect the quality of the product.

2.9   Besides the basic training on the theory and practice of Good Manufacturing Practice, newly recruited personnel should receive training appropriate to the duties assigned to them. Continuing training should also be given, and its practical effectiveness should be periodically assessed. Training programmes should be available, approved by either the head of Production or the head of Quality Control, as appropriate. Training records should be kept.

2.10 Personnel working in areas where contamination is a hazard, e.g., clean areas or areas where highly active, toxic, infectious or sensitising materials are handled should be given specific training.

2.11 Visitors or untrained personnel should, preferably, not be taken into the production and quality control areas. If this is unavoidable, they should be given information in advance, particularly about personal hygiene and the prescribed protective clothing. They should be closely supervised.

2.12 The concept of Quality Assurance and all the measures capable of improving its understanding and implementation should be fully discussed during the training sessions.

Comment

References to the "4 Ms" (men, materials, machinery, and methods) or to the "4 Ps" (personnel, premises, plant, and procedures) as the essential elements in any quality-orientated industrial enterprise are commonplace, and such generalizations serve to focus attention on basic requirements. There can be no doubt that, of these elements, it is the people (the "men" — the human species, not the gender — or the personnel) that are the most important factor in the assurance of quality. This is true of all levels within an organization, from company president and managing director to the most-junior employee. It may well be possible (if not altogether desirable) for high-quality, well-trained, dedicated personnel to compensate for a lack or deficiency in the other elements . Nothing, not even the finest premises, equipment, materials, or procedures can compensate for the quality hazard represented by low-standard, ill-trained, or poorly motivated staff. In a manner matched in only a few other industries (e.g., food), personal hygiene is significantly more important than for mere social acceptability.

Certainly, most GMP Guidelines place early and emphatic emphasis on *people*, and their management, organization, and training. It is worth stressing that good staff — well trained, managed, and motivated — will work more effectively and more productively (i.e., there are *quantity* as well as *quality* benefits).

ORGANIZATIONAL STRUCTURE

Many and varied statements have been made about the objectives, or purpose, of any commercial business. The essential elements that may be extracted from such statements are, generally, that a business exists to deliver goods or services for which there is a demand (or for which a demand can be created), and thus to make a profit.

Although some of the products with which we are concerned may be manufactured without any specific aim or desire to yield a profit, the vast majority (in terms of type, product, or number of units) are manufactured in situations where profit is a prime motive. To some, this has presented an ethical dilemma, and they have seen the profit motive as inimical, as representing an opposing force to the special dedication to quality required in the production of medicines. Others consider that making a profit from the manufacture of pharmaceuticals and the like is in no way different from profiting from the manufacture and provision of other basic human needs, such as food, shelter, and clothing. Whatever one's stance on these politico-philosophical issues, it needs to be noted that even if ethical considerations do not provide a sufficiently powerful impetus, it would still be a foolhardy manufacturer who did not allocate sufficient resources (including, most notably, a sufficient number of suitably trained and qualified people) to the prevention of the production of poor quality or defective products. If this is not done, he or she will run a risk of damaging or killing (quite literally) his market, losing his profit, and ultimately (for one reason or another) going out of business. Ethical imperatives are thus reinforced, rather than opposed, by practical and regulatory considerations. As the EC GMP Guide states, the attainment of quality "requires the participation and commitment by staff in many different departments and at all levels within the company." This spirit of involvement, this commitment, must first embrace top management and then diffuse throughout the organization as a whole.

In establishing an organizational structure for the manufacture and quality assurance of medicinal and similar products, the most generally accepted view is that there should be two separate persons, each with overall responsibility for Production *or* for Quality Control, neither of whom is responsible to the other. This organizational concept is explicitly stated, or is implied, in a number of official GMP publications, including (probably as implicit) the US cGMPs.

The World Health Organization, in its GMP Guidelines, also indicates the need for the supervision of Quality Control by an appropriate expert, reporting directly to top management and independent of other departments. As we have seen, the EC GMP Guide states:

2.3. The heads of Production and Quality Control must be independent from each other...

Aside from any statutory requirements, the separation of Quality Control from Production has on occasions been challenged by some quality experts (or "gurus") on the grounds that it removes from production personnel the healthy sense of responsibility for product quality that they rightly should have. This is to miss the point and to misunderstand the special nature of medicines manufacture and of Good Pharmaceutical Manufacturing Practice.

The more generally accepted view is that it is sound sense that the person ultimately responsible for Quality Control should be freed from the need to consider, or be influenced by, questions of *quantity* of production, meeting production schedules, sales estimates etc., all of which are, quite properly, the province of a Production Manager. The Head of Quality Control should thus be able to make decisions regarding quality standards and procedures and to approve or reject materials and products entirely unbiased by such pressures. This is by no means to say that the Production Manager and his staff do not have a responsibility for implementing quality policies and procedures and for Quality Assurance. Perhaps the misguided challenge to this view arises from differing conceptions of the precise meanings of the terms "Quality Control" and "Quality Assurance," which have been the subject of considerable confusion. If Quality Control is concerned with sampling, specification and testing, and with release/reject systems and decisions, and if Quality Assurance means *all* activities and systems concerned with the attainment of the appropriate product quality, then this sort of objection to the managerial separation of Production and Quality Control largely disappears. That is, it does if it is understood that attainment of the required quality is *everyone's responsibility*, even though some specific quality responsibilities may be assigned chiefly to the Quality Control Manager, and others to the Production Manager.

While there is fairly general agreement about the way many of the differing responsibilities for Quality Control and Production should be allocated, there are, in practice, some differences in detailed approach. In addition, there are certain functions that can be, and indeed are, seen as being joint responsibilities of both Production and Quality Control. It is essential that these various responsibilities are defined and understood. As we have seen, the EC GMP Guide (2.5 to 2.7) deals explicitly with the individual and joint responsibilities of the Head of the Production Department and of the Head of the Quality Control Department in a manner not paralleled by the US cGMPs.

The way in which the technical management of different manufacturing companies is structured can (and does) vary considerably on points of detail. It is essential that all concerned are fully aware of both their functional and reporting responsibilities, by means of organization charts and written job descriptions, and these are European legal requirements under Directive 91/356/EEC.

Although organization charts are a requirement of European law, they make good business and quality sense in all types of manufacturing enterprise anywhere in the world. It makes sense that all persons in an organization should be clearly aware of who reports to them and who their bosses are. This simple and obvious truth seems, unfortunately, to escape the notice of a number of manufacturing companies. The need for job-descriptions, in addition to being legally required (at least in Europe) for all managerial and supervisory staff, is another simple and obvious truth. A job description should, as a minimum, state:

- To whom the person reports
- The main purpose and objectives of the job
- Tasks and responsibilities

A job description should bear a reference number, current date, and date for review. It should be signed and dated by the jobholder, and by his or her immediate superior. There should be no doubt on any side what the boundaries of the job responsibilities are.

The question arises of the level in the organization down to which formal job descriptions should be prepared. Here, the EC GMP Directive 91/356, may be taken as giving sound advice as well as laying down the law. It requires that "job descriptions should be available for all managerial and supervisory staff." For example, it could be argued that, at operator level, the applicable Batch Manufacturing Instructions plus the relevant Standard Operating Procedures fill the place of job descriptions. Certainly, operators should be trained in the understanding and application of these instructions and procedures. It also makes sound quality sense to maintain records of this training, signed by the operator to confirm that they have received and understood the training given. As with job descriptions, this should not be seen as a policing exercise. The object is to ensure there is complete agreement on all sides on what, precisely, the job is all about.

To return to organization charts (or "organograms"), although the formal drawing-up of organizational structures is essential to the clear understanding of responsibilities and reporting relationships, it is important that it is also realized that in the pursuit of quality, such divisions are not to be regarded as water-tight compartments. It is, for example, difficult to see how there could be true assurance of quality in an organization where there was failure to communicate between the Production Manager and the QC Manager

This need for cooperation among various functional units, however, goes wider and deeper. The setting up of the Master Formula and Processing Method for a new product (or for an established product to be made by a modified method) must surely be a cooperative effort between Production, QC *and* Research and Development. There can also be a significant impact on Quality by Engineering and, perhaps not so obviously, Marketing. The personnel, recruitment, and training policies of a company are crucial to the attainment of quality.

MANAGEMENT AND MOTIVATION

The manufacture of quality products cannot be expected from a mismanaged, poorly motivated work force. Mismanagement may occur at all levels, from the top (e.g., chairman, chief executive) to first-line supervision (team leader, charge-hand, foreman). Mismanagement from the top is wider ranging, deeper reaching, and more insidious in its ill effects. First-line mismanagement may have more immediate and more immediately noticeable, but usually more readily curable, effects. A not entirely inappropriate analogy may be drawn with a chronic, as compared with an acute, disease state.

Self-responsible, motivated activity is more efficient than commanded activity. Other things being equal, well-motivated staff will produce more goods, with a greater assurance of the quality of those goods, than will poorly motivated staff. Conversely, in the special context of medicines manufacture, poorly motivated staff can represent a hazard to themselves, to the public, and to company profits.

Motivation and engendering of high morale in healthcare workers should be a relatively easy task. Indeed it is difficult to understand why in any pharmaceutical manufacturing operation there should be ill-motivated workers. Work of any type, far from being a curse and a punishment to man for his sins, is itself a motivator. By and large people *want* to work. Work defines a person's status. It places them in society. It provides a major source of social interaction. A person is very much what they do. We all tend to ask of a new acquaintance, what do you do?

It has been argued by some management theoreticians (who may never have walked around a factory in their lives) that pay and conditions are the less important, even insignificant, motivational factors. It is, however, difficult to see how a feeling that one is being paid less than one's worth (or less than others who are doing comparable jobs), or how a sense that the working environment and facilities are of a standard lower than the job requires, could be other than *de*motivating factors. In any event, while remuneration is a matter of company policy, legislation and basic GMP requirements should guarantee that generally, apart from a few "dirty jobs," the working environment will be more congenial than many. Even dirty jobs are acceptable when it is realised that the "dirt" is an inevitable part of the job (coal miners do not usually complain about coal). It is when the working environment is seen to be more dirty or unpleasant than it need be, that it becomes demotivating. A source of low morale, which needs to be noted, and which can arise in the most modern, clean, air conditioned, windowless, immaculately finished pharmaceutical factory, is the sense of isolation, segregation, and lack of contact with others and with the outside world, which can arise in such circumstances.

In addition to the basic need to work and the social satisfactions gained from working (plus the *possibly* debatable contributions of rewards and conditions), the other main motivating factors may be summarized as sense of purpose, sense of pride and sense of belonging.

The stimulation of such senses in the worker should be particularly easy in the healthcare industry. Medicines serve a recognized, significant social purpose. Workers will easily understand this and should readily take pride in that purpose. From induction training onward, they should be encouraged to see that they have a role to play, however marginal, in achieving the socially useful purpose of supplying products to cure, alleviate, or prevent illness. They should be made to feel that they belong to a team (from top management to most-junior shop floor worker), the aim of which is to achieve that purpose. The most important word is "communication," not just of facts but also of ideals, attitudes, and objectives. Motivation and maintenance of morale must be easier in the Pharmaceutical Industry than in most others. There is such a good peg to hang it all on.

Where morale and motivation are low, the blame must be laid squarely at the door of senior management. That said, the road to motivation and communication is *not* via "quality" posters bearing slick slogans of the type "Object 2000 — Total Quality Attainment" or "2002 — Quality Commitment Year" or "Our Objective — World Class Quality." Such trite burblings are rarely understood by the workers, achieve no purpose, and can, and do, become objects of derision. More will be achieved in terms of motivation, and two-way communication, by a more visible senior management prepared to take the trouble to walk around the factory, to see what is really going on and also *to be seen*. It is important, however, to avoid giving the impression that such a walk-around is just a high-level policing exercise, and it is also important for managers to be prepared to listen sympathetically and to comment amiably.

RECRUITMENT

Clearly, senior, supervisory, or managerial staff in Production and Quality Control must have the education, experience, and professional or technical qualifications appropriate to the jobs they perform. In many countries this is mandatory, and not just a "guideline" recommendation. A vital point that can be overlooked is that they should also have the *ability* to do the job. Specifically, they should have the ability to *manage*. In assessing a prospective senior employee's admirable paper qualifications and proven technical ability, it is perhaps a little too easy to forget that a major part of his or her prospective role will be to manage — to lead, to direct, and to motivate those people reporting to him or her. Top management should bear in mind that while the possession of outstanding technical ability and exemplary professional qualifications by no means excludes the ability to manage, neither does it guarantee that ability.

In all but the very smallest organizations, junior unqualified staff (production operators and maintenance men; cleaning, stores and service personnel) and their first-line supervision form the bulk of the work force. It is these people and the way they are trained, directed, and motivated that are perhaps the key elements in the assurance of product quality. By far the largest proportion of reported defective medicinal products results from simple human error or human misunderstanding, and not from failure at a level of high technology. Advanced qualifications are not normally required at, for example, basic operator level. Nevertheless such personnel must, at a minimum, have the education and intelligence to read and fully understand written instructions and carry them out. They must also be able to respond effectively to the very special challenges of medicine manufacture and to understand the nature and purpose of Good Manufacturing Practice. They must also have innately good standards of personal hygiene. Management needs, therefore, to be in a position to exercise some degree of selectivity in the recruitment of such staff. It follows that it should tailor its policy regarding monetary and other rewards, conditions of employment, and prospects of advancement, accordingly. It should also aim to create a working environment in which staff turnover is reduced to a minimum, since it is difficult, if not impossible, to maintain a well-trained and motivated workforce where there is rapid staff turnover. The Pharmaceutical/Healthcare Industry is not really a job for those who are just passing through, nor is it an industry where a policy of "paying peanuts and getting monkeys" makes any sort of sense at all.

TRAINING

It has been said that "People who have been trained by example that high production is more important than extreme care are going to make mistakes." (Edmond Fry, former senior official, US FDA)

Few would disagree that sound training of staff, at all levels, is of preeminent importance in the assurance and maintenance of quality. Official guidelines and regulations recommend and require it. Regulatory inspectors aim to enforce the requirement. Yet it is sadly true that training, even in some of the most distinguished and richly endowed manufacturers, is often something that is indifferently, even poorly, done and regarded as an irritant to which only occasional, makeshift resource needs to be allocated. Training must be approached with the seriousness of purpose that it needs and deserves.

The necessity for training arises whenever there is any deficiency in the knowledge, understanding, attitudes, and specific skills possessed by a person as compared with the knowledge, understanding, attitudes, and specific skills required for the successful performance of any task assigned to that person. The new recruit, or a person newly transferred to a department, is almost certain to display a deficiency in one of more of these areas, and is thus a prime candidate for training.

This need for training can, and does, emerge at all levels in an organization from senior management to most junior employee. The balance of skills training in relation to general or orientation training will vary among different jobs and different levels, both as to nature and extent. New employees may have already acquired the basic skills required from previous employment. They still need to learn how to exercise those skills in the new working environment and to satisfy their new employers that they can do so. They will still need to be made familiar with that new environment and with company background, traditions, attitudes, and policies. Crucially, they will need training in GMP or to have any previous GMP training reinforced.

The concept of sound training implies a formal, systematic approach. Merely "sitting with Nellie" on grounds that this constitutes on-the-job training is just not good enough. Nor does merely viewing a video or tape and slide show (and there are precious few good examples of these) constitute anything approaching adequate training. There is no objection to the use of videos, tape and slides, etc., per se. But to use such things in isolation and not just for illustrative and reinforcement purposes (within in the context of a well-planned interactive training presentation) is to make no more than a feeble pretense at training. No better are the "training" sessions, where the "instructor" merely stands up and reads in inexpressive parrot-fashion from someone else's prepared text. Indeed, this sort of thing is probably worse than no training at all. The trainees are not fooled (never underestimate the intelligence of the ordinary person), and the whole exercise becomes the object of demotivating derision.

Who Needs Training?

In general terms, training needs to be directed at three categories of employees:

1. Employees new to the company
2. Existing employees — when the nature or content of their job changes
3. Existing employees whose performance at a particular task declines below required standards

Basic training should also be reinforced from time to time by ongoing training programs designed to ensure that employee performance, skills, knowledge, and attitudes remain up to standard. That is, the aim should be to ensure that category 3 above does not arise. As in so many other things, prevention is better than cure.

Costs/Benefits of Training

Training of manufacturing staff takes up time and costs money. It is not outrageous, and is indeed relatively conservative, to suggest that somewhere between 2% and 5% of overall available hours need to be absorbed by training. The precise amount of time will depend upon the size and complexity of the manufacturing operation and upon (a) how far behind a company is in meeting its training obligations and (b) the extent to which a company is recruiting new staff or expanding into new products and technologies. A company manufacturing and packaging, say, just one simple liquid product will need to spend a relatively small percentage of its time on training.

In the context of more complex manufacture, a company with a long-standing, well-established, sound, and effective training program, with a relatively stable workforce, and which is not significantly expanding into new manufacturing areas, will need less time for training. On the other hand, in a company that has historically allocated inadequate resource to training, has an unstable workforce, but which is also expanding its range of activities, the time required for training will be at the upper end of the range. As training programs bed down and become increasingly effective, the overall time required will reduce. In any event, it needs to be clearly understood that training is costly. In addition to the cost of providing the training resources (trainers, in-house or consultants, training facilities and equipment, documentation), there is the unproductive time spent by the trainees. There are, however, benefits. Although these may be difficult to quantify, as has been observed, good staff, well trained and motivated, will work more effectively and productively. They will make fewer of the sorts of mistakes that result in rejects, scrap, and rework, and consequent production delays. The majority of product recalls are the result of human error. Reduce the potential for error by good training, and the possibility of encountering a recall situation correspondingly diminishes. Any company that has experienced the overall expense of a full-scale recall, with the consequent loss of consumer confidence and the possibility of regulatory action, will acknowledge that the benefits of training outweigh its cost.

In any event, training is a regulatory requirement, and failure to comply will give a manufacturer further grief. It is worth recalling that the European Directive on GMP (91/356/EEC, Chapter II art.70) states:

4. Personnel shall receive initial and continuing training including the theory and application of the concept of quality assurance and good manufacturing practice.

and that the US cGMPs require that:

211.25 (a) Each person engaged in the manufacture, processing, packing, holding of a drug product shall have education, training and experience ... to enable that person to perform the assigned functions... Training shall be in the particular operations that the employee performs and in current good manufacturing practice...

The Major Training Elements

The areas in which training needs to be given may be considered as forming three basic training elements:

1. Introductory, background (induction or orientation) training for new employees
2. GMP training, including training in hygienic practices
3. Specific skills training

No one of these elements should be neglected.

Background Training

All new employees, in any job, in any industry, need to acquire a certain basic knowledge of the company in which they will work; its pay, personnel, and promotion policies; its physical layout (they need to find their way around); its supervisory and management structure; and so on. In the pharmaceutical industry, it is also important that employees are aware of the ethical, social, and legal significance of the company's activities. The task of impressing upon employees the importance of the end product, and of the *quality* of the end product, should be relatively easy, since treatment by medicine is an inherently human-interest subject. Every opportunity should be taken to stress that the company is making medicines for patients. Not only is it a powerful factor in employee motivation, it is indisputable that people learn quicker and better, and function more effectively, when they can relate what they are learning and what they are doing to a wider context. The employee in the pharmaceutical industry needs to be able to relate his or her job not only to the work of the factory as a whole, but also to the social — the human — significance of what their company is doing. This is not just high-minded ethics. It is also, like so much of GMP, good hard-headed business sense.

GMP Training

GMP training is necessary for all Production and Quality Control workers, and also for any other personnel whose duties take them into production or control areas, or in any way bear upon the nature and quality of a company's products.

To provide a context, to stimulate interest, and to supply motivation, this training should be related to the general background of the history and use of medicines and their role in society, briefly and simply explained. The following is a brief checklist of topics that should be considered:

- Brief history of medicine — the "therapeutic revolution"
- Background of the healthcare industry and the company's place in it
- Benefits and risks of modern medicines
- Cost of medicines: R & D
- The need for GMP
- Quality Assurance and Quality Control
- Pharmaceutical dosage-forms and packs
- Company product lines
- Problems of faulty batches and recall
- The need and reasons for documentation, records, and written procedures
- The company's documentation system — and the importance of abiding by it
- Cleanliness, hygiene, and simple microbiology
- Personal hygiene
- Plant and equipment cleaning methods and schedules
- Nature and problems of microorganisms
- Microbial and cross contamination
- Clothing
- Effects of legislation, regulation, and inspection
- The overall company manufacturing and control cycle

The emphasis placed on different topics may vary in different organizations. Always, the main motivating thrust should be that the reason for being GMP-minded is for the sake of the sick patient who needs the products. A secondary motive can be the fact that if the company produces significantly defective goods it will rapidly cease to be able to pay employees. While staff must obviously have an understanding of relevant statutory and regulatory matters, and of the need to abide by the law, an approach based solely on "we have a quality system, and we follow GMP, and we do it this way because the man from the Regulatory Authority says so, and if we do not we will be in trouble" is likely to have a much lower motivating force.

Skills Training

A wide range of skills is required. Some jobs are simple, routine, and repetitive. Others may require a considerable level of expertise, concentration, and judgment. It is neither possible nor appropriate to dwell here in detail on training in all possible specific skills. Suffice it to say that skills (e.g., operating a machine, servicing equipment) cannot be satisfactorily acquired in the classroom. This is preeminently a case for showing, for demonstrating, and for hands-on practice.

In no other industry can the need for sound training be more obviously apparent, yet it is sadly true that, across the industry as a whole, there has been a wide spectrum of personnel policies and training approaches, from well-organized, efficient enthusiasm to lip-serving, indifferent inadequacy. (For a further consideration of training and of training techniques see Annex 2 to this chapter.)

Assessment of Training

The effectiveness of training should be assessed during each session by oral questions and answers, and at the end of each by a simple, largely multiple-choice, question paper. In the longer term, assessment should be by observation of operator performance and adherence to systems and procedures, as noted during periodic company self audits etc.

Retraining

Retraining and/or refresher training should be given whenever:

(a) Assessment of post-training effectiveness, or
(b) Changes in Company organization, systems or technologies

indicate the need.

Training Records

There should be two types of training records:

1. The personal file of each member of staff should contain a record of the training received, indicated by module reference number (see Figure 2.1).
2. Departmental training records should be maintained, indicating in tabular form the training received by each member of staff (see Figure 2.2).

Personnel — Hygiene and Clothing

Issues of personal hygiene and protective clothing are treated in the US cGMPS under the subheading "Personnel responsibilities," thus:

US cGMPS

Subpart B — Organization and Personnel

Sec. 211.28 Personnel responsibilities

(a) Personnel engaged in the manufacture, processing, packing, or holding of a drug product shall wear clean clothing appropriate for the duties they perform. Protective apparel, such as head, face, hand, and arm coverings, shall be worn as necessary to protect drug products from contamination.
(b) Personnel shall practice good sanitation and health habits.

Phantazpharm Inc.

Personal Training Record

Name .... Date Joined Company...............

Job Title:

1..

2..

3..

Training Received (module ref. and title)	Date

Figure 2.1 Personal Training Record

(c) Only personnel authorized by supervisory personnel shall enter those areas of the buildings and facilities designated as limited-access areas.

(d) Any person shown at any time (either by medical examination or supervisory observation) to have an apparent illness or open lesions that may adversely affect the safety or quality of drug products shall be excluded from direct contact with components, drug product containers, closures, in-process materials, and drug products until the condition is corrected or determined by competent medical personnel not to jeopardize the

Name	GI/1	GI/2	GMP/1	GMP/2	GMP/3	GMP/4	GMP/5	GMP/6	STT/1	STT/2	STT/3	STT/4	STT/5	STT/6	STT/7

Figure 2.2 Company Training Record

safety or quality of drug products. All personnel shall be instructed to report to supervisory personnel any health conditions that may have an adverse effect on drug products.

Chapter 2 (Personnel) of the EC GMP Guide has a subsection on personnel hygiene, thus:

EC GMP GUIDE

Personnel Hygiene

2.13 Detailed hygiene programmes should be established and adapted to the different needs within the factory. They should include procedures relating to the health, hygiene practices and clothing of personnel. These procedures should be understood and followed in a very strict way by every person whose duties take him into the production and control areas. Hygiene programmes should be promoted by management and widely discussed during training sessions.

2.14 All personnel should receive medical examination upon recruitment. It must be the manufacturer's responsibility that there are instructions ensuring that health conditions that can be of relevance to the quality of products come to the manufacturer's knowledge. After the first medical examination, examinations should be carried out when necessary for the work and personal health.

2.15 Steps should be taken to ensure as far as is practicable that no person affected by an infectious disease or having open lesions on the exposed surface of the body is engaged in the manufacture of medicinal products.

2.16 Every person entering the manufacturing areas should wear protective garments appropriate to the operations to be carried out.

2.17 Eating, drinking, chewing or smoking, or the storage of food, drink, smoking materials or personal medication in the production and storage areas should be prohibited. In general, any unhygienic practice within the manufacturing areas, or in any other area where the product might be adversely affected, should be forbidden.

2.18 Direct contact should be avoided between the operator's hands and the exposed product as well as with any part of the equipment that comes into contact with the products.

2.19 Personnel should be instructed to use the hand-washing facilities.

2.20 Any specific requirements for the manufacture of special groups of products, for example sterile preparations, are covered in the annexes.

Discussion

Most official guidelines or regulations on GMP stress the importance of personal hygiene, although in some of these publications references to it appear under such headings as "sanitation," which cover cleanliness of plant and equipment as well as of people.

The requirements of the two major regulatory documents are broadly similar. There are some differences. For example, the US cGMPs do not specifically require medical examinations on recruitment, as does the EC GMP Guide. The US cGMPs require that:

(d)   Any person shown at any time (either by medical examination or super-
       visory observation) to have an apparent illness or open lesions that may
       adversely affect the safety or quality of drug products shall be excluded
       from direct contact with components, drug product containers, closures,
       in-process materials, and drug products until the condition is corrected
       or determined by competent medical personnel not to jeopardize the
       safety or quality of drug products. All personnel shall be instructed to
       report to supervisory personnel any health conditions that may have an
       adverse effect on drug products.

By comparison, the EC GMP Guide offers only the weak, compromise statement:

2.15   Steps should be taken to ensure as far as is practicable that no person
       affected by an infectious disease or having open lesions on the exposed
       surface of the body is engaged in the manufacture of medicinal products.

This is indeed a feeble compromise. If it is right (as it surely is) that persons with infectious diseases or exposed lesions should not be engaged in the manu-facture of pharmaceutical products, this should be an *absolute* requirement. There should be no potential "get-out" clause.

The US cGMPs do not seem to specifically prohibit eating, drinking, chewing, smoking etc. The EC GMP Guide does.

The Need for Personal Hygiene

High standards of personal hygiene are clearly necessary for all involved in the manufacture of pharmaceuticals and the like. This necessity is most crucial where the product is exposed, particularly when that product is intended to be sterile. Nevertheless, these high standards should be demanded and achieved at all stages and with all types of product. The primary reasons are to control contamination of product, materials, or environment by that vigorous dispenser of microorganisms and general dirt, the working human being — and also to prevent cross-contam-ination through transfer of dusts and other materials via workers hands, etc. There is a very good secondary reason. Persons with high regard for matters of hygiene will more readily be able to adopt that special attitude of care and attention that the manufacturer of medicines requires.

Microorganisms can abound on body surfaces and in the nose, throat, mouth, and intestines. They may be transferred by shedding from body surfaces, generally in association with inanimate particles (e.g., skin flakes), via sneezing and cough-ing, or by direct contact with contaminated hands. The total number of microor-ganisms on the skin varies from person to person, and in accordance with their personal hygienic practices. It also varies in different parts of the skin surface. It can vary from less than 100 organisms to several millions per square centimeter of skin surface. The largest concentrations of organisms are generally to be found on the head and neck, armpits, hands, feet (and beard, if worn). Saliva can contain up to 100 million organisms per ml, and nasal secretions up to 10 million. The number of coliform bacteria alone per gram of human faeces can be of the order of 100 million.

The human body continuously sheds inanimate particles, largely consisting of skin fragments. Depending on skin type and level of activity (the more vigorous the activity, the greater the shedding), the rate of shedding is of the order of 5 to 15 grams of particles every 12 hours. Skin microorganisms are most frequently shed in association with skin particles. The extent and the hazard of microorganism dispersal increases where there is infection, especially of the skin, the respiratory system, or the alimentary canal. Steps need to be taken to control these hazards to product and environment through the observations of hygienic practices by staff and through the provision of suitable protective factory wear. Let it not be thought that the need to guard against bacterial contamination applies only to the manufacture of products intended to be sterile, or to certain creams, emulsions, suspensions, syrups, etc., which might "grow." There is evidence, for example, of the adverse clinical significance of microbiologically contaminated tablet products.

Other possible objectionable effects of microorganisms, in addition to the obvious hazards of their presence in parenterals, ophthalmic products, and products for application to wounds or broken skin include:

- possible effects on product stability
- breaking of emulsions by microbial activity
- effects on container/closure integrity (through fermentation and consequent evolution of gases)
- effects on bioavailability
- interference with analytical testing due to the products of microbial metabolism

HYGIENE MEASURES

The first step toward ensuring hygienic practices among personnel is, of course, to recruit the right sort of people in the first place (that is, people who already observe high standards of personal hygiene). It then becomes a matter of providing training that emphasizes the special risks and requirements of pharmaceutical manufacturing and of providing the necessary facilities.

There should be a general medical examination prior to employment, the extent of which may vary according to the nature of the work to be performed by the new employee. At one end of the risk spectrum is the handling of exposed sterile product, at the other, the handling of already packaged products. No person with a communicable disease or with open lesions on exposed body surfaces should engage in the manufacture of medicinal products, most certainly not where the product may be exposed. Further, staff should be instructed to report any such conditions and supervisors to look out for them. Steps should be taken to encourage such reporting, and no person should suffer any loss, e.g., of remuneration, for doing so.

In this context, it is interesting to note the recently reaffirmed policy of the US FDA[1] regarding the employment of HIV infected workers in pharmaceutical manufacturing areas:

> ... a person infected with the AIDS virus should not be restricted *a priori* from working in a pharmaceutical ... manufacturing facility. We are not aware of any epidemiological data that suggest any increased product safety risks associated with the employment of persons with AIDS under the conditions which would exist in drug ... manufacturing, based on the fact that all ... evidence ... indicates that blood-borne and sexually transmitted infections like AIDS would not be transmitted under normal conditions in the workplace.

Although there is general regulatory and industrial agreement that there should be some form of initial and ongoing medical examination of persons working in, or entering, pharmaceutical manufacturing areas, it is difficult to find any regulatory instruction or official guidance that is any more detailed or specific than a vague general suggestion that it is a good thing. However, in 1972 the Association of Swedish Pharmaceutical Industry (LIF) published a set of "Hygiene Recommendations."[2] What now follows is largely based on that Swedish publication.

Operator Hygiene — Basic Guidelines

1. It needs to be understood that good bodily hygiene and a high level of general cleanliness are necessary in those working on the manufacture of pharmaceuticals and similar products.

2. Hands, including nails, should be kept clean, always be carefully washed after visits to the toilet, before meals, and before work commences, or recommences after a break. There is considerable risk of infection being passed on by contaminated hands. It is therefore extremely important to maintain good hand hygiene in the manufacture of pharmaceuticals and the like. To reduce the risk of infection through hand contact, the following should be required of all operators:
   - Do not touch the product, nor objects that may come in contact with the product, with unprotected hands.
   - Keep the hands well groomed with short, clean nails. Hands must be free of any lesions, wounds, cuts, boils, or any other sources of infection.
   - Wrist watches, rings, or other jewelry should not be worn on the job.
   - Hands should be washed before work and as often as the job requires.
   - Protective gloves should be worn when working with open products and when handling objects that come in direct contact with the product. (Working with gloves presupposes scrupulous care and control of the gloves themselves; failing that, the use of gloves will cause a hygienic risk rather than protect against it. When gloves of rubber or plastic material are worn, they soon become very damp inside from sweat, which contains high levels of microorganisms. It is therefore essential for the gloves to be tight-fitting and not torn or punctured. Rubber gloves for multiple uses should be cleaned and disinfected at regular intervals, after the end of each task or as often as the job requires.)

3. Persons with infectious diseases or with open lesions on the body surfaces should not work in production areas. Employees should be encouraged, and indeed required, to report if they are afflicted in this way. They should not suffer any financial or other form of loss in doing so.

4. A program for health checkups should operate for all production personnel. It should provide for regular checkups in addition to a general medical examination prior to employment. Its scope and direction should be adapted to the risks attached to individual jobs.

5. The necessity to observe good oral hygiene should be stressed. Eating, chewing gum, and the ingestion of sweets or the like should be prohibited in the working area. The risks of contamination from nose and mouth can be controlled by:
   - Not talking, sneezing or coughing in the vicinity of exposed products
   - Wearing masks

6. The following routine should be followed to achieve the best effects from a mask, if used:
   - It should cover both nose and mouth and must not be touched while in use.
   - It should be replaced as soon as risks arise that its effect has been reduced, e.g., after a certain period of use, after sneezing, or when (or before) it becomes soggy.
   - It should be thrown away after use in a receptacle provided for this purpose.

7. When working with dusty material, it may be necessary to wear protective masks of the multiple-use type. Such masks should be personally adjusted to the individual using it and should be cleaned and disinfected after each use.

8. Good care of the hair, including regular washing and cutting, reduces the risk of contaminating the product. Wearing beards, moustaches, or whiskers will require the most careful grooming if they are not to pose a hygienic risk. A hair cover should be worn in all production work areas. It should cover all hair and be replaced at regular intervals.

9. Working clothes (i.e., special protective clothing) should be worn only within the designated work areas. The material used to make protective clothing should be dirt repellent and have a tightness of weave, which makes it an effective barrier to the passage of microorganisms and particles from the body to the local environment. The material must not be fiber shedding. Working clothes should be kept separate from street clothes. Overalls, or alternatively trousers and jackets, are preferable to smocks. Disposable protective clothes might be preferable for certain operations.

10. The nature of the protective clothing provided should be appropriate to the nature of the work carried out and should be put on in accordance with written changing procedures. Dirty working clothes should be handled away from the production process. Laundered clothes should be dried and stored under conditions that preclude recontamination as far as possible.

11. Special working shoes, or overshoes, should be worn. The shoes should be cleaned and disinfected at prescribed intervals and be worn only within the work area. The commonly seen plastic (they always seem to be blue) shoe covers, which readily rip, are of limited value and are little more than a cosmetic gesture.

12. Visitors, engineers, contractors, and others who have access to and enter production areas for the performance of certain tasks, should be furnished with the same type of protective clothing as used by the personnel employed in the relevant production area.

Additional Points

A few further points are worth making:

1. If high standards of cleanliness and hygiene and the proper wearing of protective clothes are to be observed by operators, it is necessary for supervisors and managers to set an appropriately good example.
2. If hand-washing facilities are to be used, not only must they be available, they must be *conveniently* available.
3. From the GMP angle, the protection of the operator and his or her "normal" clothing is only a secondary consideration. The primary purpose of protective clothing, in this context, is the protection of the product, and thus the patient.
4. Protective garments are of no value if they are damaged, dirty, or permitted to become vehicles of contamination or cross-contamination. Suitable changing rooms should be provided, and the protective garments should not be worn outside the controlled factory environment or in any area where they could collect or distribute potential contaminants.
5. Medical checks should include sight testing, including checks for color blindness. This is something that is often neglected, and it can assume great significance in jobs where visual acuity and distinction of colors are important.
6. The question of headwear can be difficult. In aseptic production areas there can be no question. All hair on the head and face, as well as on the bodily surfaces, must be completely covered. In other, less-critical areas, the importance of head covering is both variable (in accordance with the operations carried out), and, it must be said, debatable. There is, however, a powerful argument in favor of sound head covering, even in areas where the product is not exposed (e.g., in the labeling of filled and sealed containers), since it helps to engender a salutary attitude among the workforce that they are engaged in "medical" work and they are part of a team making medicines. However, the minuscule paper objects that may on occasions be seen perched precariously atop operators' elaborate hairdos would seem to have little direct practical or psychological value. The selection of suitable headwear and the peaceful persuasion of operators to wear it properly can call for considerable management skill. But then again, *all* matters concerned, directly or indirectly, with the quality of pharmaceutical products call for considerable management skill.

REFERENCES

1. Young, Frank E., US Commissioner of Food and Drugs — letter to the 3M Company, October 17 1986, quoted by Paul Motise of FDA in "PDA Letter," April 1998.
2. Association of the Swedish Pharmaceutical Industry, "Hygiene Recommendations," Stockholm, 1972.

ANNEX 1 TO CHAPTER 2

The Qualified Person

It is important to note that a person who is qualified, even in a relevant discipline, is not necessarily a "Qualified Person" (QP) in the specialized European legislative sense.

Every company manufacturing pharmaceuticals in the EU, or importing pharmaceuticals into a member state, must have at least one Qualified Person. The QP's functions are as follows:

In the case of manufacture, the QP is legally required to:

> "Secure that each batch of product has been manufactured and checked in compliance with the law in the member state … and in accordance with the Marketing Authorization." and he or she must certify to that effect "in a register or equivalent document." In this context, an appropriately signed Batch Manufacturing Record, for example, is considered to be an "equivalent document."

In the case of the importation of pharmaceutical products into the EC from a non-EC source:

> "The QP must ensure that each imported batch undergoes a complete re-analysis in the importing state, even if certificates of analysis of confirmed reliability are available, and he must certify in writing to that effect. (In some specific cases a waiver may be permitted.) No such requirement for re-analysis applies to export/import between EU member states."

> *The above quotations are paraphrases of EC Directive 75/31, Chapter IV, Article 22, (a) and (b).*

What is the difference, it may be asked, between these requirements and the release of product by a QC. Manager, or Unit, independent of Production, as has long been the practice in the U.S. and U.K.?

The QP concept derives from long-standing traditional arrangements in some European countries, notably France and Belgium, where it was a legal requirement for a manufacturer to appoint a "Responsible Pharmacist," who could be in charge of both Production *and* QC, and not necessarily a full-time employee at all. In the relevant EC Directive, the Responsible Pharmacist was converted to the Qualified Person, and now in Britain, for example, there is this somewhat strange legalistic graft of a traditional Gallic approach on to the preexistent requirement for separate QC and Production Managers. A British Manufacturer's License (equivalent to the EC Manufacturing Authorization) has to name, as it always has since the introduction of the UK Medicines Act 1968, a Production Manager and a QC Manager. In addition, one or more Qualified Persons also must be named, who may or may not be the same persons as the QC Manager or the Production Manager.

Under "grandfather" transitional provisions (now virtually expired), it was relatively easy for anyone already in post to be accepted as a QP. Under the

Permanent Provisions, the requirements in terms of education, and professional and academic qualifications, knowledge and experience are somewhat more rigorous. The requirements (Directive 75/319/EEC) for a QP include:

    a. "Formal Qualification...after recognized course of Study, bearing at least upon":

        Applied physics
        Organic chemistry
        General and inorganic chemistry
        Analytical chemistry
        Pharmaceutical chemistry
        General and applied biochemistry
        Physiology
        Microbiology
        Pharmacology
        Pharmaceutical technology
        Toxicology
        Pharmacognosy

    b. Practical Experience For 2 years...or 1 year...or 6 months (depending on the nature of the "formal qualification")

In the U.K., health ministers have delegated to the relevant professional bodies (The Royal Society of Chemistry, The Royal Pharmaceutical Society, and The Institute of Biology) the responsibility for maintaining and publishing a register of those considered, in terms of all the criteria, acceptable as Qualified Persons. While the UK Medicines and Healthcare Products Regulatory Agency (MHRA, formerly MCA) has the last say on the acceptability of an applicant to be named on a license, it is not usual for them to reject a person who is listed in the joint professional register.

The three professional bodies have produced a joint statement on knowledge and experience requirements, and a Code of Practice for Qualified Persons.

It is believed that the intention is that there should be a reciprocity of Qualified Person status across the community, but that has yet to be fully tested, and it could be that some member states will not be inclined to accept that any person other than a pharmacist could be a Qualified Person.

ANNEX 2 TO CHAPTER 2

Training and Training Techniques

The Learning Process

It may be useful first to discuss briefly both the learning process and the main points to be considered by anyone (i.e., a trainer or instructor attempting to impart knowledge or aid the acquisition of skills) aiming effectively to activate that process.

In any consideration of approaches to training, there are two essential factors, which need constantly to be borne in mind. These are that people learn quicker, more thoroughly, and are better able to use the knowledge or skill acquired if:

  a. The specific skill or knowledge is placed in a wider or more general context, so the trainee is aware of "how it all fits," and is not expected to acquire information as isolated, disconnected scraps.
  b. The reason *why* something is done or required is explained (and understood), as well as what is to be done and how to do it.

The first stage in any learning process is the *reception* of new information. This information is initially received via the senses, and then transmitted to the brain. All senses can, or may, be involved (sight, hearing, touch, smell, taste). Of these various senses, it is generally accepted that sight provides the strongest stimulus to the learning process. It is also true that the simultaneous stimulation of more than one sense has a powerful reinforcing action. Thus, trainees will learn quicker, better, and more retentively if they are not merely *told*, but also *shown*, and, in practical skills training, allowed to touch and try out (i.e., get the feel) for themselves. Even in more theoretical, nonskills training, the learning process is facilitated if a verbal presentation is reinforced by pictures, slides, diagrams, charts, and by concrete examples. The stimulation of the senses of smell and taste generally have a much more limited application in the industrial training process.

A distinction may be drawn between the mere *reception* of sensory stimuli, and their useful and effective *perception* in the brain. There can be few people who have not sat musing, or dozing, with half an eye on the television screen. In such circumstances, clearly some stimulus is being applied to both eye and ear, albeit that the level of organized *perception* by the higher sensory centers may indeed be very low. A number of factors, such as fatigue, can affect levels of perception (and, indeed, reception). Trainees who are so fatigued that they fall asleep are unlikely to receive any relevant sensory stimuli, and thus will fall at the first hurdle. Perception-affecting factors include:

  1. Inherent interest (or otherwise) of the subject or its manner of presentation
  2. Health and fatigue of trainees
  3. General physical condition of trainees (hunger and overrepletion both inhibit the learning process, as does the need to attend to other natural functions)

4. Environmental conditions in the training room — heating, lighting, and ventilation
5. Presence or absence of distractions
6. Trainee familiarity, or perhaps overfamiliarity, with the subject
7. Motivation and the will, or desire, to learn

We cannot learn without sensory stimuli, but the mere reception and perception of such stimuli does not constitute learning. The brain must be able to organize and make judgments on the perceived sense-data, relate them to other relevant, previously learned information and file them away in memory for later retrieval. This has been termed the *cognitive* or *assimilation* phase of learning.

The final, clinching phase in the learning process has been termed the *effector* phase, although perhaps in our present context the smarter (not to say "cooler") term might be the *validation* phase. This is where trainees, either verbally (by the spoken or written word), by practical demonstration, or both, reveal or attempt to put into practice what they have learned. This phase serves both to monitor progress and, most powerfully, to reinforce the overall learning process.

Further Factors Influencing the Overall Learning Process

In addition to the factors already mentioned, which bear particularly on the reception and perception phases of learning, other factors that influence the learning process include:

1. Length and frequency of training sessions
   Although it varies from person to person, there is a limit to the amount of new information with which the brain can cope (that is, receive, perceive, assimilate, and use) at any one time. Just as muscles need time for recovery following physical effort, so brains need time to recover after hard learning effort. Shorter, periodic learning sessions are better than occasional long ones.
2. Planned, structured training
   Training that proceeds in accordance with a preconceived, well-thought-out, and structured training program is more effective than unplanned, casual *ad hoc* training. This does not mean that a training plan should be so inflexible as to allow no scope for adjustments to meet individual needs, different persons' ability to absorb knowledge, or to concentrate over different periods of time.
3. Sequence
   Information that is presented in a logical sequence is far more readily and effectively absorbed than information that is random. This applies both to the sequence within a given learning session, and also to the logical sequence from one session to another.
4. Feedback
   People learn better if they are made aware how well (or how badly) they are doing. This is, of course, a two-way benefit. The instructor needs to be in touch with how well his message is getting across, in order that suitable adjustments in content, style, and presentation can be made. For trainees

to know that they are progressing satisfactorily is both motivating and re-inforcing. The revelation of where they are going wrong will enable further assistance to be given and suitable corrections to be made. Overemphasis, at too great a frequency, should, however, not be placed on failure. This will only serve to demotivate trainees.

5. Vividness of original reception and perception
   Mere exposition in a dull and lifeless manner inhibits learning and fails to impress the memory. Interesting learning sessions, presented by instructors who are able to project enthusiasm and some degree of character, and who can both communicate and also encourage active trainee involvement (as distinct from just sitting and listening), all backed up by memorable visual images or direct hands-on contact with any relevant hardware, stimulate learning and fix things in the memory.

6. Repetition
   That repetition is a major factor in the learning process is well known and acknowledged; practice makes perfect. It needs also to be noted that overrepetition is counter-productive, and that it is just as easy to learn bad habits by repetition as it is good habits. One should not practice one's mistakes.

7. Frame of reference
   Learning is faster and more effective — and retention is better — if the learner can relate new knowledge or skill to that already possessed.

People do not all learn with the same facility or at the same rate. There is considerable variation in the ability of different people to absorb and retain new information, although skills once acquired tend to be retained more readily than factual information. (Does anyone ever completely forget how to swim or ride a bicycle?)

Training — Approach and Technique

The above excursion into the learning process has been taken in order to provide a basis for a discussion of the way training should be organized, approached, and presented.

As a preliminary, we may isolate some key factors that influence the effectiveness of training:

1. Training should not be haphazard. It should be conducted in accordance with a planned, structured program, with records maintained of the training given, and assessments made of trainee progress. Training records should be kept in two forms: (a) a record of the training received by each individual employee, and (b) a tabular record, where the names of employees are listed and the training modules are ticked off as they are received (see Figure 2.1 and Figure 2.2).

2. Training is not a "one-off" business. All training should be reassessed from time to time, and augmented and reinforced as necessary.

3. Understanding and retention of information are greatly assisted by:
   a. An understanding of *context* — of "how it all fits in"
   b. An understanding of the *whys*, as well as the *whats* and *hows*
4. Training presentations should be made as interesting and as attractive as possible, with the engagement of as many of the trainee's senses (particularly sight), other than just hearing, as possible. In other words, do not just talk, *show*.

Preparing for a Training Session

When planning to give a specific training session, or a series of interlinked sessions, a trainer needs to pay attention, before the event to:

a. Defining in his or her own mind what the *objectives* of the training are
b. The *preparation* for the training

During the actual training session, although guided by preparations made in advance, the trainer will need to consider:

c. How he or she is *transmitting* the prepared material, in order to best achieve the desired objective(s)
d. How the transmitted information is being *received, perceived,* and *assimilated* by the trainees

Objectives

Before attempting to train, the trainer must have an absolutely clear understanding of the objective(s) of the training that is to be given. This applies whether the session is of the more formal training-room type, or is less formal "on-the-job." That is, the training should proceed in accordance with a preconceived, carefully considered training program. It should be intended and designed to impart a definite amount of specified knowledge, understanding or skill, with both trainer and trainee having a clear appreciation of how the specific session relates to the latter's overall training needs. This does not mean that a training schedule should be a rigid, inflexible, dehumanized affair. But it also means that merely leaving it to Nellie or Fred to show the new guy around to give them an idea of what to do is not good enough.

Careful mental framing of objectives will provide a salutary concentration of the trainer's mind on the job in hand and provide a measure for the later determination of how well those objectives have been achieved.

Preparation

Sound preparation is obviously essential. The trainer's own training, knowledge, and skills (both job skills and skill as a trainer) form the bedrock of the preparation. In addition, a trainer should take the trouble to prepare properly for each session. The effort required for this preparation will vary widely according to circumstances. It could be extensive, where, for example, the trainer is starting from scratch, solely on the basis of his own store of knowledge of the subject. Then it will be necessary to collect thoughts, organize, and concentrate knowledge, plan a logical sequence for the presentation, prepare speaking notes (or at least jot down headings to maintain the proposed presentation properly sequenced and on track), prepare or obtain visual aids, and so on. On the other hand, where the basic preparation of a training module has already been done (either by the trainer or by someone else), and the trainer is used to giving the particular presentation, the effort required in preparation may be considerably less. It may involve little more than a consideration of the numbers and abilities of the trainees, with an appropriate mental adjustment in accordance with that consideration, and a check that all necessary materials, training aids, demonstration equipment, and the like will be available at the right time and in the right place.

At the very least, any person about to impart any knowledge, or teach any skill, should keep firmly in mind that although this may be the 50th time for them, it might be the 1st time for the trainees. Every attempt should be made to keep the approach fresh, and the way to do that it *not* merely to stand up and read a previously prepared text. Even if such a text is available, it should serve only as a basis for the trainer's own presentation, in his or her own individual style. To simply read out someone else's text is to court disaster. Trainees are rarely fooled and may treat the whole exercise with cynical contempt. However experienced the trainer is in relation to the topic in hand, it makes good sense to review notes, consider sequence, to think if an updating of information is required, and to consider if there are any topical examples (good or bad) that might usefully be cited.

With practice, a trainer will become more and more able to function without constant reference to notes. Broad general headings, however, should keep things in sequence, and help to ensure that what should be dealt-with is, indeed, covered. One of the spin-off benefits of slides, or overhead transparencies, is that they also serve as sequence-headings for the trainer. Provided, that is, that the trainer carefully checks in advance that they are all in the correct order. Many a potentially good presentation has been ruined by slides or overheads that have got out of order either during, or since, they were last used. The effect on the hapless trainer's composure can be dire.

Training Room and Equipment

Having considered the trainer's mental preparation (the "software"), mention of slides and overheads leads naturally to a consideration of the "hardware" preparation. Here is a checklist:

- Training room available at time required?
- Adequate capacity?
- Sufficient seating (if required)?
- Tables, or other writing surfaces, available (if trainees will be required to write, take notes, answer question papers, consult documents)?
- If so, writing implements available, or trainees told to bring them?
- Transparencies or slides prepared? Undamaged? In correct order?
- Overhead projector available? Does it work? Spare bulb?

- 35mm projector available? Does it work? Spare bulb?
- Remote control? Trainer familiar with remote control? If no remote control, who will operate projector?
- Chalkboard, white board, or flip chart?
- Chalk or felt pens?
- Videos?
- Video player and screen? Do they work?
- Trainer knows how to start (and stop) the equipment?
- Demonstration material (equipment, samples, documents, etc.) available as and when needed?

Careful preplanning to ensure that all these aspects have been considered in advance can be crucial to the success, or otherwise, of training. Nonavailability, or failure to function, of these physical components of training can lead to distracting gaps in a training session, and perhaps worse, can so discompose a trainer as to turn a professional presentation into a display of apparent bumbling amateurism.

The writer has more than once seen trainers or lecturers virtually self-destruct as a result of failure of visual aid equipment. A little time spent in attention to these things in advance will be amply repaid.

Transmission

The way the training message is transmitted and comprehended depends, to a significant extent, on the personal qualities of the trainer. For example:

a. The trainer's knowledge
   The trainer's own knowledge of the subject and ability in the skill concerned are self-evidently significant factors. A trainer who does not know his subject is clearly in no position to teach others about it. Most important is the recognition by the trainees that the trainer speaks or demonstrates with authority. Respect for that authority will stimulate attention and, hence, learning. If that respect is not granted, attention will wander and all that the trainees will learn is a new level of contempt for the poseur.

b. Appearance
   An expensive three-piece suit, or upper-set garden party ensemble, is not necessarily required for an instructor in good practices in the manufacture of, say, tablets, but a scruffy overall and unkempt appearance should be avoided. A smart, well-groomed appearance inspires confidence and helps command attention.

c. Voice, manner, and approach
   It is difficult to train successfully if the voice is dull, monotonous, or very quiet. Such things *may* be improved by practice, but very quietly spoken persons can, realistically, only be expected to train groups of no more than two or three people, if they are able to do it at all. A trainer's manner should be outgoing, firm but friendly, aimed at establishing contact, and putting trainees at their ease. Trainers should, however, be themselves and not attempt to act a part. There is nothing wrong with laughter. Indeed a

little humor can be a positive aid to learning. It can ease any tension and make the learning memorable. It should not, however, be overdone. Full-time comedy is for comedians. In any event, trainers should carefully evaluate their talents as mirth-provokers. To some it comes easy, even too easy. Others, who are not naturals probably will do best to avoid the gags. Jokes that bomb, cause embarrassment, or provoke low groans are likely to impede the attainment of the desired objective.

Feedback from the trainees should be encouraged and obtained. Can they see and hear properly? Are they comfortable, not too hot or too cold? Do they understand? Have they any questions? Every effort should be made to turn a one-way lecture into a live (even lively) two-way event. On no account should a slow learner be ridiculed, even lightly. It does that person, and the trainer, no good, and it can provoke resentment in the other trainees. Difficulties should be treated sympathetically and sarcasm avoided.

In addition to the personal qualities and approach of the trainer, other factors that affect the transmission (and the reception and perception) of information include:

a. The training environment
   The questions that need to be asked are: Is it neither too hot nor too cold? Is it well ventilated? Can all the trainees see and hear? Far too many training sessions are ruined by a dark, hot and stuffy training environment, where the general tendency is to doze. The keen trainer will check on the comfort of the trainees and on whether or not they can see and hear properly as the session proceeds. A little advance attention to these factors will pay dividends.

b. Presence or absence of distractions
   Random sights and noises can distract the attention of trainees. Given the choice, there are usually many other things that a trainee would rather be hearing, seeing, doing, or handling. Efforts should be made by the trainer to ensure, as far as possible, concentration on the matter at hand. What can be done will, to an extent, be dictated by the place available for training. Some things can only be taught and practiced on the factory floor, where inevitably there will be distractions. Yet in even a well-appointed, dedicated, training room there may be distractions. It is a strange, but nevertheless real, aspect of human behavior that what is going on outside the window, or materials left over from, or intended for, other training sessions (e.g., diagrams, charts, writing on flip charts or blackboards, etc.) always seem more interesting than the matter under discussion. Such distractions should be removed as far and as quickly as possible — and that includes pictures, diagrams, writing on flip charts, and so on that the trainer has used earlier in a session. Anything that is no longer immediately relevant should be removed or erased. It is a false economy to attempt to cram as much as possible on one sheet of a flip chart. Use one side to explain one point, and turn it over. External distractions, such as persons passing by windows, may be more difficult to control. Use of some form of blind may be helpful.

c. Quality and impact of visual aids

As already discussed, visual images both reinforce and facilitate the reception and assimilation phases of learning. This statement should, perhaps, be modified by saying that *relevant* visual images, properly used, can have this effect.

No doubt stimulated by the often-repeated, and generally accepted, advice to illustrate and emphasize by the use of slides and the like, some trainers tend to overdo the visual aids, to an extent that training sessions become more like picture shows. Visual images could be regarded as rather like the jam on the bread. They certainly help the bread go down, and assist in its assimilation, but a trainer should not offer all jam and too little of the solid bread of training. Utterly to be avoided are irrelevant images, the only justification for which is their easy availability in some clip-art package.

Common visual-aid materials, with comments on their use, include:

i. Black- (or green- or white-) board and chalk (or erasable pen)
   ■ Yellow chalk usually shows up better than white on black or green.
   ■ Write, or draw in sufficient size to be clearly visible to all.
   ■ Right-handed trainers should have board on their left (as they face their audience). This way they will obscure less of the board as they write or draw.
   ■ The temptation to talk to a board or screen, rather than the audience, needs to be fiercely resisted.
   ■ The board should be cleared, as soon as the trainer has finished with what is on it. To leave old material can distract trainees from the new topic to which the trainer has turned.

ii. Flip charts (or other pads of large white paper sheets)
   ■ A common fault is too small drawing or writing. It must all be bold and simple.
   ■ As above, material that is finished should be removed.
   ■ Trainers should ensure, in advance, that sufficient felt pens (that are not exhausted or dried-out) are available. Again, the temptation to talk to the chart, rather than the audience, needs to be fiercely resisted.

iii. 35mm slides and overheads
   These are perhaps the most commonly used form of visual aid. Opinions differ as to which of the two forms is to be preferred. In the writer's view, based on his own experience as a trainer and in observing others, the great advantage of the use of an overhead projector, and films or acetates, is that it gives the speaker greater and more immediate control of things. Given that the trainer has taken the trouble, in advance, to assemble the overheads in the right order, little should go wrong. If a 35mm projector, without a remote control, is used, the trainer has to rely on a projectionist to show the right slide at the right time. Even with a remote control, operated by the speaker, there is always the problem of mechanical (or electrical) failure of the projector, or jamming of

slides. A remote control, hand held by the speaker, although it eliminates divided control, does not diminish the potential for such problems. And it adds the problem, which all to frequently surfaces, of the trainer in full flight, pressing the backward instead of the forward button, and then panicking and taking some embarrassing time to get back on track. Reversed, or inverted 35mm slides are more difficult to correct (and can result in spilled slides) than overheads. It is also much more difficult to refer back to previously displayed images with 35mm slides.

There are a number of failings common to the use of both 35mm slides and overheads:

- Projected words, rather than pictures or diagrams, are of debatable value.
- Words that cannot be read, or images that cannot be clearly seen and interpreted, are a positive, distracting menace. (The all too commonly heard "I don't know if you can all read or see this" should be taken as an abject admission of failure. If it cannot be clearly seen and interpreted, then it should not be shown.)
- The ever-increasing availability of computer graphics and clip-art packages has led to the blossoming of irrelevant images on slides and overheads. Some trainers and other presenters seem to be obsessed with the thought that any image is better than no image at all. They are very wrong. No matter how charming, arresting, seductive, or amusing the image, if it is not relevant, then it is a distraction.
- All too often, slides and overheads are too wordy, too complex, or too crowded. All these faults should be avoided, and things made simple, bold, relevant, and direct.
- Once a slide or overhead has made its point, it should be removed or it will become a distraction. Again, this is probably easier with overheads. If a break is intended in the showing of a series of 35mm slides, then a blank should be inserted or the projector switched off.
- Another commonly encountered failing is where a speaker displays a complex and detailed slide, and says words to the effect that "....this is all clearly (!) illustrated by this slide," and then goes on, without explanation, to talk about something entirely different. The slide will neither teach nor illuminate, nor will the trainees be giving full attention to the new topic.

iv. Computer-generated slide shows, operated by the speaker from a laptop keyboard *can* be very effective, but *only* if the speaker is absolutely assured of what they are doing. There have been many presentations that have gone into destruct mode because the presenter has pressed a wrong key and has been unable to find their way back on-course.

The technique of pointing and the use of pointers also need to be considered:

- When pointing at a board (or a screen or a flip chart) the trainer should point with the hand that is on the same side as the board, so as to remain facing the audience.
- If a pointer is used, it should not be allowed to wander around the board (or screen). It should be pointed directly at the item to be stressed, held motionless for a sufficient time for the point to sink in, and then taken away. Laser pointers are a particular problem for all but the steadiest of hands. For those with the slightest inclination to wobble, some additional support should be found. A little red spot, dashing about, can be very distracting.
- A pointer should be used only as a pointer, not as a walking stick, a swagger stick, a conductor's baton, or as a device to tap out interesting rhythms on the floor or table.

Handout Material

Longer-term retention of information will be greatly assisted if trainees are encouraged to make a few notes or are handed out some form of printed material, written and presented in a readily absorbable and attractive fashion, to take away and read at leisure. This can take the form of material prepared in house. Alternatively there is, for example, the "Quality Rules," widely used series of basic GMP training texts written by the author of this book.

Summary/Checklist

To conclude this review of training, the following is a summary/checklist for actual and potential trainers.

1. Know your subject. It is usually necessary to know more than the bare bones of the topic to be presented in order to be able to deal with questions. A trainer lacking knowledge of his subject will not be able to fool all his audiences all the time.
2. Be absolutely clear about what your objectives are.
3. Try to approach each session as something new, no matter how many times you have spoken on the subject.
4. Check the availability of the training room, and on the heating, lighting, ventilation, and sound or noise level.
5. Make sure that all required equipment, materials, handouts, slides, etc. are ready to hand. Check that equipment works. (Spare bulb for projector? Felt pens? Chalk?)
6. Look smart and sound bright.
7. Speak up and check up on how your message is being received.
8. Avoid and eliminate distractions. (Note that few things are more distracting than hunger, thirst, a need to visit the toilet, and a stuffy atmosphere.)
9. Establish contact with audience. Be friendly, but do not let the session become just a genial chat.
10. Use humor (if it works for you), but be careful not to overdo it or cause embarrassment.

11. Encourage trainee participation (questions, discussions, workshops). Training should be a two-way, not just a one-way exercise.
12. Have sympathy for those in difficulties. *Never* ridicule. By ridicule you "lose" not only the unfortunate victim, but also most of the other trainees as well.
13. Present the subject in a logical, preplanned sequence.
14. Place the subject in its wider context.
15. Explain the *whys* as well as the *whats* and *hows*.
16. Avoid distracting mannerisms (tapping, shuffling, and the like).
17. Engage as many senses as possible — especially sight, but remember to remove slides, drawings, etc. as soon as they have made their point.
18. Don't carry on for too long. Several short sessions, with breaks for mental recovery, are better than one long one.
19. Look at your audience (not at screens, boards, or notes), but do not stare at anyone in particular, no matter how beautiful she or he may be.
20. Hammer home important points by repetition and summary, but do not bore by overdoing it.
21. Do not act. Be yourself, but your best self.

Selection and Training of Trainers

The easy options are to employ consultants to present on-site training courses or to send trainees to external courses, seminars, or symposia. The former tend to be expensive; the latter *very* expensive. They both can be of variable quality and utility. It is essential to establish, before engaging consultants or electing to send trainees on external courses, the credentials of those who will be presenting the training. When considering the use of a consultancy body to present on-site training, it is essential to agree on, and to tightly specify, in advance a program of topics to be covered, and in what degree of depth. If this is not done beforehand, it is both pointless and too late to complain after the event that the required topics were not covered, that subjects not required were presented, or that the level was pitched too high or too low for the trainees. All this should be agreed, in writing, in advance, as should the price structure. A number of companies have had a nasty shock at the final invoice, listing expensive extras that they had not anticipated. Training managers need to be aware that many of the "essential" or "intensive" seminars that feature so prominently in junk mail are not conceived and presented as a coherent, integrated whole, but are given by a loose assemblage of itinerant orators who roll up, say, their well-worn piece, collect their fee, and then move on to say it all again at some other time and place, and in another context.

All this is not to say that training given by persons from outside the company is of no value. To the contrary, outside experts and specialists can present a more profound or wider view. When it comes to training in Quality Assurance, GMP, or regulatory matters, for example, they can ensure that the requirements in such areas are seen by the trainees, not as company quirks, but as important issues of wider application and relevance. That all said, any company of any size, is well advised to provide much of its training needs from its own internal resources. This is not just a question of economics. This way, the company has full control over the training provided, its content, and when and where it will be held. It is salutary for trainees to feel that colleagues, or more-senior personnel, have the knowledge, skill, and ability to teach them what they need to know. To permanently opt out and use external people could be seen as an admission of failure or inadequacy. To opt in, it will, in addition to devising training programs and syllabi, be necessary to select and train trainers.

To throw people in at the deep end just because they know (or are believed to know) about a subject, without any consideration of their actual or potential ability to teach that subject, is just not good enough. A prime consideration is willingness, interest, and enthusiasm to be a trainer. No person who is unwilling should be forced into the job. This is very definitely a case of 1 volunteer is worth infinitely more than 10 pressed men.

It is worth setting down some basic concepts:

■ It is a mistake to think of training as something that is easy to do. It is not. To be a good trainer is hard work, although it can be very rewarding and useful for personal development, particularly in terms of confidence building.
■ Some people are naturally brilliant trainers, public speakers, and the like. On the other hand, there are a few who, no matter how they try, will never make it as trainers. Between these two extremes, there is the great mass who, with a little effort and practice, and with appropriate practice and encouragement, can become more than adequately good at it.
■ Potential trainers who are nervous at the thought of standing up and speaking to a group of people need to be assured that everyone has nerves, even experienced speakers, and particularly at the start of a presentation. With a little practice, trainers learn to control, or cope with those nerves, and far from letting it all spoil their presentation, they can in fact energize themselves on the adrenaline flow.

Any organization intending to set up a training scheme based on its own corps of internal trainers, having determined what are its training needs and at least roughed-out training programs and basic syllabi, should then identify potential trainers. The first criterion is the obvious one that a trainer should know his or her subject. Other judgments will have to be based on personal attributes and qualities. Can the person under consideration be expected to be able to explain and teach that subject in a clear, confident, and interesting manner? Those who seem to be suitable should then be approached and asked if they are interested in taking on a training role in addition to their normal function. Acting as a trainer should be made to appear as attractive as possible, and a little encouragement given, but absolutely no pressure should be applied to force anyone to do anything they do not want to do. It may be possible to offer financial incentives, but in any event, the job of trainer must be presented as a positive move, leading to greater job satisfaction and personal development, and certainly not as some additional chore that the company regrets having to ask anyone to take on. An alternative is to announce generally the establishment (or reestablishment) of a training scheme, and call for volunteer trainers. Any company that, through either approach, fails to recruit an adequate body of potential trainers must ask itself what is wrong with its motivational and personnel policies and practices. In any company where there is a good, well-motivated sense of belonging, and where there is a healthy relationship between management and labor, there should be no great problem in finding a sufficient number of people ready and willing to be trained to be trainers.

An approach that has been found to work well on a number of occasions is first to call a meeting of all those selected or who have volunteered. At this meeting the proposed training program should be outlined, and the basic principles and techniques of training, as discussed earlier in this chapter, should be explained. It would be useful to distribute handouts of the major points. The trainee trainers should then be allocated topics (with a summary of the points to be covered) upon which they are requested to develop a training module (outline script and slides, overheads, or other visual aids, complete with any other demonstration material).

It is generally better to allocate any given topic to more than one person — say, to two or three — so they can assist each other in the preparation. All should then be asked to return in, say, two months in order to present to the other members of the group the training module they have prepared. In the interim, each member of the group should be asked to prepare a short talk, on any subject they care to select (hobbies, interests, anything) to be given to the rest of the group in, say, two to four weeks. This is essentially a confidence-building exercise, and it needs careful management to prevent it from becoming the reverse. While it is serious in its intention, attempts should be made to make it fun, rather than an ordeal. At the end of each talk, the course leader should call for *constructive* comments from the other trainee trainers on the content, interest, and presentation of the talk, and then give his or her own views. Every effort should be made to be encouraging. Depending on the number of trainee trainers, it may be necessary to spread this phase over more than one day.

The real crunch comes when the newly prepared modules are presented. This, too, may extend over more than one day. This time, the comments requested, while still *constructive*, should be required to be more detailed and specific:

- Content — adequate coverage? factually correct?
- Interest
- Presentation
- Voice — clear? interesting or boring? sound level?
- Contact and interaction with audience?
- Questions, feedback, discussion.
- Any mannerisms or other distractions?
- Clarity and impact of visuals

As before, the course leader should add his or her own comments on each presentation. Where, say, two persons have prepared a given module, they could either present half each or preferably each separately present the whole module, but on different days. This will have the added advantage of affording an opportunity to compare different approaches to the presentation of the same topic. (It is, of course, assumed that no presentation will consist of just a slavish reading out of a previously prepared text. The script written during the preparation of a module should be considered as neither more nor less than notes upon which speakers base their own words.)

Almost certainly, the first time around, there will be some corrections and polishing up needed for most modules. Any presentations that are judged to be first-class presentations of exemplary material may be considered to have yielded a well-formed trainer and a training module for future use. This is unlikely to apply in many cases. It will usually be necessary to repeat the process after a month or two, during which time any necessary correction and polishing can take place. After this second session, it should be possible to settle on the corps of company trainers and to eliminate, as gently and sympathetically as possible, any no-hopers. The company will also have acquired a set of training modules for future use and for use in the later training of further trainers.

This may seem a lengthy and time-consuming process, but it is one with which the author has been successfully involved on a number of occasions, and there is no doubt that, if well managed, it does work well.

3

PREMISES/BUILDINGS AND FACILITIES

Note: The US cGMPs have one subpart, C, on Buildings and Facilities and another one, D, on Equipment. The EC GMP Guide covers Premises and Equipment in a single chapter, 3.

REGULATORY STATEMENTS

US cGMPs

Subpart C — Buildings and Facilities

Sec. 211.42 Design and construction features

(a) Any building or buildings used in the manufacture, processing, packing, or holding of a drug product shall be of suitable size, construction and location to facilitate cleaning, maintenance, and proper operations.

(b) Any such building shall have adequate space for the orderly placement of equipment and materials to prevent mix-ups between different components, drug product containers, closures, labeling, in-process materials, or drug products, and to prevent contamination. The flow of components, drug product containers, closures, labeling, in-process materials, and drug products through the building or buildings shall be designed to prevent contamination

(c) Operations shall be performed within specifically defined areas of adequate size. There shall be separate or defined areas or such other control systems for the firm's operations as are necessary to prevent contamination or mix-ups during the course of the following procedures:

(1) Receipt, identification, storage, and withholding from use of components, drug product containers, closures, and labeling, pending the appropriate sampling, testing, or examination by the quality control unit before release for manufacturing or packaging;

(2) Holding rejected components, drug product containers, closures, and labeling before disposition;

(3) Storage of released components, drug product containers, closures, and labeling;

(4) Storage of in-process materials;

(5) Manufacturing and processing operations;

(6) Packaging and labeling operations;

(7) Quarantine storage before release of drug products;

(8) Storage of drug products after release;

(9) Control and laboratory operations;

(10) Aseptic processing, which includes as appropriate:

    i.   Floors, walls, and ceilings of smooth, hard surfaces that are easily cleanable;

    ii.   Temperature and humidity controls;

    iii. An air supply filtered through high-efficiency particulate air filters under positive pressure, regardless of whether flow is laminar or nonlaminar;

    iv. A system for monitoring environmental conditions;

    v.   A system for cleaning and disinfecting the room and equipment to produce aseptic conditions;

    vi. A system for maintaining any equipment used to control the aseptic conditions.

(d) Operations relating to the manufacture, processing, and packing of penicillin shall be performed in facilities separate from those used for other drug products for human use.

Sec. 211.44 Lighting

Adequate lighting shall be provided in all areas

Sec. 211.46 Ventilation, air filtration, air heating and cooling

(a) Adequate ventilation shall be provided.

(b) Equipment for adequate control over air pressure, microorganisms, dust, humidity, and temperature shall be provided when appropriate for the manufacture, processing, packing, or holding of a drug product.

(c) Air filtration systems, including prefilters and particulate matter air filters, shall be used when appropriate on air supplies to production areas. If air is recirculated to production areas, measures shall be taken to control recirculation of dust from production. In areas where air contamination occurs during production, there shall be adequate exhaust systems or other systems adequate to control contaminants.

(d) Air-handling systems for the manufacture, processing, and packing of penicillin shall be completely separate from those for other drug products for human use.

Sec. 211.48 Plumbing

(a) Potable water shall be supplied under continuous positive pressure in a plumbing system free of defects that could contribute contamination to any drug product. Potable water shall meet the standards prescribed in the Environmental Protection Agency's Primary Drinking Water Regulations set forth in 40 CFR part 141. Water not meeting such standards shall not be permitted in the potable water system.

(b) Drains shall be of adequate size and, where connected directly to a sewer, shall be provided with an air break or other mechanical device to prevent back-siphonage.

Sec. 211.50 Sewage and refuse

Sewage, trash, and other refuse in and from the building and immediate premises shall be disposed of in a safe and sanitary manner.

Sec. 211.52 Washing and toilet facilities

Adequate washing facilities shall be provided, including hot and cold water, soap or detergent, air driers or singles-service towels, and clean toilet facilities easily accessible to working areas.

Sec. 211.56 Sanitation

(a) Any building used in the manufacture, processing, packing, or holding of a drug product shall be maintained in a clean and sanitary condition, Any such building shall be free of infestation by rodents, birds, insects, and other vermin (other than laboratory animals). Trash and organic waste matter shall be held and disposed of in a timely and sanitary manner.

(b) There shall be written procedures assigning responsibility for sanitation and describing in sufficient detail the cleaning schedules, methods, equipment, and materials to be used in cleaning the buildings and facilities; such written procedures shall be followed.

(c) There shall be written procedures for use of suitable rodenticides, insecticides, fungicides, fumigating agents, and cleaning and sanitizing agents. Such written procedures shall be designed to prevent the contamination of equipment, components, drug product containers, closures, packaging, labeling materials, or drug products and shall be followed. Rodenticides, insecticides, and fungicides shall not be used unless registered and used in accordance with the Federal Insecticide, Fungicide, and Rodenticide Act (7 U.S.C. 135).

(d) Sanitation procedures shall apply to work performed by contractors or temporary employees as well as work performed by fulltime employees during the ordinary course of operations.

Sec. 211.58 Maintenance

Any building used in the manufacture, processing, packing, or holding of a drug product shall be maintained in a good state of repair.

EC GMP GUIDE

Chapter 3 Premises and Equipment

Principle

Premises and equipment must be located, designed, constructed, adapted and maintained to suit the operations to be carried out. Their layout and design must aim to minimise the risk of errors and permit effective cleaning and maintenance in order to avoid cross-contamination, build up of dust or dirt and, in general, any adverse effect on the quality of products.

Premises

General

3.1 Premises should be situated in an environment which, when considered together with measures to protect the manufacture, presents minimal risk of causing contamination of materials or products.

3.2 Premises should be carefully maintained, ensuring that repair and maintenance operations do not present any hazard to the quality of products. They should be cleaned and, where applicable, disinfected according to detailed written procedures.

3.3 Lighting, temperature, humidity and ventilation should be appropriate and such that they do not adversely affect, directly or indirectly, either the medicinal products during their manufacture and storage, or the accurate functioning of equipment.

3.4 Premises should be designed and equipped so as to afford maximum protection against the entry of insects or other animals.

3.5 Steps should be taken in order to prevent the entry of unauthorised people. Production, storage and quality control areas should not be used as a right of way by personnel who do not work in them.

Production Area

3.6 In order to minimise the risk of a serious medical hazard due to cross-contamination, dedicated and self contained facilities must be available for the production of particular medicinal products, such as highly sensitising materials (e.g., penicillins) or biological preparations (e.g., from live micro-organisms). The production of certain additional products, such as certain antibiotics, certain hormones, certain cytotoxics, certain highly active drugs and nonmedicinal products should not be conducted in the same facilities. For those products, in exceptional cases, the principle of campaign working in the same facilities can be accepted provided that specific precautions are taken and the necessary validations are made. The manufacture of technical poisons, such as pesticides and herbicides, should not be allowed in premises used for the manufacture of medicinal products.

3.7 Premises should preferably be laid out in such a way as to allow the production to take place in areas connected in a logical order corresponding to the sequence of the operations and to the requisite cleanliness levels.

3.8 The adequacy of the working and in-process storage space should permit the orderly and logical positioning of equipment and materials so as to minimise the risk of confusion between different medicinal products or their components, to avoid cross-contamination and to minimise the risk of omission or wrong application of any of the manufacturing or control steps.

3.9 Where starting and primary packaging materials, intermediate or bulk products are exposed to the environment, interior surfaces (walls, floors and ceilings) should be smooth, free from cracks and open joints, and should not shed particulate matter and should permit easy and effective cleaning and, if necessary, disinfection.

3.10 Pipework, light fittings, ventilation points and other services should be designed and sited to avoid the creation of recesses which are difficult to clean. As far as possible, for maintenance purposes, they should be accessible from outside the manufacturing areas.

3.11 Drains should be of adequate size, and have trapped gullies. Open channels should be avoided where possible, but if necessary, they should be shallow to facilitate cleaning and disinfection.

3.12 Production areas should be effectively ventilated, with air control facilities (including temperature and, where necessary, humidity and filtration) appropriate both to the products handled, to the operations undertaken within them and to the external environment.

3.13 Weighing of starting materials usually should be carried out in a separate weighing room designed for that use.

3.14 In cases where dust is generated (e.g., during sampling, weighing, mixing and processing operations, packaging of dry products), specific provisions should be taken to avoid cross contamination and facilitate cleaning.

3.15 Premises for the packaging of medicinal products should be specifically designed and laid out so as to avoid mix-ups or cross-contamination.

3.16 Production areas should be well lit, particularly where visual on-line controls are carried out.

3.17 In-process controls may be carried out within the production area provided they do not carry any risk for the production.

Storage Areas

3.18 Storage areas should be of sufficient capacity to allow orderly storage of the various categories of materials and products: starting and packaging materials, intermediate, bulk and finished products, products in quarantine, released, rejected, returned or recalled.

3.19 Storage areas should be designed or adapted to ensure good storage conditions. In particular, they should be clean and dry and maintained within acceptable temperature limits. Where special storage conditions are required (e.g., temperature, humidity) these should be provided, checked and monitored.

3.20 Receiving and dispatch bays should protect materials and products from the weather. Reception areas should be designed and equipped to allow containers of incoming materials to be cleaned where necessary before storage.

3.21 Where quarantine status is ensured by storage in separate areas, these areas must be clearly marked and their access restricted to authorised personnel. Any system replacing the physical quarantine should give equivalent security.

3.22 There should normally be a separate sampling area for starting materials. If sampling is performed in the storage area, it should be conducted in such a way as to prevent contamination or cross-contamination.

3.23 Segregated areas should be provided for the storage of rejected, recalled or returned materials or products.

3.24 Highly active materials or products should be stored in safe and secure areas.

3.25 Printed packaging materials are considered critical to the conformity of the medicinal product and special attention should be paid to the safe, and secure storage of these materials.

Quality Control Areas

3.26 Normally, Quality Control laboratories should be separated from production areas. This is particularly important for laboratories for the control of biologicals, microbiologicals and radioisotopes, which should also be separated from each other.

3.27 Control laboratories should be designed to suit the operations to be carried out in them. Sufficient space should be given to avoid mix-ups and cross-contamination. There should be adequate suitable storage space for samples and records.

3.28 Separate rooms may be necessary to protect sensitive instruments from vibration, electrical interference, humidity, etc

3.29 Special requirements are needed in laboratories handling particular substances, such as biological or radioactive samples.

Ancillary Areas

3.30 Rest and refreshment rooms should be separate from other areas.

3.31 Facilities for changing clothes, and for washing and toilet purposes should be easily accessible and appropriate for the number of users. Toilets should not directly communicate with production or storage areas.

3.32 Maintenance workshops should as far as possible be separated from production areas. Whenever parts and tools are stored in the production area, they should be kept in rooms or lockers reserved for that use.

3.33 Animal houses should be well isolated from other areas, with separate entrance (animal access) and air handling facilities.

DISCUSSION

Regarding Buildings and Facilities, or Premises, the requirements of the two documents are, with a few exceptions, closely similar. Here, as elsewhere, the mode of expression is markedly different. As befits a series of *regulations*, the US cGMPs are positive, direct, and state, usually unequivocally, what *shall* be done. Chapter 3 of the EC GMP Guide presents a number of the ambiguities and limp equivocations to which reference has been made in the "Note on the genesis of the EC GMP Guide" in the Introduction. A particular example here is:

3.7 Premises should *preferably* be laid out in such a way as to allow the production to take place in areas connected in a logical order corresponding to the sequence of the operations and to the requisite cleanliness levels. (writer's emphasis)

Surely premises should *always* be laid out in such a manner. In what possible circumstances would it be acceptable for premises to be laid out in an illogical, nonsequential fashion, and in disregard of the "requisite cleanliness levels?" Compare the US cGMPs no-nonsense statement, 211.42 (b):

"Any such building shall have adequate space for the orderly placement of equipment and materials to prevent mix-ups between different components, drug product containers, closures, labeling, in-process materials, or drug products, and to prevent contamination. The flow of components, drug product containers, closures, labeling, in-process materials, and drug products through the building or buildings shall be designed to prevent contamination."

There are a number of other examples of this sort of feeble imprecision in the EC GMP Guide that readers will no doubt be able to note for themselves.

Significant differences in *content* include:

US cGMPs	EC GMP Guide
Location to be suitable to "facilitate cleaning, maintenance and proper operations …"	" … must be located to … suit the operations carried out." (Allows the modifying effect of "measures taken to protect the manufacture.")
Include, in this subpart, relatively brief requirements for "aseptic processing" areas, which *maybe* are intended to refer to sterile products manufacturing areas generally.	Brief, nonspecific coverage in EC Chapter 3. Very considerably more detailed coverage in Annex 1 "Sterile Products Manufacture."
Specifically require "smooth, hard surfaces that are easily cleanable" for walls, floors, and ceilings in "aseptic processing" areas only.	Requires that "… interior surfaces (walls, floors and ceilings)" should be "smooth, free from cracks and open joints," nonparticle-shedding, and permit "easy and effective cleaning" in *all* areas "where starting and primary packaging materials" and "intermediate and bulk products are exposed …" (3 .9).
"Operations relating to the manufacture, processing and packaging of penicillin shall be performed in facilities separate from those used for other drug products for human use." "Completely separate air-handling systems" also required for penicillin manufacture, etc.	"Dedicated and self-contained facilities" required, not only for "penicillins," but also for "biological preparations (e.g., from live micro-organisms)" for "certain antibiotics," for "certain hormones" and for "certain cytotoxics." Which antibiotics, hormones, etc. are "certain" and which are not is not made clear.

-- continued

US cGMPs	EC GMP Guide
Specific requirements for the safe disposal of sewage, waste, and rubbish.	No corresponding specific mention.
No specific requirement to prevent entry of unauthorized persons, or to prevent production, storage, and QC areas being used as a right-of-way.	Specific requirement to prevent "entry of unauthorised persons," and to "prevent production, storage and QC areas being used as a right-of-way by persons who do not work in them."
Require "adequate washing facilities ... and clean toilet facilities" to be "easily accessible to working areas."	Requires that "rest and refreshment rooms should be separate from other areas" (3.30), and that facilities for changing and for washing and for toilet purposes should be easily accessible," but "toilets should not directly communicate with production or storage areas" (3.31).
Sec. 211.48 of the UScGMPs (Plumbing) requires the supply, under continuous positive pressure, of potable water, and *appears* to suggest that this is the grade of water to be used in manufacture of drug products.	See below. [a]

---

[a] This question of water quality is not covered in Chapter 3 of the EC GMP Guide, although water *treatment* merits a paragraph in the EC GMP Guide Annex 1 on Sterile Products. Water is, however, covered in considerable detail by the "Note for Guidance on Quality of Water for Pharmaceutical Use," issued by the EC Committee for Proprietary Medicinal Products (CPMP) and the EC Committee for Veterinary Medicinal Products (CVMP) and published by The European Medicines Evaluation Agency (EMEA) in October 2001 (in operation from June 1, 2002). In line with the European Pharmacopoeia, this guidance note distinguishes four grades of water: potable water, purified water, highly purified water, and water for injections. In this guidance note potable water is not considered suitable for use as an ingredient of pharmaceutical/medicinal products, but only for the earlier stages in the manufacture of some active pharmaceutical ingredients (APIs) and as an initial rinse in the cleaning and rinsing of equipment

Although the most important single factor in the assurance of the quality of pharmaceutical/medicinal products is the quality of the people who manufacture them, the premises in which they are manufactured will also have an important bearing on the quality of those products.

A manufacturing facility is a building into which are fed:

1. Raw and packaging materials (or part processed products)
2. People ready to work
3. Services (air, heat, light, power, water, etc., plus any additional support systems for 2 above)

From it will emerge:

1. Finished products (or part processed products)
2. People leaving after their workday
3. Waste, scrap, rubbish, and effluent

A primary consideration, therefore, for the siting and building of the facility is that it must be possible (and, preferably, conveniently possible) to feed materials, people, and services to the site, and then to distribute products issuing from it and to dispose of waste, effluent, etc. Thus, factors influencing the selection of a location for the construction of a manufacturing factory include:

- Ease of access *to* the site for and by:
    - People (i.e., proximity to centers of human habitation, and thus of labor availability, is an important factor)
    - Materials suppliers
    - Services (water, electricity, gas, etc.)
- Climate — prevailing wind direction (and thus the potential for airborne contamination), extremes of temperature (which can bear upon product stability), and rain, snow, and fog, which will affect ease of access
- Local building restrictions, and restraints on use and disposal of toxic, flammable, or explosive materials
- Local fire safety regulations
- Availability of development grants, which may on the one hand mean that more money can be spent on building and equipping the factory to high quality standards, yet on the other may be negated by lack of availability of suitable labor

Within the facility there will be various flow-patterns. These flows will be principally of materials and products, and of personnel. Materials will be received, held pending test, released for use, held in store, dispensed for manufacture, and processed into products that are then packaged, tested, and held in quarantine pending release, and then stored pending distribution.

Working along with material — and product — flow patterns, and indeed allowing or causing them to happen, are personnel-flows, as people arrive for work, change into suitable protective clothing, carry out work, take breaks, change back to outdoor clothes, and leave for home.

In addition to the material/product and personnel flows there will be flows of air of differing qualities (plain, conditioned, filtered), the flow of various services through pipework, ducting and conduit, and the disposal-flows of waste, defective or contaminated material, and of rubbish, sewage, and effluent.

The basic factors (all of which have quality implications) bearing upon the design, structure, and layout of a manufacturing facility may be summarized as:

- Location
- Structure
- Internal surface finishes
- Size, scale, and complexity of manufacturing operations

- ▪ Protection (from weather, pests, dust, dirt)
- ▪ Security (*not* just an economic issue. Break-ins can cause contamination and mix-up)
- ▪ Space — sufficient for *orderly* manufacture and storage and to avoid congestion and chaos
- ▪ Internal layout — smooth work-flows, that are (ideally):
  - ▪ Unidirectional, with
  - ▪ Minimum of crossing-over and with
  - ▪ Minimum of backtrack
    (All to reduce the potential for contamination and mix-up)
- ▪ Segregation of different types of operations and products
- ▪ Grouping together of similar operations and products
- ▪ Lighting, heating, ventilation
- ▪ Installation of services and fittings
- ▪ Drains and waste disposal
- ▪ Buildings maintenance

Other significant factors will be the company's marketing strategies, and its inventory and physical distribution policies.

LOCATION

Many established manufacturing facilities are situated where they are for historical, and long-forgotten reasons. Sites for new factories have been selected for many different, and not always the best, reasons. Factors truly relevant are:

1. Is the site suitable for the erection of a building of the size, shape, and height proposed? Will the existing terrain allow the insertion of foundations, which will support such a structure?
2. Is the site of sufficient area to accommodate not only the building, but also access roads, parking areas, hard standing for delivery and dispatch vehicles and, perhaps, a certain amount of pleasing external landscaping and planting?
3. Do national and local regulations permit a building of the size, type, and shape proposed?
4. What are the risks of water damage, flooding, pollution, pest or /vermin infestation, or contamination and/or objectionable odours from other nearby activities? What control (e.g., via the local authority) will the manufacturer have over any possible future development of such activities?
5. Will it be possible to attract suitable staff?
6. Will local and personal transport allow convenient staff-access to the site?
7. The convenience and economics of getting materials to the site, and distributing products from it.
8. The logistics and geographical relationships between the site and any other company-owned facilities, its subsidiaries, warehousing, and distribution agents, wholesale and retail outlets.

9. Availability of services — water, power, electricity, fuel oil, telecommunications, waste, and effluent disposal.
10. Potential for future expansion.

Other relevant factors might include altitude (at high altitudes physical characteristics, e.g., boiling point, of some materials are altered), climate (extremes of temperature can affect products and materials, as may excessive rain or flooding, snow, and fog — or at least they may necessitate more extensive ventilation and temperature control systems), prevailing wind (risk of airborne pollution), local noise, and tax incentives or development grants (possible extra money available for higher standard plant and equipment).

All these considerations (although some may seem to be more of an economic and commercial nature) do indeed bear upon product quality, directly or indirectly. Some are obvious. What if the local terrain, building regulations, and area available do not permit the construction of the splendidly quality-preserving or -enhancing building that has been conceptually planned? It is no good constructing a factory in an area where the required numbers, and quality, of workers cannot be recruited. Crucial consideration must be given to the availability of staff able to rapidly grasp and respond to the special disciplines, and indeed the culture, appropriate to the manufacture of pharmaceuticals. Even noise can affect quality, at least indirectly, by affecting the powers of concentration of operators on the matter at hand. Anything that causes chaos and confusion (late arrival of ingredients and packaging materials, "panic production," overstuffed stores, trying to expand or re-layout buildings that resist it) can result in error, mix-up, and contamination.

SITE SECURITY

Protection of the site against intrusion, theft, and vandalism is not just an economic issue. It is also an important quality issue. Break-ins, for whatever purpose, can result in mix-ups and contamination, which may escape immediate detection. All but the smallest facilities should, ideally, be surrounded by a secure perimeter fence. Access to the site (staff, contractors, visitors) must be strictly by authorized persons only, using some form of personal identification badges.

STRUCTURE AND FINISHES

There are many ways to build factories. The most widely adopted approach is based on a steel or reinforced concrete frame with fill-in external walls of brick, building block, coated steel panels, or combinations of these. Such structures provide a degree of flexibility in arranging internal non-load-bearing walls, which can be constructed from structural blocks rendered, made smooth, and finished with a hard drying, smooth, impervious surface finish, or from prefabricated partition panels of various types. Internal wall surfaces (and indeed all surfaces in processing areas) should be impervious, nonporous, nonshedding, and be free of cracks, dirt retaining holes, and flaking paint. They should be washable and able to resist repeated applications of cleaning and disinfecting agents. Internally, there should be no recesses that cannot be cleaned, and a minimum of projecting ledges, shelves, fixtures, fittings, and the like.

Services, pipework, ducting, and conduit should be installed so as not to create unclean able dust-traps, preferably within walls or above ceiling voids. If pipework etc. must pass through walls, it should be thoroughly sealed-in on both sides.

In certain highly critical processing areas (e.g., sterile products), walls need to be smoothly coved to floors and ceilings.

Floors should be even-surfaced, be free from cracks, and allow for easy cleaning and removal of any spillages. They should conform to the requirements similar to those indicated for walls above. They need also to be tough. Expansion joints should be flush-sealed with a suitably resilient compound.

Where drains or drainage gullies are installed, they should be easily cleanable (and clean) and trapped to prevent reflux. Floors should fall to drains, not vice versa.

Overhead ducts, pipes, and roof joists should be avoided. A common approach is to employ suspended or false ceilings, with the void above the ceiling being used for pipework and services. Ceiling panels or tiles should be close fitting and sealed or clamped together at joints. The entire ceiling should have a smooth impervious surface, easy to keep clean. Acoustic tiles are generally inappropriate in processing areas, except perhaps where product is not exposed.

Lighting should be fitted flush, or suspended from the ceiling in such a manner that the fittings may be kept clean.

Doors and window-frames should all have a smooth, hard, impervious finish, and should close tightly. Window and door frames should be fitted flush, at least on sides facing inward to processing areas. (That is, and for example, a door or window between a transit corridor and a processing room may need only to be flush fitted — no window ledges, etc. — on the processing side. A door or window fitted between two processing rooms should be flush fitted on both sides).

Any windows from production areas to the outside should be tightly sealed and not normally openable.

Doors, except emergency exits, should not open directly from production areas to the outside world. Any emergency exit doors should be kept shut and sealed, and designed so as to be openable only when emergency demands.

Despite the space-saving advantages, sliding doors should be avoided because of the difficulty of maintaining the sliding gear in a clean condition.

BASIC DESIGN AND LAYOUT

As previously stated, fundamental to good pharmaceutical factory design are the concepts of:

- Segregation of different types of operation
- Grouping together of related types of activity or product
- Smooth, mainly unidirectional, flows of materials (starting, packaging, and in-process), intermediates and products, with minimal crossing over of work flows, or backtracking. Similar considerations apply also to personnel flows.

In any but the simplest facility, manufacturing only one product (or a small range of closely similar products), it is perhaps not possible to achieve ideal segregation, grouping, and flow, but the objective is clear. It is, essentially, to avoid mix-up and contamination and additionally to create and maintain an orderly, efficient working environment in which supervision, rapid appraisal of just what is going on, and communication are all facilitated. In addition to the Quality aspects, there are, of course, the economic benefits of more efficient production and higher productivity. The current trend among the larger multi-product companies to "rationalize" manufacturing sites to single product-type (e.g., tablets and capsules, or liquids, only) factories is doubtless driven by quality, as well as economic considerations.

The immediate surrounds of the building should be such that they may be, and are, maintained in a clean, tidy, and orderly condition. Around the entire perimeter there should be a width of concrete, tarmacadam or similar material, which should fall to drains and prevent water seepage into the building. All outside walls should be sealed to prevent entry of dust, damp, and insects through cracks and gaps, as should all cutouts for windows, piping, and duct work. All external loading and unloading points should be provided with protection from the weather.

The building, and the site as a whole, should be secure, with access restricted to authorized personnel at specified times only. This, too, has quality as well as economic implications. Intruders, even if their intent is no more than petty pilfering, could be the cause of contamination and mix-up. The access of vermin, birds, insects, and pests should also be prevented.

The general external appearance ought not be neglected. It might be argued that aesthetics have nothing to do with product quality, but a good, clean, attractive external (and internal) appearance does help to encourage desirable operator attitudes.

In practice, *total* realization of the ideal layout is rarely found. Nature of the site, local conditions, and availability and placement of services all tend to dictate modifications, and as business and product range change and develop, and premises expand, supplementary flows tend to be grafted onto the original pattern. Nevertheless, the aim should always be to remain as close to the ideal as possible.

In fact, with the current general trend among larger companies to rationalize toward factories producing just a single product type, the achievement of ideal factory layouts is probably becoming somewhat easier.

Naturally, internal building requirements vary according to the nature of the operations carried out or type of product produced within the various departments, sections, or rooms. Not surprisingly, there are rather special requirements for the design, finishes, layout, and environmental control of premises for sterile products manufacture. These are matters that we will come to later (Chapters 13 and Chapter 14).

PLANT SERVICES, SYSTEMS, AND UTILITIES

A manufacturing facility, built and finished as designed, still requires various other inputs, in addition to people, equipment, and materials, before the manufacture of products can begin. These can be referred to collectively as "plant services, systems, and utilities."

First and foremost, air must be provided for those working in the factory to breath. That air will need to be at a comfortable temperature so the workers' innate enthusiasm for work will remain unimpaired. Providing ventilation by simply opening the windows and doors is not acceptable. Some form of forced, conditioned air supply is required, and this in turn has further implications for the control of potential airborne contamination. Attention is also necessary to the other needs of the factory and the people in it.

Water is needed — for drinking and washing, cleaning, rinsing, producing steam — as a major ingredient.

Services are needed to permit equipment to function, as are systems for dust control and collection, and systems for the disposal of waste and effluent. Cooling systems may also be required.

So, plant services, systems and utilities requirements include the following:

- Heating, ventilation, and air conditioning (HVAC)
- Lighting
- Water (of various grades)
- Steam
- Compressed air
- Various other gases
- Vacuum
- Electricity
- Cooling systems
- Dust control and collection systems
- Effluent and waste disposal systems and drainage
- Bulk solvent and other bulk liquid supply systems
- Lubrication services

To this list may be added the provision of cloakroom, toilet, canteen, and communication systems.

It would be a mistake to think that these things are all "just engineering" issues. All have strong quality implications. If these services are not adequately provided, and operating satisfactorily, then the quality of the product will suffer, either directly or indirectly, and as a further consequence, patients may also suffer as a result of failure of product quality. Furthermore, if some of these services are inadequately provided, factory personnel may suffer (from discomfort or inconvenience or worse) and if they suffer, so will product quality.

HVAC

"Natural" ventilation (via doors and windows) is not acceptable because of the risk of product contamination from the outside world (particulate matter, dust, dirt, microorganisms, insects, etc.). Control of humidity is also important for a number of products, particularly effervescent products. Noneffervescent tablets, gelatin capsules, and tablet coatings, packages and packaging materials, and various other medicinal and similar products can also be adversely affected by humidity, which can also encourage microbial growth.

Windows from production areas to the outside world should thus normally remain shut, and preferably not be openable. External doors should be air locked, or only openable in an emergency.

Therefore, some form of forced, conditioned, usually filtered, air supply is required. The nature and quality of that air supply will depend on the nature of the process being performed and the products produced in the area concerned. Perhaps the most critical requirements are for the manufacture of sterile products.

Air systems can, therefore, have a positive effect on product quality if they are properly designed, installed, operated, and maintained. If not, they can have an adverse effect.

Care must be take to ensure that an air supply system is doing what it is supposed to do. That is, supplying air of a higher quality than the outside environmental air at an appropriate temperature and level of humidity. The purpose is immediately defeated if the system is drawing air from a source that contains contaminants of a nature, and at a level, that the conditioning and filtration systems are not adequate to deal with. Siting of the air intake is, therefore, critical.

Maintenance of the system to ensure that it continues to operate to designed standards is also critical. Damaged or holed air ducts can cause more contamination than no system at all, by drawing in contaminated air and dust from service voids, lagging material, etc., as a result of venturi effects. Air filters should be functioning to the standard required and as specified. If they are damaged, they too can become a source of contamination. As they become blocked and their efficiency decreases, they need to be changed. The system should be designed, and the changing operation carried out, so that changing filters does not have the effect of merely spreading around all the dust, etc., that has been collected on the filter over a period of time.

Air supply ducts must be installed, preferably in "voids" or above false ceilings, but, in any event, where it does not create uncleanable surfaces or recesses. It is essential that HVAC systems are subject to a formalized program of planned preventative maintenance (PPM). This will need close liaison between the manufacturing and maintenance departments to ensure that production does not continue in ignorance of a "down" HVAC system.

LIGHTING

Lighting levels should be adequate to permit operators to do their work properly, accurately, and attentively. Too little light may cause operators to miss things they should be noticing, or to work without the necessary precision. It will cause eye strain and fatigue, which in turn can have indirect adverse effects on product quality. Too bright light, producing glare and dazzle, can also be fatiguing and have similar ill effects on quality.

Although daylight is preferable from a number of aspects, it needs to be noted that a number of pharmaceutical products and materials are affected by UV light. The design and layout of a modern pharmaceutical factory also usually make artificial lighting inevitable. It should be installed so as not to create uncleanable dust traps, e.g., preferably flush-fitted to the ceiling, or with smooth easily accessible and cleanable surfaces.

WATER

Water is, overall, by far the biggest single usage item in pharmaceutical manufacturing. It is used:

- As an *ingredient* (many liquid products consist mainly of water)
- As an *in-process material* used at some stage in manufacture, but which does not appear in the final product. (An example is the water that is used in granulating and coating solutions in tablet manufacture. Little of it remains in the end-product — except, of course, any non-volatile impurities, or even microorganisms it may have contained)
- For drinking
- For washing (people, floors, walls, equipment, containers)
- For rinsing
- For cooling
- As a source of steam

Basic considerations, all of which have significant product-quality implications, are:

- The quality of the feed water to the plant
- The uses to which the water will be put
- The standards to which the waters used for different purposes must comply
- The water treatment methods that must be applied to ensure that water used for various purposes complies with the appropriate standard
- The design and installation of water treatment systems
- The control and monitoring of the quality of the output water

Put simply, and rather obviously, the nature and extent of the water treatment will depend on the quality of what is available as source water and on what is needed as the output from the treatment process.

Water may be originally obtained from a number of sources. Water from wells or bore-holes, given suitable treatment, has been used to manufacture pharmaceuticals. In many countries, the most usual source is normal mains, or town, water of potable (drinkable) quality.

For pharmaceutical purposes, it may be considered that there are three basic grades of water:

- Potable Water
- Purified Water
- Water for Injections

As noted earlier, a recent EC Guidance note (2001), following the lead of the European Pharmacopoeia, distinguishes a fourth grade of water — Highly Purified Water. (See also Santora and Mani.[1])

Potable Water

Potable water, quite simply, is water that is fit and safe to drink — that is, it is the stuff that comes from the mains and out of the faucets (or taps).

As far as it is possible to ascertain, no detailed monograph on potable water appears in any pharmacopoeia. Some pharmacopoeias make reference to it in terms such as "suitable water freshly drawn from the public supply" and "palatable and safe to drink," but no monograph. An international standard was published by the World Health Organization (WHO) in 1971 and an EC Guideline (1978) set standards for appearance, pH, limits for toxic substances and microbial contamination, and so on, but the precise definition of quality standards for water tends to vary with location.

To produce potable water, the primary source material (from rivers, lakes, wells, etc.) needs some form of treatment (flocculation, settling, filtration, chlorination, etc.). Potable water can contain a range of dissolved organic and inorganic substances, suspended colloidal matter, and relatively low levels of microorganisms. Although it has been suggested that potable water can be used as an ingredient in the manufacture of some nonsterile pharmaceutical products (creams, ointments, and tablet granulations for example), informed opinion holds that potable water should only be used for drinking, personal washing, and also for the initial washing and rinsing of equipment and containers, provided (in the case of surfaces in contact with product) this is followed by rinsing with either purified water or water for injections, as appropriate and relevant.

Although it may seem strange that water that is fit to drink is not considered fit to be used as an ingredient of pharmaceutical products, this is indeed generally considered to be so. "Mains water" (or "city water") usually (dependent on location and original source) contains small, but not insignificant, quantities of dissolved, and possibly suspended, impurities. Some of these, although harmless to normal fit people when swallowed, can cause harm to those who are ill and weak or when administered by other routes than by mouth. They can also adversely affect formulations, for example by causing precipitation, or through ionic solutes disturbing the delicate balance of some emulsions. Potable water will also contain at least some level of (ever-increasing) microorganisms. These can cause infection in patients and break down some formulations (e.g., emulsions).

Particular care is necessary when potable water (and indeed any water) is held in a storage tank, where microbial growth could be prolific.

Purified Water

Purified water is potable water that has been treated so as to conform with defined official standards. That is, for example, the monographs that appear in the United States Pharmacopoeia (USP), the European Pharmacopoeia (EP), and the British Pharmacopoeia (BP). These monographs set down tests and limits for *chemical* purity, based on specific limit tests, and more general techniques such as electrical conductivity and residue on evaporation, but do not specify allowable microbial levels.

It is usual, however, for manufacturers to define their own in-house limits, with a limit of not more than 100 organisms per ml being common. Commonly adopted "warning limits" vary from 10 to 50 organisms per ml. Often, the complete absence of particular types, or groups, of organisms (e.g., coliforms, pseudomonads) is specified.

Purified water is produced from potable water by distillation, ion exchange, reverse osmosis or other suitable means.

It is used for "general" manufacturing purposes (that is, generally as an ingredient of nonsterile and certainly *not* of injectable products) and as a final rinse for washing of containers and other primary packaging components, and in final rinses when cleaning equipment (in both cases when these are intended to be used only for nonsterile products).

Water for Injections

Here it is necessary to draw an important distinction — between water for injections, which is in bulk (e.g., in a bulk holding tank or circulating in a ring-main distribution system) and water for injections that has been sterilized and is in fact sterile.

Various pharmacopoeias make this distinction in different ways. The EP/BP defines water for injections as, in effect, water that complies with the requirements for purified water, with the additional requirement of not more than 0.25 IU of bacterial endotoxin.

The EP/BP further distinguishes two subgrades of water for injection: "water for injections in bulk" (in effect, the water that is used in the preparation of bulk solutions intended ultimately for injection, and which will be sterilized at a later stage in the process) and "sterilized water for injections." Sterilized water for injections is defined as water for injections that has been filled and sealed into "suitable containers" and then "sterilized by heat in conditions which ensure that the product still complies with the test for bacterial endotoxins." Thus, this definition is specifically directed at the water in sealed ampoules or vials, which is used to dissolve or suspend sterile powders immediately prior to injection.

Note that the EP/BP requires that water for injections should be produced by distillation. Some other countries permit the use of reverse osmosis. The USP also distinguishes between "water for injection," "sterile water for injection," "sterile water for irrigation" and "bacteriostatic water for irrigation."

Water for injections (not necessarily sterilized if the product is later to be sterilized, but most certainly sterilized, and *sterile*, if it is not) is used for the manufacture of injections, ophthalmic products, and other sterile products intended for critical clinical applications. Here we encounter a matter of fundamental importance, and although it will be encountered again when we turn later to consider sterile production in more detail, it is so important that it is well worth stressing now:

Although "water for injections in bulk" is not required to be sterile, this does not mean that it may contain an abundance of organisms, and it is usual for manufacturers to set their own in-house limits. Opinions tend to vary on what these should be, but not more than 500 cfu (colony forming units) per liter, with not more than 100 cfu per liter as a "warning limit" and a complete absence of specified organisms (e.g., coliforms), is commonly suggested.

Other Waters

Other waters include water used for cooling and as boiler feed for the production of steam.

Cooling water used for cooling equipment does not have any defined standard — nor does it need any, provided that it is retained within a sealed system and does not come into contact with product or the production environment. It has been suggested that it is prudent to add chemicals to such water in order to minimize microbial growth. However, in the accidental event of contact with product or environment, it would need to be recognized that, while the microbial risk may have been reduced, the chemical contamination risk has been increased.

The water used following the sterilization cycle in some types of autoclaves to cool the sterilized load is a different matter altogether. It should be sterilized water for injections quality to protect against the potentially hazardous consequences of water (or residues from it) remaining on the load or, say, entering a vial or ampoule through a faulty seal or a crack.

The quality of water used to feed boilers is, from a pharmaceutical point of view, of no importance — *provided* there will be no contact (direct or indirect) between the steam produced by the boiler and the products manufactured, or with the contact surfaces of the equipment used to manufacture them. Where the steam *will* come into contact with products, containers, or the contact surfaces of manufacturing equipment, the water used to produce it should not contain volatile additives like amines or hydrazines. If the steam is intended to be used for sterilization (e.g., in autoclaving, "live-steaming," or sterilize in place [SIP]) then it must be "clean steam" (or "pure steam"), produced from deionized (or reverse-osmosis water) water by a well-designed clean steam generator, which will yield a condensate that complies with the requirements for water for injections.

Water Treatment and Supply Systems

The EC GMP Guide Annex 1 on sterile products (and the statement is relevant to water used for other purposes) states (paragraph 35):

> Water treatment and distribution plants should be designed, constructed and maintained so as to ensure the reliable production of water of an appropriate quality. They should not be operated beyond their designed capacity. Water for injection should be produced, stored, and distributed in a manner which prevents microbial growth, for example by constant circulation at a temperature above 70°C .

(Many would argue that the temperature at which water should be held and circulated should be not less than 80°C.)

In the treatment of water to produce the required quality grade(s), it is not merely a question of the correct selection, installation, and maintenance of the major items of equipment (e.g., stills, de-ionizers). It is a matter of viewing the whole water production, supply, and distribution process as an integrated system, and controlling it and monitoring it to ensure the consistent supply of water of the required quality. This requires consideration of the source-water arriving at the plant, its nature and quality, and what sort of settling, coarse filtration, scavenging, or other pretreatment it may require; through to the deionization equipment, its installation, monitoring, maintenance and regeneration; on to the still itself, its installation, control and monitoring, via any holding vessel (with provision for elevated temperature storage, and with vent valves protected by hydrophobic bacteria-retentive filters) and the recirculation system; to final delivery to production areas.

The overall concept should be what has been termed a "sanitary design" — a system that aims at minimizing microbial growth, at minimizing chemical and particulate contamination arising from the system itself, and that permits cleaning and sterilization "in place." Except in the very smallest systems, where water is taken direct from the still as required, water should be distributed to the required production outlets via holding tank(s) and a recirculating loop, in all of which the water (at 80°C) is maintained in constant turbulent motion. Tanks and pipework should be constructed of 316 stainless steel, with internal surfaces (including all welds) highly polished to prevent minipockets of stagnant water where organisms can flourish. The following sources of contamination should be avoided or kept to a minimum:

- Excessive length pipe runs
- Too many valves
- Nonsanitary valves and joints
- Threaded joints
- Dead-legs
- Undrainable loops and bends
- Unprotected vents
- Pumps
- Tanks

It is pointless to install a system that works well just for the first week or so. It needs to be monitored and maintained to ensure that it continues to work well. Here there is a vital need for close cooperation between microbiological quality control which will perform the microbiological monitoring, and the engineers who will need to service and maintain the system to ensure it remains capable of supplying the quality of water required. Care needs to be taken to ensure that in the very act of sampling for microbiological and chemical testing, the system itself is not contaminated.

The following, extracted from EC Guidance Note on the Quality of Water for Pharmaceutical Use (2001) will serve as a summary of the various grades of water and their usage:

4. Requirements of the European Pharmacopoeia

The European Pharmacopoeia provides standards for the following grades of water:

- Water for Injections
- Purified Water
- Highly Purified Water

4.1   **Potable Water** is not covered by a pharmacopoeial monograph but must comply with the regulations on water laid down by the competent authority. Testing should be carried out at the manufacturing site to confirm the quality of the water. Potable water may be used in chemical synthesis and in the early stages of cleaning pharmaceutical manufacturing equipment unless there are specific technical or quality requirements for higher grades of water. It is the prescribed source feed water for the production of pharmacopoeial grade waters.

4.2   **Water for Injections** (WFI) is water for the preparation of medicines for parenteral administration when water is used as a vehicle (water for injections in bulk) and for dissolving or diluting substances or preparations for parenteral administration before use (sterilised water for injections).

Production

Control of the chemical purity of WFI presents few major problems. The critical issue is that of ensuring consistent microbiological quality with respect to removal of bacteria and bacterial endotoxins. Distillation has a long history of reliable performance and can be validated as a unit operation, hence it currently remains the only official method for WFI.

WFI in bulk is obtained from water that complies with the regulation on water intended for human consumption laid down by the competent authority, or from purified water, by distillation in an apparatus of which the parts in contact with the water are of neutral glass, quartz or suitable metal and which is fitted with an effective device to prevent the entrainment of droplets. The correct maintenance of the apparatus is essential. During production and storage, appropriate measures are taken to ensure that the total viable aerobic count is adequately controlled and monitored.

WFI complies with the tests for Purified Water with additional requirements for bacterial endotoxins (not more than 0.25 IU of endotoxin per ml), conductivity and Total Organic Carbon

4.3   **Purified Water** is water for the preparation of medicinal products other than those that require the use of water which is sterile and/or apyrogenic. Purified Water which satisfies the test for endotoxins may be used in the manufacture of dialysis solutions.

Production

Purified Water is prepared by distillation, by ion exchange or by any other suitable method, from water that complies with the regulations on water intended for human consumption laid down by the competent authority.

4.4   **Highly Purified Water** is intended for use in the preparation of products where water of high biological quality is needed, except where Water for Injections is required.

Production

Highly Purified Water is obtained from water that complies with the regulations on water intended for human consumption laid down by the competent authority. Current production methods include, for example, double-pass reverse osmosis coupled with other suitable techniques such as ultrafiltration and deionisation. Highly Purified Water meets the same quality standards as WFI, but the production methods are considered less reliable than distillation and thus it is considered unacceptable for use as WFI."

Steam

Possible uses of steam include:

- General factory heating
- Production process heating (steam-jacketed vessels, heating coils)
- Steam cleaning
- Sterilization (autoclaving, "live-steaming" of vessels and pipes, sterilization in place [SIP])

Where steam is not associated with product manufacture, and does not come into contact with product or manufacturing materials (or with surfaces that will contact product or materials), then, pharmaceutically speaking, the quality of that steam is not particularly relevant. Where there is any such contact, then the steam should be of such a quality that, when condensed, the water thus produced would comply with the requirements for purified water. When used as the sterilizing medium (e.g., in autoclaves, SIP systems) the steam should be clean steam. That is, steam that, when condensed, will form water for injections quality water.

Gases/Compressed Air

Various gases may be used for a variety of purposes, for example, inert gases used as a protective "blanket" or to displace air in an ampoule head-space, as propellants in aerosol products, as sterilants (e.g., ethylene oxide), as a source of flame in glass ampoule sealing.

Any gas that may come into contact with a product (or product contact surfaces), or that is used in the manufacture of a product, must be treated as if it were a raw material and must therefore be subject to standard quality control procedures to ensure that it conforms to predetermined quality standards. A number of gases are used in laboratory test procedures. If these are not of the required or specified quality, then the reliability of the test results may suffer.

Gases supplied in cylinders should be properly color coded in accordance with the relevant national or international standard, and additionally identified as to lot or cylinder number. Cylinders should be stored under cover, without exposure to extremes of temperature. Storage conditions should ensure that their markings remain clearly visible. Pressure gauges should be regularly checked and calibrated. Gas pipelines, from cylinders or from bulk gas storage, should be clearly marked as to contents. It should not be possible to switch pipelines and connections and thus to supply the wrong gas. Dedicated, pin-indexed valves and connections, as (one hopes) used in hospital gas supply lines, should be employed where possible.

Gases (including compressed air) may need to be filtered when supplied to production areas generally. Gases (including compressed air), when supplied to sterile products manufacturing areas (and other controlled environments), will certainly need to be filtered (as close to the point of use as possible) to ensure that they conform to the particulate and microbial standards for the area.

Electricity

Continuity of electricity supply is essential for a number of systems or processes (air supply and extraction, particularly for sterile manufacture; fermentation plants; incubators) and thus backup systems should be available in the event of mains failure. Ideally, there should be automatic changeover and reset from mains to emergency generator supply. Just what the needs and priorities are for emergency backup should be a matter of discussion and agreement between the Production, Engineering, and Quality functions. The actions to be taken on mains failure should be agreed and set down as a Standard Operating Procedure (SOP), with out-of-hours contact telephone numbers of relevant key personnel (Engineering, Production, Security).

Certain equipment (computers, microprocessor control systems, some analytical instruments) may need voltage stabilization in order to operate reliably.

Solvents and Other Bulk Liquids Supplies

Large manufacturing organizations may well receive a number of liquid materials in bulk (solvents, sugar syrup, "liquid glucose," glycerin, etc.), which are pumped from the supplier's delivery tanker to storage tanks. Often it is not a practical proposition to keep the tanker waiting in the yard while full quality testing is carried out, and the material may thus be provisionally accepted and pumped to storage on the basis of passing, perhaps, just one or two tests (including a specific identity test), with full testing to follow later. It is therefore important that (if and as necessary) storage tanks, pipework, and valve systems are installed so that liquids held pending full test cannot be used before formal release by Quality Control. Bulk liquid storage and pipework systems should be installed and maintained so as to prevent mix-up, cross-contamination, and inadvertent switching of pipelines. All pipes should be marked clearly to identify their contents and the direction of flow.

Lubricants and Lubrication

To say that moving parts need lubrication is to state the obvious. What should be just as obvious, but does not always seem to be so, is that there is a world of difference between the lubrication of a piece of equipment that is used to fabricate, say, machine parts, and one that is used to manufacture pharmaceuticals, etc. Lubricants should not be allowed to come into contact with starting materials, products, or product containers. Care needs to be taken to avoid hazarding product quality through contamination from leaking seals, lubricant drips, and the like.

Gland packing materials should be inert and nonreactive, and, wherever possible, food-grade lubricants should be used. All those concerned with lubrication should be aware that pharmaceutical manufacturing is a different world — lubricants are potentially serious contaminants.

Waste Disposal and Drainage

Careful control of waste material and its disposal is important for a number of reasons. If reject or scrap product or material is allowed to accumulate in an uncontrolled fashion, it can represent a cross-contamination hazard. If it is a vehicle of microbial growth, then it could also become a viable contamination hazard.

Clearly, product scrap or waste can also represent an environmental and public toxic hazard and must therefore be disposed of in accordance with all national and local legal requirements. Chemical and solvent wastes may require agreement with the local authority as to their disposal, and any necessary pretreatment. Emitted gases and vapors may also need treatment, not only to avoid environmental pollution, but also to prevent product contamination.

Similar care and control is necessary over the disposal of scrap, rejected, or discarded packaging materials. All waste printed packaging materials (printed containers, tubes, labels, cartons, leaflets, etc.) must be kept under secure control and destroyed, under close supervision, as soon as possible, to prevent unauthorized (either inadvertent or deliberate) reuse.

Drains (internal and external) should be sufficient in size, number, and location to do the job intended. They should not be, or allowed to become, vehicles of contamination. They should have trapped gullies, with air breaks as necessary, to prevent back-siphonage. Internally, open drainage channels should be avoided, if possible. If they are necessary, they should be shallow to facilitate cleaning and disinfection. There should be written cleaning and disinfection procedures for internal drains. These procedures should be strictly implemented.

In the critical areas of sterile products manufacturing facilities, drains should not be installed.

Cloakroom, Toilet, Canteen, and Communication Facilities

All these have product-quality implications. If operators are not comfortable (e.g., hungry, thirsty, need to use the toilet), their work will suffer — and so will product quality. They should not eat or drink in production areas. They need to be properly dressed in the correct standard protective clothing. They need to observe hygienic personal practices. Supervisory staff, at least, will need to be able to communicate with each other over distances; hence, the quality-significance of the provision of good, standard cloakrooms, toilets (*not* opening directly to production areas), canteens, and communication systems. The good, or alternatively bad, effects on operators' morale, motivation, and attitude can be considerable. No management can reasonably expect operators to respect the ideals of high quality standards when they provide them only with ugly, wretched, and inadequate canteens and dark, dismal, and dirty cloakrooms and toilets.

REFERENCES

1. Santora, M. and Mani, C., What water for pharmaceutical use? *Eur J. Pharm & Parent Sci.*, 8(1), 15–20, 2003.

4

CONTAMINATION AND CONTAMINATION CONTROL

This chapter marks a break in the pattern, so far established in this book, of basing each chapter on a subpart of the UScGMPs, with which the corresponding section(s) of the EC GMP Guide are compared. In the earlier chapters we have encountered a number of requirements, in different subparts and sections, that are concerned with contamination control. The aim of this, more general, chapter is to consider the rationale of those requirements and to discuss compliance measures.

An issue that dominates thinking about pharmaceutical facillities, their design and layout, and how they are operated, to an extent matched in few other industries, is the need:

    a. To avoid contamination of materials and products, either one by another ("cross contamination") or by extraneous matter
    b. To prevent mix-ups of ingredients ("starting materials" or "components"), products, and packaging materials

These dominant, almost overriding, concerns bear powerfully upon the issues of siting, design, structure, layout, surface finish, ventilation, and waste disposal.

CONTAMINATION — TYPES AND SOURCES

The word "contamination" covers a range of different substances. Simply, it is stuff in the wrong place, or where it should not be. The various possible forms of contamination can be classified into two main types:

- Living (or Viable)
- Nonliving (or Nonviable)

The hazard that any contaminant represents will depend on its precise nature and where it is found.

Living, or Viable, Contamination

While it would be quite reasonable to regard a frog, or a shark, swimming about in a tank of liquid product as a viable contaminant, the term is generally taken to refer not to such macroorganisms, but to microorganisms — such things as bacteria, molds and fungi, yeasts, and viruses.

A major problem with microorganisms is that there can be many millions of them present on a surface, or in a liquid, without there being any obvious indication that they are present. As an illustration, picture a one-liter bottle or bag of intravenous infusion fluid. When made, it should look clear and bright. Even if only a few microorganisms are present at first, under the right conditions (that is, right conditions for them) they could grow and multiply very rapidly. Yet even if there are 1 million microorganisms present in every ml (that is, 1000 million in the liter bottle), then only the very keenest eye will be able to detect the very, very faint cloudiness caused in the liquid. For the average pair of eyes to be able to detect just a very faint milkiness, 10 million microorganisms would need to be present per ml, or 10,000,000,000 (ten thousand million) in the whole liter.

Like all living things, microorganisms grow, feed, and reproduce. Many have no built-in means of locomotion, but can be transferred from one place to another by air or liquid currents — or by the movement of a host organism, for example, a person. Some can move themselves about in liquids, or on wet surfaces, by the beating of short cilia or whiplike flagella. Some aerobic microorganisms respire using oxygen in a manner analogous to that of mammals. In contrast, some (the anaerobes) cannot grow at all in the presence of oxygen. On the other hand, most microorganisms can survive without oxygen for quite a long time. Three things they all must have in order to grow and reproduce are moisture, food, and warmth. A lack of these three essentials will not necessarily kill microorganisms; they just will not be able to flourish, grow, and reproduce without them. Many are remarkable survivors under the most trying conditions, but if moisture, food, and warmth are removed (that is, if things are kept dry, clean, and cool or cold), a good step will have been made toward controlling the spread of microorganisms, even if they have not been killed or completely removed. Even extremes of cold will not kill them; it just keeps them under control. That is, it stops them from growing and multiplying.

As for what microorganisms can use for food, although some individual types (or species) are very selective about what they can feed on (for example, a specific sugar or a specific protein), across the range of microorganisms as a whole, they use an amazing variety of substances for food. Obviously, they can live on things like meat, fruit, milk, and bread. Many feed on what might be termed just plain dirt. Some have been known to use the most unlikely things as food — like aero-engine fuel and dilute disinfectants.

Bacteria reproduce by the simple process of each individual dividing itself in half. Under good conditions (that is, when they have moisture food and warmth), they can divide in this way once every 20 minutes. So, in 20 minutes 1 bacterium becomes 2 bacteria, in 40 minutes 4, in an hour 8, and so on. In 12 hours there will be more than 2 million million descendants of the original organism.

Some bacteria can cause disease — from the minor to the very serious — as can some molds, yeasts, and viruses, but by no means are all microorganisms harmful. In fact, in normal circumstances the great majority of them are quite harmless to healthy people. A number are very useful to us. But others are indeed pathogenic — causing diseases, from the most minor illnesses to those that cause death. Others can spoil food and break down things like medicinal and cosmetic creams, lotions, and other liquids. Even microorganisms that are normally harmless can be a danger if administered to people who are already ill, and normally quite innocuous organisms can be lethal if administered to patients, in sufficient quantity, by injection.

Thus, as far as possible, microorganisms need to be kept out of, and off, pharmaceutical products of all types. Products that are intended to be injected, or used in the eye, on open wounds, or inserted into body cavities, tissues, or blood vessels must be sterile, that is, completely free from all living organisms.

Contaminated liquids, intended to be taken by mouth, in addition to possibly being "spoiled" by microorganisms, can also infect a patient swallowing the liquid. Liquids, creams, and ointments, intended for application to the skin surface, if contaminated with microorganisms, can (in addition to the spoilage risk — growing microorganisms can cause breakdown of emulsions) cause skin infections. (Note: (a) Skin diseases are not necessarily just slightly irritating, trivial matters. They can be very serious, even lethal, and (b) a number of the active substances that are used in skin preparations to treat inflammation, rashes, etc. can have the effect of suppressing the normal immune response to bacterial infection. The presence of organisms in such products could thus represent a doubly serious patient hazard.)

Even with dry products taken orally, such as tablets and capsules, there have been cases of serious illness in patients taking products infected with bacteria. Some molds (which can grow on tablets) produce some very toxic substances.

It is important always to remember that people taking medicines are usually doing so because they are already ill, and thus their resistance to infection may well be lower than normal.

It is thus crucially important that manufacturing premises are built, laid out, surface-finished, serviced, maintained, and drained so as to minimize the harboring and proliferation of microorganisms.

Nonliving Contamination

In addition to those living (viable) forms of contamination, there are also the nonliving forms. These can further be classified into two main groups:

■ Active contamination
■ Inert (or inactive) contamination

By "active" is meant chemically, or physiologically active, or having some activity when introduced into the human (or other animal) body. So, contamination can be classified, overall, as follows:

    a. Living (e.g., microorganisms)
    b. Nonliving
        Active
        Inert (or Inactive)

In addition, there is another form of contamination that must be guarded against when making products for injection. These are pyrogens (or bacterial endotoxins — for practical purposes the terms may be regarded as virtually synonymous).

Active Nonliving Contamination

Examples are:

    a. Powder, dust, or crystals from other batches of product, or residues of other solutions, suspensions, or creams "left over" in containers, vessels, or items of equipment that have not been properly cleaned and dried

    b. Ingredient materials left in containers that are then reused without proper cleaning, or powders that have been spilt (for example, in a dispensing operation)

    c. The uncontrolled release of dust, gases, vapors, sprays, or organisms from materials and products in process

This type of contamination tends to be called "cross-contamination." It may also be caused by the dust on the clothes and shoes of people who have been using or weighing bulk chemical substances. Since many chemical and biological substances, like many of the active ingredients used in modern medicinal products, can have very powerful effects in and on the body, even in very small amounts, the potential hazards of this sort of contamination hardly need emphasizing. It would seem obvious that the dangers will vary widely with the nature of the contaminant, and that is largely true. The most dangerous will be highly potent substances that are taken in low doses; things like steroid hormones, cytotoxic substances, and sensitizing agents, such as certain antibiotics. But problems can be caused by other, apparently less potent, substances.

There are many medicinal substances that, although they are well tolerated and safe in normal doses when taken by most people, can cause severe reactions in a sensitive minority. The classic case is penicillin and similar antibiotics. For the majority of patients, there is no problem — but in some, the antibiotics can cause serious reactions, even in minute amounts. Moderate doses of aspirin, taken as recommended, do not cause harm to most people. In a few, it causes marked sensitivity reactions. In general, the safest thing is to assume that somebody, somewhere, could react to traces of nonliving active contaminants.

The presence of any chemical or microbiological contaminant in a medicinal product of such a nature and in such a quantity as may have the potential to adversely affect the health of any patient or impair the therapeutic activity of the product is clearly unacceptable. Particular attention should be paid to the problem of cross-contamination, since even if it is of a nature and at a level unlikely to affect health directly, it may be indicative of unsatisfactory, and potentially dangerous, manufacturing practices.

Many regulatory authorities take a distinctly strict view of cross-contamination. Various suggestions have been made about quantitatively specifying acceptable levels of cross-contamination, based upon the potency, or activity of the contaminant. For example:

> No more than 1/1000th of minimum daily dose of X (the contaminant) in maximum daily dose of Y (the contaminated product).

But there is no universal agreement on this point. Some cynics have argued that acceptable levels vary in accordance with the sensitivity of the analytical methods available to detect them, and that as analytical chemical technology advances, and methods become ever more sensitive, the levels considered to be acceptable get lower. Nevertheless, the dangers of this active contamination are fairly obvious. Perhaps not quite so obvious is the significance of inert nonliving contamination.

Inert Nonliving Contamination

The concern here is with particles, fibers, flakes, dusts, and the like, that do not have any specific chemical or biological activity in or on the body. Contamination by the inert excipients (or "fillers") that are used in some medicinal products could be regarded as "inert cross-contamination." Other commonly used terms are "particulate contamination," "particulate matter," or just "particulates." Particles, particulate matter, particulate contamination, and particulates all mean the same thing — little bits floating about in the air or in liquids, or deposited on surfaces, or in products.

The list of such particles is almost endless and includes: atmospheric and house dust, fine soil, sand, ash, smoke, dandruff, skin flakes, pollen, fibers (from natural and artificial textiles or from paper), flaking paint, powdering plaster or masonry, metal particles from moving machine parts or from drilled or ground metals, rubber or composition particles from belt drives in machines, inert powders from other products, and so on.

There has been much discussion and argument over the dangers (or otherwise) of fine particles contaminating medicinal products, particularly those that are injected or inserted into the body. Clearly, it is important to avoid hard particles in eyedrops, and it is well known that excessive inhalation of a wide range of dusts causes serious lung problems. Particles in products intended for application to body surfaces can abrade the skin and give rise to infections. The dangers of inert particles in products taken by mouth will vary with the nature and level of the contamination, but even small amounts of relatively harmless materials can spoil the look of tablets or liquids that are meant to be clear.

Perhaps surprisingly, some researchers have argued that the injection of inert particles is not as dangerous as it may at first seem. Others have claimed that there are hazards, and that injected particles can block small blood vessels, or pass to the lungs and block the fine bronchial tubes, or lodge in the liver and causes damage, and so on. So, in spite of some of the arguments, it is generally considered necessary to control the level of particles in injections (and other sterile products), and the pharmacopoeias specify the levels that are permitted for particles in certain injections and other sterile products. This in turn means controlling the numbers of particles in the rooms (in the air and on surfaces) in which such products are made.

There is another very good reason for keeping down contamination by inert particles. It is that most airborne bacteria and other microorganisms do not merely float around in the air on their own. They are usually associated with particles. So, if the level of particles in general is controlled, a step has been made toward controlling the level of microorganisms.

One form of possible contamination so far not mentioned includes oil, grease, and other lubricating materials. At best, it can spoil the look of products. At worst, it can be toxic or harbor microorganisms.

Pyrogens

Although it is nonliving, this form of contamination is produced by living organisms that may be present, for example, in water or in a solution. It may also be present on the surfaces of containers, vessels, instruments, devices, and other materials that have been in contact with liquids that contained organisms, or that have been left wet so that microorganisms could develop. These bacteria-produced contaminants are the pyrogens. (or bacterial endotoxins). Pyrogens are polyliposaccharides produced from the outer cell walls of certain gram-negative bacteria. When injected or otherwise inserted into patients, they can cause a rapid rise in body temperature (hence, "pyro" "gen" — "giving rise to heat"), with chill, shivering, vasoconstriction, pupillary dilation, respiratory depression, and an increase in blood pressure. There may also be pains in joints and back, headache, and nausea. In seriously ill patients, the effects of pyrogen can be very serious. Since most sterilization processes do not necessarily remove or destroy pyrogens, it is very important to guard against the development and growth of microorganisms, and the formation of pyrogens. The sterilization process alone cannot (usually) be relied on to ensure that the product is both sterile *and* free from pyrogens. Very careful control over the entire manufacturing cycle is essential to ensure that pyrogen-producing organisms are excluded, or at least kept to a minimum, throughout the process.

Sources of Contamination

Contamination by active chemical substances can be caused by dust and powder, spilt or released during processing, which are floating in the air, or which have settled on surfaces, or in vessels or equipment. Contamination can also arise from residues left over in or on containers, vessels, and equipment that have been used for other products or materials, and that have not subsequently been properly cleaned. Other possible sources include traces of materials that have been used for cleaning and disinfection.

There are many possible sources of what we call inert nonliving contamination (or particulate contamination). For example:

- ■ Buildings
    - ■ Unsealed stonework, brick, mortar, plaster, flaking paint, etc.

- Sawdust, brick chippings, metal filings generated during repair, maintenance, installation, and restructuring
- Raw Materials
  - Ingredients that might not be highly active in themselves can nevertheless be a big source of highly undesirable particles.
- Equipment
  - Dirty equipment
  - Moving machine parts, and belt drives
  - Materials used to lubricate equipment
- General environment, which can deposit a wide range of contaminants such as dust, dirt, soil, sand, smoke, and ash
- Containers, packages, paper, and cardboard, which can all produce considerable amounts of fibrous contamination

Filtered air supplies to the rooms in which products are manufactured, although intended to reduce particulate contamination, can have the reverse effect if the filters are damaged or if the system is not properly maintained. Air extraction systems, if badly designed, installed, and maintained can in fact be a *source* of contamination by withdrawing it from one location only to blow it over another.

There is thus a very wide range of possible sources of contamination. A further major source is people and their clothing.

By the use of rooms with special filtered air supply, quite a high level of control over particles can be achieved. Put people in a room, and it is a different story. The human animal sheds thousands of millions of dead skin cells and fragments per day. This amounts, it has been claimed, to a total weight of somewhere between 5–15 grams per day per person (Association of the Swedish Pharmaceutical Industry, 1972).[1] The more we move, and the more vigorously we move, the greater the shedding becomes. We shed 3 or 4 times more particles when we move about than when we are at rest. Depending on the type of cloth, we also disperse large numbers of fibers from our clothing. This amount also increases as we move about.

Personnel moving from one location to another can carry powders, dusts, fibers (and microorganisms) with them as they go — on their bodies, clothes, and shoes.

The human male sheds approximately 1000 bacteria-carrying particles per minute. People are, indeed, a major source of both living and nonliving contamination.

Sources of Microorganisms

To the question "where do microorganisms come from?" the simple answer is that, like all other organisms (including people), they do not just "happen." They come from parent microorganisms. The big difference is, of course, that in favorable conditions they multiply so much more rapidly than, for example, we do.

Microorganisms are almost everywhere. Some have managed to flourish in strong acids, some in hot springs, and some in certain disinfectants. They are found, by the millions, in or on:

- The environment around us — that is, in the air (indoors and out), on the ground, in the soil, on walls, floors, and surfaces in general — almost everywhere.
- Water — they exist in water from the mains, in rivers, seas, and lakes; in puddles, on wet surfaces and wet floors; in damp surfaces of containers and equipment that has not been properly dried. In general, bacteria will be found in all forms of water, whether it is in large quantities or just light surface films, except water that has been specially sterilized and sealed in against any recontamination.
- Raw materials — used for making products.
- Containers and closures — used for packaging products.

All these are sources of contamination that something can be done about. It is not always easy, but it is possible. One source of contamination, and it is a major one, is rather more difficult to deal with. It is, of course, people. In spite of increasing automation, it is still necessary to involve people in the manufacture of medicinal products, and people generally object to being sterilized (in the microbiological sense), treated with strong disinfectants, or eliminated altogether.

CONTROL OF CONTAMINATION

The control of contamination is a major issue in the design, construction, and layout of a manufacturing facility, and indeed in QA/GMP as a whole. Much depends on people and the way they behave, and the protective clothing they wear. (Note: the reference here is to clothing worn to protect products and materials from contamination by people.) Other important control measures are the application of well-planned and proven cleaning and disinfection procedures. Crucial factors are also the design, structure, surface finishes and layout of factories, the design, installation, and maintenance of equipment, and the design, installation, efficiency, and maintenance of factory services such as ventilation, heating, lighting, and water supply. Proper factory and equipment design and layout can also reduce the risk of what can perhaps be regarded as extreme cases of contamination — the complete mix-up of one product with another, of one ingredient with another, or of one packaging material (especially printed materials) with another. An understanding of these problems is crucial to an appreciation of the quality-influencing aspects of buildings and equipment.

Cleaning and Disinfection

Cleaning is quite simply the removal of dust, dirt, debris, and residues. The more difficult question to answer is "how clean is clean?" and inevitably the answer is "it all depends." One normally expects domestic dwelling places, kitchens, etc. to be clean. It is generally expected, with good reason, that areas used for the manufacture of medicinal and other healthcare products should be cleaner than mere "domestically clean." How *much* cleaner will depend on the nature of the product being manufactured, its intended route of administration, and the potential hazards of any contamination of the product. It is thus reasonable to suggest that the highest conceivable level of cleanliness is required for the manufacture of sterile products intended for injection. On the other hand, a significantly lower level may well be acceptable for, say, the manufacture of foot-dusting powders. This dust, dirt, debris, and residues can arise from a number of sources:

- Airborne dust, dirt, and particles
- Particles, fibers, hairs, and exudates shed by humans
- Spillages and breakages
- Particles from friction in machines
- Oil and grease from lubricated moving parts
- Residues from previous products
- And, just plain dirt (it is one of the fundamental laws of the universe that things that are not regularly cleaned get dirty)

For obvious reasons, areas, surfaces, and equipment in and on which products are made must be kept clean. Dirt, and the microbes that it can harbor, must not get into or on products. But there is another good reason for regularly and scrupulously cleaning away this dirt.

Floors, walls, ceilings, and work surfaces often need to be disinfected. Disinfectants can be inactivated by dirt. Dirt (particularly oily or greasy films, and proteinlike matter) can also protect microorganisms against the action of disinfectants. So, before disinfection, it is important to first *clean* surfaces.

Where gross amounts of dirt are present, it may be necessary to first remove most of it by scrubbing. Then surfaces may be cleaned by the application of a cleaning agent, followed by rinsing. In most normal circumstances all that is needed for the cleaning of floors, walls, and work surfaces is clean water with the addition of detergent, followed by a clean water rinse. The quality of the water used will depend on the nature of the operations carried out on, or near, the surfaces in question. Obviously, it must be microbiologically clean, and it may be appropriate (for example, in sterile product manufacture) for at least the final rinse water to be of high quality "water" for injection standard.

Manufacturing tanks, pipelines, and associated equipment need also to be cleaned and rinsed after use, and before any sterilization that may be necessary. This may be done by simple manual methods, or by Clean in Place (CIP). Here, cleaning is accomplished by automatically pumping cleaning agents and by rinsing liquids, under pressure, around the entire system without necessarily dismantling it.

Disinfection

A disinfectant is a chemical substance, or combination of substances, which, when applied to surfaces will kill microorganisms, with the exception of some bacterial spores. Disinfection is not the same as sterilization, which is the destruction or removal of all microorganisms, and indeed of all organisms generally.

When something is sterilized, if it is done properly, *all* living organisms are destroyed, or removed. A disinfectant is something that cannot quite achieve that. It is not possible to be certain that, by use of chemical solutions alone, *all* living organisms will be destroyed, particularly bacterial spores.

However, usually the aim with walls, floors, ceilings, and work surfaces is not necessarily that they should be rendered sterile, but that they should be as clean as possible, with any microbiological contamination kept to a minimum.

Other words that mean more-or-less the same as "disinfectant" are "germicide," "bactericide," and "biocide." An "antiseptic" is a milder substance that can, for example, be used on skin surfaces and wounds to control or prevent infection without harming the patient. Antiseptics cannot be used to disinfect premises. The term "sanitize" has been used in so many different senses (for example, after the bombing at the Atlanta Olympic Games, the local police chief declared on international TV that the area was now safe as it had been sanitized) that it is virtually devoid of meaning, and therefore its use in any scientific or technological context should be abandoned.

Types of Disinfectant

A wide range of substances are used as disinfectants. They may be single substances, like alcohols or phenols, and there are a number of commercially available mixtures. It is usually best not to make "do it yourself" mixtures. It could be dangerous, and some disinfectants can neutralize each other's activity.

Disinfecting agents vary in the range of their activity and in the concentrations at which they are effective. All have their own special advantages — and disadvantages. For example, alcohols are inflammable, phenols and chlorine compounds can be dangerous and corrosive, iodine compounds can stain some surfaces, and so on. Some examples of disinfectants, with their range of effects, etc. are shown in Table 4.1. This is a very simplified table. The important message is that it cannot be said that all disinfectants are the same. They all have different activities and ranges of effect, and are effective in different concentrations.

Disinfectants should always be used in accordance with instructions and at the right dilution (instructions as given either in the supplier's literature or in company procedures). Since some microorganisms can grow readily in dilute disinfectants, dilutions of disinfectants should not be stored unless they are sterilized. Otherwise, dilutions should be made freshly each time they are needed.

Another, traditional method of disinfecting clean rooms is by fumigating or "gassing," usually with formaldehyde gas, although this can present problems due to the unpleasant, choking, and toxic nature of the gas.

Rotation of Disinfectants

Many manufacturers use different disinfectants over a period of time, on an alternating, or rotating, basis. The reasoning behind this is to prevent the development of disinfectant-resistant strains of microorganisms. Although there have been some discussions in the literature about whether or not it has this effect, alternation of disinfectants remains a recommendation of a number of experts (and some regulatory inspectors), and it is probably a worthwhile practice.

Table 4.1   Disinfectants for Premises — Types and Applications

Substance	Suitable Concentration	Effect on Bacteria	Effect on Spores	Effect on Vegetative Fungi	Advantages	Disadvantages
Ethanol	70%	Good	Fair	Fair	Quick acting; evaporates rapidly, leaving no residues	Limited range of effect; flammable
Phenols	0.5–3%	Excellent	Good	Excellent	Broad range of effect; may be combined with surfactants	Corrosive on some surfaces (including skin)
Formaldehyde		Excellent	Good	Good	Broad range of effect; used for "gassing"	Premises not accessible during treatment; can be corrosive; short- and long-term human toxicity problems
Isopropanol	70–90%	Good	Good	Good	Quick acting; evaporates, leaving no residues	Not the most effective
Iodine and iodophors	75–150 ppm	Excellent	Good	Excellent	Quick acting; effective in low concentrations	Can be corrosive; stains some surfaces
Chlorine compounds (hypochlorite, chloramines, etc.)	1–4%	Excellent	Good	Excellent	Broad range of effect	Corrosive
Quaternary ammonium compounds	1–5%	Good	Fair	Fair	Some cleaning effect; odorless	Limited effect; inactivated by soap detergents

Cleaning and Disinfection in Processing Areas

Routine cleaning and disinfection in clean rooms and other processing areas should be regularly carried out in accordance with an established program, following a standard written procedure. That is, cleaning and disinfection is not something to be done just when it seems like a good idea, or when time permits.

Written programmers and procedures will, naturally, vary in detail from manufacturer to manufacturer, and in accordance with the type of product being manufactured. Some important general points on cleaning and disinfection of rooms, areas, and surfaces may be set out as follows:

1. There should be an approved written program and procedure, which must always be followed exactly.
2. It is necessary to *clean* thoroughly first, before disinfecting.
3. It is important to ensure that the cleaning and disinfecting process does not, in fact, create more contamination.
4. All cleaning and disinfecting agents and materials should themselves be clean and not shed fibers or particles. (It is not possible to clean, using muddy water and dirty, hairy cloths.)
5. Cleaning implements and wiping cloths, having been applied to a surface, should not be rewetted by direct return to the container of cleaning or disinfecting agent, but first rinsed (and squeezed-out) in a second bucket of clean water.
6. Nonshedding materials should be used for wiping surfaces, and dry, dust-creating brushes should not be used. If it is necessary to remove significant quantities of powdery materials, then wet or vacuum methods are preferable.
7. All cleaning and disinfection of a room should start at the part of the room furthest from the entrance, otherwise there is a danger of the cleaner "painting himself or herself into a corner" and having to cross the cleaned area in order to get out.
8. When cleaning walls and other vertical surfaces, work should always start at the top and work down — again, to avoid recontamination of parts already cleaned and disinfected.
9. It is vital that the right cleaning and disinfecting agents are used, in the right dilutions, as directed in the company's written procedure. Remember, dilutions of disinfectants should be made up fresh, in clean containers. They should not be stored for later use unless they are sterilized.
10. All cleaning equipment and implements must themselves be thoroughly cleaned after use and stored in a clean, dry condition.
11. All spilt materials (liquids or powders, or breakages) should be cleaned up in a way that will minimize the possibility of creating further contamination. Again, dry brushing should be avoided and wet or vacuum methods employed. If there is a risk of microbial contamination, the cleaned-up area or surface should then be disinfected. Any spilled material that represents a microbiological hazard should be placed in a container, immersed in disinfectant, covered, and removed from the room.

REFERENCES

1. Association of the Swedish Pharmaceutical Industry, *Hygiene Recommendations,* Stockholm, 1972.

5

EQUIPMENT

REGULATORY REQUIREMENTS

US cGMPs

Subpart D — Equipment

Sec. 211.63 Equipment design, size, and location

Equipment used in the manufacture, processing, packing, or holding of a drug product shall be of appropriate design, adequate size, and suitably located to facilitate operations for its intended use and for its cleaning and maintenance.

Sec. 211.65 Equipment construction

(a) Equipment shall be constructed so that surfaces that contact components, in-process materials, or drug products shall not be reactive, additive, or absorptive so as to alter the safety, identity, strength, quality, or purity of the drug product beyond the official or other established requirements.

(b) Any substances required for operation, such as lubricants or coolants, shall not come into contact with components, drug product containers, closures, in-process materials, or drug products so as to alter the safety, identity, strength, quality, or purity of the drug product beyond the official or other established requirements.

Sec. 211.67 Equipment cleaning and maintenance

(a) Equipment and utensils shall be cleaned, maintained, and sanitized at appropriate intervals to prevent malfunctions or contamination that would alter the safety, identity, strength, quality, or purity of the drug product beyond the official or other established requirements.

(b) Written procedures shall be established and followed for cleaning and maintenance of equipment, including utensils, used in the manufacture, processing, packing, or holding of a drug product. These procedures shall include, but are not necessarily limited to, the following:

(1) Assignment of responsibility for cleaning and maintaining equipment;

(2) Maintenance and cleaning schedules, including, where appropriate, sanitizing schedules;

(3) A description in sufficient detail of the methods, equipment, and materials used in cleaning and maintenance operations, and the methods of disassembling and reassembling equipment as necessary to assure proper cleaning and maintenance;

(4) Removal or obliteration of previous batch identification;

(5) Protection of clean equipment from contamination prior to use;

(6) Inspection of equipment for cleanliness immediately before use.

(c) Records shall be kept of maintenance, cleaning, sanitizing, and inspection as specified in Secs. 211.180 and 211.182.

Sec. 211.68 Automatic, mechanical, and electronic equipment

(a) Automatic, mechanical, or electronic equipment or other types of equipment, including computers, or related systems that will perform a function satisfactorily, may be used in the manufacture, processing, packing, and holding of a drug product. If such equipment is so used, it shall be routinely calibrated, inspected, or checked according to a written program designed to assure proper performance. Written records of those calibration checks and inspections shall be maintained.

(b) Appropriate controls shall be exercised over computer or related systems to assure that changes in master production and control records or other records are instituted only by authorized personnel. Input to and output from the computer or related system of formulas or other records or data shall be checked for accuracy. The degree and frequency of input/output verification shall be based on the complexity and reliability of the computer or related system. A backup file of data entered into the computer or related system shall be maintained except where certain data, such as calculations performed in connection with laboratory analysis, are eliminated by computerization or other automated processes. In such instances a written record of the program shall be maintained along with appropriate validation data. Hard copy or alternative systems, such as duplicates, tapes, or microfilm, designed to assure that backup data are exact and complete and that it is secure from alteration, inadvertent erasures, or loss shall be maintained.

Sec. 211.72 Filters

Filters for liquid filtration used in the manufacture, processing, or packing of injectable drug products intended for human use shall not release fibers into such products. Fiber-releasing filters may not be used in the manufacture, processing, or packing of these injectable drug products unless it is not possible to manufacture such drug products without the use of such filters. If use of a fiber-releasing filter is necessary, an additional non-fiber-releasing filter of 0.222 micron maximum mean porosity (0.45 micron if the manufacturing conditions so dictate) shall subsequently be used to reduce the content of particles in the injectable drug product. Use of an asbestos-containing filter, with or without subsequent use of a specific non-fiber-releasing filter, is permissible only upon submission of proof to the appropriate bureau of the Food and Drug Administration that use of a non-fiber-releasing filter will, or is likely to, compromise the safety or effectiveness of the injectable drug product.

EC GMP Guide

Equipment is covered in the EC GMP Guide in a subpart of Chapter 3, Premises and Equipment. It reads as follows:

Equipment

3.34   Manufacturing equipment should be designed, located and maintained to suit its intended purpose.

3.35   Repair and maintenance operations should not present any hazard to the quality of the products.

3.36   Manufacturing equipment should be designed so that it can be easily and thoroughly cleaned. It should be cleaned according to detailed and written procedures and stored only in a clean and dry condition.

3.37   Washing and cleaning equipment should be chosen and used in order not to be a source of contamination.

3.38   Equipment should be installed in such a way as to prevent any risk of error or of contamination.

3.39   Production equipment should not present any hazard to the products. The parts of the production equipment that come into contact with the product must not be reactive, additive or absorptive to such an extent that it will affect the quality of the product and thus present any hazard.

3.40   Balances and measuring equipment of an appropriate range and precision should be available for production and control operations.

3.41   Measuring, weighing, recording and control equipment should be calibrated and checked at defined intervals by appropriate methods. Adequate records of such tests should he maintained.

3.42   Fixed pipework should be clearly labelled to indicate the contents and, where applicable, the direction of flow.

3.43   Distilled, deionized and, where appropriate, other water pipes should be sanitised according to written procedures that detail the action limits for microbiological contamination and the measures to be taken.

3.44   Defective equipment should, if possible, be removed from production and quality control areas, or at least be clearly labelled as defective.

Comparison

In general, the requirements of the two documents are broadly similar, although the US cGMPs are somewhat more detailed.

DISCUSSION

Manufacturing equipment should be capable (and more than that, be *demonstrably* capable) of producing products, materials, and intermediates that are

US cGMPs	EC GMP Guide
Very reasonably requires that lubricants, coolants, etc. should not come into contact with products	Not mentioned specifically here; could be considered to be covered by "production equipment should not present any hazard to products"
Equipment, in general, is required to be cleaned, manintained and "sanitized"	Equipment is required to be cleaned and maintained, but only water pipes are required to be "sanitized"; "neither document offers a definition of this term
Both require written cleaning procedures; US cGMPs are more detailed	Both require written cleaning proced ures. EC less detailed than US
Requires "removal or obliteration" of "previous batch identification"	Not mentioned here
Requires protection of cleaned equipment from contamination prior to use	Clean equipment to be "stored only in a clean dry condition"
Hazards of repair and maintenance not mentioned here	"Repair and maintenance operations should not present any hazard to… products"
Control over computer or related systems required	Computers not mentioned here; covered in Annex 11 Computer Systems
Labeling of fixed pipes not mentioned	Fixed pipework to be clearly labeled to indicate contents and direction of flow
Requirement for use of nonfiber releasing and nonasbestos filters for injectable products	Not covered here; Annex 1, on Manufacture of Sterile Medicinal Products requires, rather weakly, that "Fibre shedding characteristics of filters should be minimised"
Calibration is required here for "automatic, mechanical or electronic equipment," but not specifically for measuring devices generally. Calibration of laboratory instruments *is*, later required. (See 211.160 b (4) and 211.194.d)	Calibration required for "measuring, weighing, recording and control equipment"; records to be maintained

intended and that conform to the required or specified quality characteristics. In other words, not only should products be fit for their intended purpose, but so should the items of equipment used to produce them.

Furthermore, the equipment must be designed and built so that it is possible (and relatively *easily* possible) to clean it thoroughly. Surfaces that come into contact with products should have smooth, polished finishes, with no recesses, crevices, difficult corners, uneven joints, dead-legs, projections, or rough welds to harbor contamination or make cleaning difficult. Equipment must also be capable of withstanding repeated, thorough cleaning. Traces of previous product, at levels that might be acceptable in other industries, are totally unacceptable in the manufacture of pharmaceuticals. It may also be necessary for equipment to be sterilized before use. It then becomes important that it is capable of withstanding the sterilization treatment — for example, the stress of steam at elevated temperature, under pressure.

Lubrication of moving parts should be designed and performed so that the product cannot be contaminated by the lubricant — or by, for example, metal particles from parts that have not been properly lubricated.

As far as the properties of the materials of construction of the equipment are concerned, there are two major concerns:

1. The possibility of contamination, or degradation, of the product by the material from which the equipment is constructed
2. The action of the product, or material in-process, on the material from which the equipment is constructed

Contamination of product can arise from shedding or leaching of contaminants from the equipment into the product or from reaction between the product and the material of the equipment. Product could otherwise be degraded by this sort of interaction or by ab- or adsorption of components of the product onto, or into, the equipment.

Corrosive action of product on equipment can damage that equipment, and in turn lead to further product contamination or degradation.

It is worth remembering that there are two aspects of the potential release of product contaminants by equipment: they could be toxic to patients, even in very small amounts, and they could cause product decomposition. As an example of the latter — penicillin can be inactivated by trace heavy metals.

This is not the place for a detailed discussion of the properties of the materials of construction of pharmaceutical plant and equipment, but it is worth noting that each case must be considered on its merits. For example, although stainless steel is widely, and generally successfully, used, there are a few examples of liquid solution products where the active ingredient can be degraded by contact with stainless steel mixing and storage vessels. In some such cases, plastic vessels have been found to be the best alternative. This illustrates the crucial importance of selecting equipment fabricated from materials appropriate to the product to be manufactured.

Fixed equipment should be installed, piped in, and supplied with services in a manner that creates a minimum of recesses, corners, or areas that are difficult to get to for cleaning. The pipework mazes beloved by some installation engineers should be avoided. Where pipework or ducting passes through walls or partitions, it should be sealed in on both sides.

In summary, equipment should be designed and located to suit the processes and products for which it is to be used. It must be shown to be capable of carrying out the processes for which it is used (that is, it should be properly commissioned, or "qualified") and of being operated to the necessary hygienic standards. It should be maintained so as to be fit to perform its functions, and it should be easily and conveniently cleanable, both inside and out. Parts that come into contact with materials being processed should be minimally reactive or absorptive with respect to those materials, and there should be no hazard to a product through leaking seals, lubricant drips, and the like, or through inappropriate modifications or adaptations. Equipment should be kept and stored in a clean condition and checked for cleanliness before each use. Washing and cleaning equipment should not, itself, become a vehicle of contamination. All measuring, weighing, recording, and control equipment should be serviced and calibrated at defined intervals according to an established procedure. Fixed pipework should be labeled as to contents, with an indication (where applicable) of the direction of flow. Defective equipment should be removed from manufacturing areas or clearly labeled as defective.

CLEANING OF EQUIPMENT

Between batches (or "campaigns") all manufacturing equipment and vessels must be thoroughly cleaned and (as necessary) disinfected or sterilized.

There should be written procedures for doing this, which must be followed exactly. Each piece of equipment has its own particular areas where there is a risk, given the right conditions, of microbial growth.

The best modern equipment is usually designed and built to reduce these risks as far as possible. It needs to:

- Be easy to dismantle and clean
- Have internal surfaces that are smooth, continuous, with no pits or rough, unpolished welds
- Have no dead-legs, or water or dirt-traps

It may be necessary to strip (or partially strip) equipment down before cleaning it. A written standard procedure should always be followed. With mobile equipment, there is the advantage that it can be removed from the manufacturing room for cleaning in a wash bay.

Once equipment has been cleaned and disinfected or sterilized, steps should be taken to ensure that it cannot become recontaminated. Care must also be taken to ensure (by labeling or segregation) that there is no possibility of mix-up between items that have been cleaned and disinfected or sterilized and those that have not.

Clean in Place (CIP) and Sterilize in Place (SIP)

The traditional way of cleaning, between batches or products, the internal contact surfaces of equipment — mixing vessels, storage vessels, and any associated pipework — was (and to a significant extent, still is) to do it by hand. This requires opening up vessels, dismantling, and stripping down, with subsequent reassembly. The efficacy of cleaning by simple manual methods will be crucially influenced by the zeal (or lack of) of the human cleaner. It will also result in long down times, and thus, poor plant utilization. It also introduces the potential hazard of recontamination of internal surfaces when the equipment is reassembled. Well-

designed clean in place (CIP) systems provide an answer to these problems —
but the crucial phrase is "well designed." For the manufacture of products intended
to be sterile (or microbiologically clean or "low count"), the concept is extended
to sterilize in place (SIP). It needs always to be remembered that, before sterilization
(or disinfection) it is first necessary to *clean*. If traces of product, or other material,
remain on surfaces through ineffective cleaning, they can build up, protecting
organisms from the sterilizing agent (e.g., steam), and thus prevent proper sterilization.

There are a number of advantages of CIP/SIP, especially if the process is
automated (as it should be for maximum efficiency and efficacy). These advantages include:

- Reduction in equipment down-time/increased plant utilization
- Reduction in labor costs
- Elimination of the variability of the human factor, thus a more consistent and
  reproducible process
- Elimination of the recontamination hazard on reassembly

The major disadvantage is the higher initial cost of plant purchase and installation.

Effectiveness of cleaning is a function of a number of factors, including time,
temperature, and rate of turbulent flow of the cleaning solution; the concentration
(and activity in relation to the soiling material to be removed) of chemical cleaning
agents in the cleaning solution; and the surface finish (smoothness or roughness)
of the surfaces to be cleaned. All these factors interact. For example, all other
things being equal, it will take a longer time to completely clean a relatively rough
internal surface as compared to a high-polish, smooth one. Higher temperatures
will need lower times and flow rates, and so on. Cleaning solutions commonly
employed contain caustic agents and detergents, and it must be remembered that,
before cleaning is complete, it is necessary to ensure removal of the cleaning
agents themselves. That is, there must be a rinsing stage, using (for aqueous
products) water of a quality appropriate to, and compatible with, the products to
be manufactured in the equipment.

To attempt to "bolt on" CIP/SIP systems to existing plant and equipment
is to court disaster. The process, and the system, must be designed and built
in from the very start; that is, at the process development, or scale-up stage.
The plant needs also to be designed and built so as to be able to withstand
(and *safely* withstand) the temperatures involved.

By the very nature of a CIP process, it is not possible to take a look to see if
the equipment is clean. Indeed, to do so would defeat the whole object of the
exercise. This makes validation of the CIP cleaning process especially important.

Cleaning Validation

It is increasingly being considered that it is not sufficient merely to apply *ad hoc*
cleaning methods and then *assume* that things (particularly equipment, manufacturing and holding vessels, and the like) are clean just because they look clean,
and as we have noted, in a CIP process it is just not possible to see if the internal
surfaces of the equipment "looks clean." It is necessary to employ fully

documented, *validated*, cleaning procedures. That is, cleaning procedures for which there is documented experimental evidence that they do, in fact, achieve the level of cleanliness that is both intended and appropriate in the given circumstances. Validation of cleaning processes will be discussed in more detail later.

Sterilize in Place (SIP)

"Sterilize in place" (SIP) is the term applied to a process of sterilizing the internal product-contact surfaces of a complete system of manufacturing and holding tanks and associated pipework (transfer lines, filling lines, etc.) while that system is assembled and in place, without having to take the system apart, separately sterilize the various elements of the system, and then reassemble it aseptically. The sterilizing agent employed is steam at sterilizing temperature, and it is thus necessary to design the system, from the outset, so it is able to withstand the temperatures and pressures required. As with CIP, "bolt-on" SIP is not a practical proposition — nor is it a safe one. Removal of air, and condensate, is crucial to ensuring that dry saturated steam, at the required temperature, for the required time, makes contact with all internal surfaces of the equipment. Evacuation is not usually possible, and air must be removed by properly designed and positioned bleed valves. In most systems, there will be significant amounts of condensate, which must be removed by drainage points placed in all horizontal and low parts of the system. Wherever possible, pipework should be angled so as to assist drainage. The process can be controlled automatically so as to maintain sterilizing conditions throughout. At the completion of the sterilizing phase, air or nitrogen is introduced through a bacteria-retentive filter, and the system is purged of any residual steam or condensate. A flow of pressurized gas is then maintained to dry the system, which should then be kept under positive sterile air (or nitrogen) pressure to maintain internal sterility before the system is used. The efficacy of any SIP system must be demonstrated by appropriate process validation.

CALIBRATION

A number of items of equipment used in manufacturing are themselves measuring devices (e.g., balances, scales, volumetric measures, metered valves) or have measuring devices (from quite complex pressure gauges, strain gauges, and load-cells to the more humble dipsticks and sight glasses) associated with them. All need to be calibrated — and maintained in a state of calibration.

Confusion sometimes exists between the two terms "metrology" and "calibration." Metrology is the science or study of measurement. In the EC GMP Guide, calibration is defined as the following:

> CALIBRATION: The set of operations which establish, under specified conditions, the relationship between values indicated by a measuring instrument or measuring system, or values represented by a material measure, and the corresponding known values of a reference standard.

This definition is only partially satisfactory, although linguistic purists might argue that is, indeed, what calibration *is*, and that nothing more is needed.

However, because the above definition gives no indication of *purpose*, or of what is the next step once the relationship has been established, the following definition (from the US National Standards Laboratory) is preferred:

> CALIBRATION: The comparison of a measurement system or device of unknown accuracy to another measurement system or device with a known accuracy to detect, correlate, report or eliminate by adjustment, any variation from the required performance limits of the unverified system.

There is no doubt that calibration involves a comparison of the unknown (or uncertain) with the known, and it is usually taken that the next step is to make any correction or adjustment that this comparison has shown to be necessary.

We measure many things — for example, time, linear dimensions (length and distance), area, volume and capacity, mass and weight, temperature, heat, pressure, velocity, electrical values (current, voltage, resistance, etc.), etc. Some of these are fundamental measures (e.g., time, linear dimensions, mass), others are derived from them (e.g., area, volume, velocity). The units used have been established in various ways, and some "absolute standards" have changed over the years. A meter, for example, was originally defined as 1/10,000,000th of the length of the polar quadrant through Paris. The definition has changed a number of times since, and the current definition (since 1983) is the distance traveled by light, in a vacuum, in 1/299,792,258th of a second.

In ordinary, routine work it is hardly necessary (or practicable) to refer each time to the ultimate, or absolute, standard for any measurement, and it is usual to make the necessary comparison with something lower in the league table, but which has in turn been reliably calibrated and certified against a higher standard. This introduces the concept of a hierarchy of standards:

ABSOLUTE STANDARDS
↓
INTERNATIONAL STANDARDS
↓
NATIONAL STANDARDS (e.g., UK NPL, US NBS)
↓
CERTIFIED REFERENCE STANDARDS
↓
WORKING REFERENCE STANDARDS

Laboratories, calibration departments and technicians, and the like will tend to have available reference standards (e.g., weights) or devices (e.g., thermometers), which have been certified (for example by the UK National Physical Laboratory or the US National Bureau of Standards) and from which their own internal working standards are derived or against which they are compared.

Terminology

The terminology of calibration tends to be akin to that used in analytical validation (see later). Thus, in the context of calibration and metrology:

Accuracy is the closeness of an observed or measured value to the true, or a reference, value. (Closeness to the truth.)

Precision is the closeness of agreement between different measurements of the same value, in a series of measurements, using the same measuring device. (Closeness to each other or togetherness.)

Range is the interval over which a device or system will operate with suitable accuracy and precision. (Is it the right tool for the job?)

Sensitivity is the degree to which the device or system can detect small differences in a measured value.

Since the quality of a product depends so much on the quality (i.e., fitness for purpose) of the measuring devices used in its manufacture and testing, it is important that the calibration of all measuring and testing equipment (whether it be intended for manufacturing or laboratory use) should not be conducted on a whim or only when a device is clearly not functioning. It should be *managed* as a well-controlled operation, run according to preplanned programs and schedules, with written, approved procedures for the calibration of each type of instrument or device, and with records maintained of calibrations carried out. Whatever the format of the documentation system employed, it should clearly signal when an instrument or device is due for calibration. The main steps to Good Calibration Practice may be set out as follows:

a. Carefully review all manufacturing and control processes to determine and record all the measurements that need to be made and to define the accuracy and precision required when making them. On this basis, select and obtain the necessary test and measuring equipment accordingly, or discard and replace any test equipment found to be unsuitable, or inadequate for the purpose.

b. Mark, or by some other means (e.g., by reference in documents or records to plant or model numbers) identify all measuring equipment to ensure it is calibrated at defined intervals against certified reference standards.

c. Prepare and implement written calibration procedures, programs, schedules, and records, that will ensure measuring devices are indeed calibrated, as intended, at the prescribed time intervals. The careful determination of the intervals between routine calibrations of a given instrument or device is critical to the success, or otherwise, of a calibration program. It should not be a general, overall figure, applicable to all instruments. Each time interval should be specifically selected for each device, after considering:

■ Type of measuring device
■ How crucial is the accuracy and precision of the device in relation to quality and, hence, consumer safety
■ Degree of accuracy and precision required
■ Device manufacturer's recommendations
■ Extent of use
■ Stress placed upon device in use
■ Any tendency of device to display drift

- Previous history, and records, of device in use
- Environmental conditions

d. Keep calibration records, detailing what calibrations have been carried-out, when, and by whom. Regularly review these records to ensure that the required calibrations are, in fact, being carried out at the specified intervals.

e. Ensure visibility of calibration status. That is, label the equipment with an indication of when it last was calibrated, and when it next is due, or record this information in an immediately accessible document or record book.

f. Ensure the calibration status of any measuring device or instrument before it is used.

g. Carry out documented retrospective assessments of the validity of previous measurements and tests whenever a piece of measuring or test equipment is found to be out of calibration. (It is irresponsible to fail to consider the potential effects of potentially false previous results when a measuring device is found to be reading incorrectly, and to act accordingly, no matter the economic consequences.)

h. Ensure that measuring equipment is handled and stored so that its accuracy and general fitness for use is not hazarded.

i. Protect the equipment and any associated software against unauthorized adjustments that would invalidate its setting.

j. Ensure that calibrations, inspections, tests, and measurements are carried out under suitable environmental conditions and that reference standards are very carefully protected against damage or deterioration. (To turn briefly to the department of the absurd, one recalls a company where the pristine condition of their set of standard balance weights was maintained by a vigorous weekly application of brass polish, and another that identified its set of standard weights each with a dab of red paint.)

The most important requirement is the need to ensure that calibration work is carried out by trained, experienced personnel who really know what they are doing and know the importance of what they are doing. It is also important that there is a formally assigned, accountable responsibility for calibration.

If calibration work is carried out under external contract, it should be subject to a formal written contract, clearly defining the nature and extent of the work required, and the content and format of the resultant test report(s).

MACHINE MAINTENANCE

All machinery is subject to the deleterious effects of wear, dirt, stress, and corrosion, acting individually or in combination with one another. To minimize these adverse effects, and the inevitable consequent decline in machine performance, efficiency and useful life, and (most importantly) in product quality, it is vital to take appropriate preventative measures. Thus, a comprehensive written maintenance program should be prepared for each piece of mechanical production equipment, setting-out each and every required maintenance activity in detail. It should include statements of the frequency with which each activity should be performed, in terms of real time (e.g., daily, weekly, monthly, yearly) or machine time (e.g., number of hours machine running time). The frequency and time base should be clearly defined in the written program(s) for each maintenance procedure to be carried out on each machine.

Machine maintenance should be carried out on a planned preventative, not on an emergency curative, basis. In the manufacture of pharmaceuticals, etc. the adage, useful perhaps on a domestic basis, "if it ain't broke, don't fix it" is definitely not applicable.

Formal maintenance records, which can be readily related to the overall maintenance program, should be compiled as each maintenance operation is performed and held on file in order to ensure, and to make it possible to demonstrate, that all required maintenance operations are indeed carried out as and when required by the program.

Matters generally to be considered in order to combat the deleterious effects mentioned above include, but are not necessarily limited to:

a. Dirt (e.g., dust, grit, and other abrasive particulate matter) can be a major cause of loss of machine efficiency and useful machine life, especially when mixed with moisture, oil, or grease. Maintenance programs should ensure that machine surfaces are kept in an appropriately clean condition. Buildup of dirt can be minimized by ensuring that surfaces (except of course those that require lubrication) are free of oil, grease, and moisture.

b. Wear between moving parts in contact is inevitable. The extent of wear, and the rate at which it occurs, can be minimized by the application, at a specified frequency, of the correct, defined lubricants. It should be noted that excessive lubrication can be almost as damaging as insufficient lubrication. Moving parts should be inspected for wear at regular, defined, intervals. Failure to monitor wear can lead to machine failure and possible serious damage.

c. Regular inspections for corrosion should be made, looking for signs such as discoloration, chemical deposition, and flaking or "bubbling" surface finishes. Inspection should not be limited to the machine itself, but should cover brackets, supports, and ancillary equipment. If corrosion is discovered, immediate steps should be taken to treat it, to discover the cause(s) and to prevent recurrence.

d. All machinery should have been constructed and installed so as to tolerate the strains to which it will be subjected when used for its intended purpose, over its expected operational life-span. It should not, however, be assumed that faults will never occur through stress and/or fatigue. Regular inspections should be made to detect any signs of this (e.g., stress cracks) on all parts of the machine under any stress.

Other more specific points that need to be covered in the maintenance program include (but are not necessarily limited to) checking and confirming the correct operation (as relevant) of:

■ Electric motors and pumps
■ Automatic valves and switches

- Any other automatic systems
- Steam traps in SIP systems
- Thermocouples and RTDs
- Any alarm systems, both visible and audible

The written maintenance program should also cover any special maintenance specified or recommended by the manufacturer of any given piece of machinery.

Formal change control procedures and documentation should be in-place that will ensure that no significant machine-engineering changes or modifications can take place without prior authorization, nor without full assessment of any potential effects on product quality.

Unless maintenance programs, maintenance records, and change control procedures are developed and implemented (to ensure that manufacturing equipment and attendant instruments and control devices remain in the same qualified and maintained state as they were during any validation studies conducted using that equipment), then any assurance hopefully derived from those validation studies could well be negated.

Requirements for equipment design, specification, qualification, calibration, and maintenance apply equally to equipment, installations, or services that are ancillary, subsidiary, or provide support to manufacturing equipment, such as:

- Electrical power supplies
- HVAC systems
- Steam generators (to ensure that the steam produced does indeed comply with the required specification)
- Air compressors (to ensure, e.g., the supply of appropriate quality, oil-free, compressed air)
- Heat exchangers
- Chillers
- Water purification and supply systems
- CIP and SIP systems

This also includes all measuring, indicating, controlling, monitoring, and recording instrumentation associated with these various items of equipment, systems, and services.

6

MATERIALS CONTROL

Notes:

a. The US cGMP Regulations devote one specific subpart (E) to "Control of Components and Drug Product Containers and Closures." This topic is not covered in any one single chapter of the EC GMP Guide. The corresponding GMP requirements are, however, set out *inter alia* in a number of paragraphs, distributed among Chapters 4 (Documentation), 5 (Production), and 6 (Quality Control). It is thus not possible to make a simple, direct US subpart/EC chapter comparison. In what follows, for the purposes of comparison, each section of the US subpart E will be considered along with the correspondingly relevant passage(s) in one or other of Chapters 4, 5, or 6 of the EC GMP Guide.

b. Non-US readers should note that the term "component," as used in the US cGMPs means "any ingredient intended for use in the manufacture of a drug product, including those that may not appear in such drug product" (Part 210, Sec. 210.3, Definitions). It is thus equivalent to the European "starting material" and is not to be confused with the common European term "packaging component."

c. In the terminology of the EC GMP Guide, a "primary packaging material" is a container or closure, or other packaging material that comes in direct contact with the product. "Secondary packaging materials" are those that do not come into contact with the product. Printed packaging materials may be primary or secondary.

REGULATORY STATEMENTS

US cGMPs

Subpart E — Control of Components and Drug Product Containers and Closures

Sec. 211.80 General requirements

(Note: This section has four subsections, (a) to (d). subsections (a) and (d) are concerned with written procedures and documentation, (b) and (c) with the physical aspects of the storage of components, etc. We will return to (b) and (c) later.)

(a) There shall be written procedures describing in sufficient detail the receipt, identification, storage, handling, sampling, testing, and approval or rejection of components and drug product containers and closures; such written procedures shall be followed.

(d) Each container or grouping of containers for components or drug product containers, or closures shall be identified with a distinctive code for each lot in each shipment received. This code shall be used in recording the disposition of each lot. Each lot shall be appropriately identified as to its status (i.e., quarantined, approved, or rejected).

EC GMP Guide

Chapter 4 Documentation

4.10 There should be appropriately authorised and dated specifications for starting and packaging materials, and finished products; where appropriate, they should be also available for intermediate or bulk products.

4.11 Specifications for starting and primary or printed packaging materials should include, if applicable:

(a) a description of the materials, including: the designated name and the internal code reference; the reference, if any, to a pharmacopoeial monograph; the approved suppliers and, if possible, the original producer of the products; a specimen of printed materials;

(b) directions for sampling and testing or reference to procedures;

(c) qualitative and quantitative requirements with acceptance limits;

(d) storage conditions and precautions;

(e) the maximum period of storage before re-examination

Receipt

4.19 There should be written procedures and records for the receipt of each delivery of each starting and primary and printed packaging material.

4.20 The records of the receipts should include:

(a) the name of the material on the delivery note and the containers;

(b) the "in-house" name and/or code of material (if different from a);

(c) date of receipt;

(d) supplier's name and, if possible, manufacturer's name;

(e) manufacturer's batch or reference number;0 total quantity, and number of containers received;

(g) the batch number assigned after receipt;

(h) any relevant comment (e.g., state of the containers)

4.21 There should be written procedures for the internal labelling, quarantine and storage of starting materials, packaging materials and other materials, as appropriate.

Sampling

4.22 There should be written procedures for sampling, which include the person(s) authorised to take samples, the methods and equipment to be used, the amounts to be taken and any precautions to be observed to avoid contamination of the material or any deterioration in its quality.

Testing

4.23 There should be written procedures for testing materials and products at different stages of manufacture, describing the methods and equipment to be used. The tests performed should be recorded (see Chapter 6, item 17).

4.24 Written release and rejection procedures should be available for materials and products, and in particular for the release for sale of the finished product by the Qualified Person(s) In accordance with the requirements of Article 22 of Directive 75/319/EEC.

Chapter 5 Production

Principle

Production operations must follow clearly defined procedures; ...

5.2 All handling of materials and products, such as receipt and quarantine, sampling, storage, labelling ... should be done in accordance with written procedures or instructions and, where necessary, recorded.

5.12 At all times during processing, all materials, bulk containers, major items of equipment and, where appropriate, rooms used should be labelled or otherwise identified with an indication of the product or material being processed, its strength (where applicable) and batch number. Where applicable, this indication should also mention the stage of production.

5.13 Labels applied to containers, equipment or premises should be clear, unambiguous and in the company's agreed format. It is often helpful in addition to the wording on the labels to use colours to indicate status (for example, quarantined, accepted, rejected, clean, ...).

Starting materials

5.25 The purchase of starting materials is an important operation which should involve staff who have a particular and thorough knowledge of the suppliers.

5.26 Starting materials should only be purchased from approved suppliers named in the relevant specification and, where possible, directly from the producer. It is recommended that the specifications established by the manufacturer for the starting materials be discussed with the suppliers. It is of benefit that all aspects of the production and control of the starting material in question, including handling, labelling and packaging requirements, as well as complaints and rejection procedures are discussed with the manufacturer and the supplier.

5.27 For each delivery, the containers should be checked for integrity of package and seal and for correspondence between the delivery note and the supplier's labels.

5.28 If one material delivery is made up of different batches, each batch must be considered as separate for sampling, testing and release.

5.29 Starting materials in the storage area should be appropriately labelled. Labels should bear at least the following information:
   - the designated name of the product and the internal code reference where applicable;
   - a batch number given at receipt;
   - where appropriate, the status of the contents (e.g., in quarantine, on test, released, rejected);
   - where appropriate, an expiry date or a date beyond which retesting is necessary.

When fully computerised storage systems are used, all the above information need not necessarily be in a legible form on the label.

5.30 There should be appropriate procedures or measures to assure the identity of the contents of each container of starting material. Bulk containers from which samples have been drawn should be identified.

5.31 Only starting materials which have been released by the Quality Control Department and which are within their shelf life should be used.

5.32 Starting materials should only be dispensed by designated persons, following a written procedure, to ensure that the correct materials are accurately weighed or measured into clean and properly labelled containers.

5.33 Each dispensed material and its weight or volume should be independently checked and the check recorded.

5.34 Materials dispensed for each batch should be kept together and conspicuously labelled as such.

Packaging materials

5.40 The purchase, handling and control of primary and printed packaging materials shall be accorded attention similar to that given to starting materials.

5.41 Particular attention should be paid to printed materials. They should be stored in adequately secure conditions such as to exclude unauthorised access. Cut labels and other loose printed materials should be stored and transported In separate closed containers so as to avoid mix-ups. Packaging materials should be issued for use only by authorised personnel following an approved and documented procedure.

5.42 Each delivery or batch of printed or primary packaging material should be given a specific reference number or identification mark.

5.43 Outdated or obsolete primary packaging material or printed packaging material should be destroyed and this disposal recorded.

Chapter 6 Quality Control

Documentation

6.7 Laboratory documentation should follow the principles given in Chapter 4 ... and the following details should be readily available to the Quality Control Department:
   – specifications;
   – sampling procedures;
   – testing procedures and records (including analytical worksheets and/or laboratory notebooks);
   – analytical reports and/or certificates;
   – data from environmental monitoring, where required;
   – validation records of test methods, where applicable;
   – procedures for and records of the calibration of instruments and
   – maintenance of equipment.

Sampling

6.11 The sample taking should be done in accordance with approved written procedures that describe:
   – the method of sampling;
   – the equipment to be used;
   – the amount of the sample to be taken;
   – instructions for any required sub-division of the sample;
   – the type and condition of the sample container to be used;
   – the identification of containers sampled;
   – any special precautions to be observed, especially with regard to the sampling of sterile or noxious materials;
   – the storage conditions;
   – instructions for the cleaning and storage of sampling equipment.

DISCUSSION

At this stage, the US cGMPs simply state the basic requirements. The EC GMP Guide has similar requirements, distributed among a number of different chapters, but in considerably more specific detail.

The US cGMPs establish as a basic general requirement that all aspects of the receipt, storage, handling, approval (or rejection) of components and drug product containers and closures shall proceed in accordance with approved written procedures. It is noteworthy that, in contrast to some GMPs that have been published in other parts of the world, the US cGMPs make it laudably clear that not only should there be these written procedures, but also that they *"shall be followed."*

It cannot be argued that a requirement to have and to follow written procedures is anything other than a sound general principle. In the manufacture of anything

Table 6.1 Comparison of requirements on receipt of components (starting materials) and packaging materials

US cGMPs	EC GMP Guide
Require written procedures for receipt, identification, storage, handling, sampling, testing, and approval or rejection which *shall be followed*	Similar requirements, but gives more detail on the *content* of the relevant procedures, specifications, and records
Both US cGMPs and EC GMP Guide require sampling of goods received, but neither offer much on number and quantity of sample to be taken	Rather more detail on the *mechanics* of sampling, but not on what constitutes a valid or representative sample
Require identity and status labeling, plus a "distinctive code"	Similar requirement, plus where appropriate, expiry or retest date; (bulk containers, major equipment, and rooms also to be identity labeled)

as important to human health and well-being as drug (or medicinal) products, every activity must be preplanned and formally defined in advance. Nothing can be left to chance. There is no room for "playing it by ear" or "by the seat of the pants." Manufacture of consistent quality drug products demands consistent, predetermined, *defined* activity.

It would certainly be a mistake to conclude that the general emphasis on "the paperwork" (or its electronic equivalent) is just one more expression of the innate bureaucratic urges of the government departments which, by and large, are responsible for the publication of GMP regulations or guidelines. Documentation is, in fact, the main structural supporting member, indeed the backbone, of any system of Quality Assurance.

In essence it is all very simple. It is about establishing written instructions for all significant activities, about following those instructions in practice, and about making records of those activities. The objectives are, in short:

1. To state clearly, in advance and in writing, what is to be done
2. To do it — in accordance with those instructions
3. To record what was done and the results of doing it

There are a number of very good practical, and patient-safety, reasons for proceeding in this way. The reasons for all this documentation are:

1. To ensure there is no doubt about what has to be done, by having formally approved written instructions for each job, and then following them
2. To define standards for materials, equipment, premises, services, and products
3. To confirm, as work proceeds, that each step has been carried out, and carried out *correctly*, using the correct materials and equipment

4. In the longer term, to keep, for later reference, records of what *has* been done, for example, manufacturing and test records, installation, commissioning, servicing, and maintenance records
5. To enable investigation of complaints, defect reports, and any other problems, and to permit observation of any drifts away from defined quality standards
6. To help decide on, and take, any necessary corrective action (including action to prevent reoccurrence) in the event of any complaint or defect report

A further very good reason for documentation is to overcome a common human failing. The great majority of us are, like Hamlet, "indifferent honest." That is, most of us are pretty honest most of the time. At opposite ends of the honesty spectrum are the few that are always totally honest, on the one hand, and the congenital liars, on the other. Both species are relatively rare. If a manufacturing instruction reads, for example, "after 15 minutes, check that the temperature is between 42°C and 47°C," most people will conscientiously check that this is so — *on the first few occasions*, but may later drift into being less careful. If the instructions require that a temperature within the required range be confirmed by ticking and initialing in a box, the average "indifferent honest" mind will be more acutely concentrated on ensuring that what is required is indeed done properly, but may well begin to lapse after the process has been performed scores, or even hundreds, of times. The best assurance is provided by an instruction that reads along the lines of: "after 15 minutes, check the temperature, which should be between 42°C and 47°C. Record the temperature reading in the box and initial. If the temperature is outside this range, report this immediately to the section head." Few people will ever be inclined to enter a completely false, or "invented" reading, and a more precisely factual record will also have been made for later investigation or review.

Documentation helps to build up a detailed picture of what a manufacturing function has done in the past and what it is doing now, and thus it provides a basis for planning what it is going to do in the future.

One common GMP recommendation is that manufacturers should, from time to time, carry out detailed reviews of their own operations — that is, perform "self inspections," or "internal quality audits." Detailed reviews of past records and documents are a great aid in doing this. Certainly, regulatory inspectors, during their inspections of manufacturing sites, often spend much time examining a company's documents and records. It has been suggested that some regulatory agencies adopt the attitude that "if there are not detailed instructions it will not be done, and if a written record has not been made and retained, it has not been done." While this may be something of an extreme position, it is a useful thought to keep in mind. Another way of looking at it is that Documentation is Quality Assurance made visible.

IMPLEMENTATION

(Note: the forms, labels, written procedures, etc. in the rest of this chapter are intended as illustrative examples only. There are, of course, many other possible designs, layouts, and styles.)

Figure 6.1 is a flow diagram illustrating the ordering, receipt, sampling, approval (or rejection), and dispensing of starting materials, thus:

The Purchasing Department orders the material on the basis of a Starting Material Specification provided to them by the Quality Control Department. Purchasing Department sends the order to an Approved Supplier, that is a company that has been approved, jointly by the Quality Control and Production departments to supply the material in question.

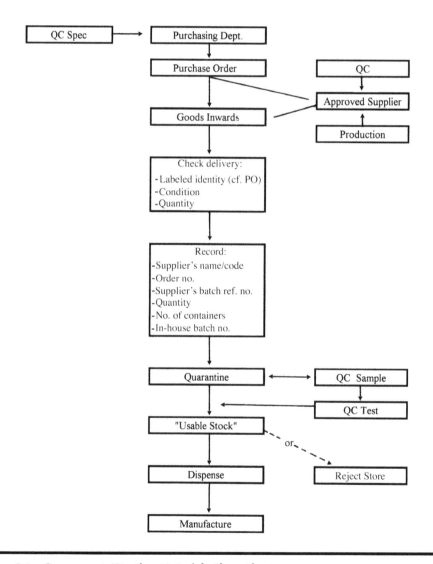

Figure 6.1 Components/Starting Materials Flow Chart

Note: The US Food and Drug Administration (FDA) investigators do inspect Active Pharmaceutical Ingredient (API) manufacturing sites. At the time of writing, the current expectation in the European Union is that requirements for compliance with GMP by, and inspection and approval of, starting materials and component suppliers will become mandatory within the next year or so. Then, depending on the form this regulatory approval or certification takes, the supplier will need to be approved in a formal regulatory sense. As guidance on what it considers to be cGMP for the manufacture of APIs, the FDA has made available a document entitled "Guidance for Industry — QA7 Good Manufacturing Practice for Active Pharmaceutical Ingredients." This guidance was prepared by the International Conference on Harmonisation (ICH), which aims to promote international agreement on regulatory requirements. The ICH consists of representatives from the United States, the European Union, and Japan. The US FDA has declared that this ICH guidance "describes cGMPs for the manufacture of APIs" (US Federal Register, September 25, 2001). The European Commission has issued an annex to the EC GMP Guide on Good Manufacturing Practice for Active Pharmaceutical Ingredients, which is also based on the ICH QA7 guidance.

To continue with the flow-diagram, at the time of placing the order, the Purchasing Department sends a copy of the purchase order to the Goods Inwards (or Receiving) Department, where it is (accessibly) retained, pending the receipt of the goods.

On receipt, the goods are carefully examined by a responsible member of the Goods Inwards Department for general condition, and to check for any signs of external damage, soiling, or dampness. At the same time, the labeled identity of the delivered material is checked and compared with the Goods Inwards copy of the purchase order, and with any supplier's delivery, or advice note, to confirm that the material delivered is, as far as its labeling is concerned, the material that was ordered. If there is any doubt about the nature or the quality of the goods delivered, the Quality Control Department is contacted immediately. A check is also made at this time on all the identity labels on the containers in a multicontainer delivery. Different suppliers' batches within one delivery are to be segregated, one from an other, with a different internal lot number for each entered on the QUARANTINE label, which is applied to each container.

If all the containers in the delivery appear to be correct and in good condition, the Goods Inwards Department then place on each container a QUARANTINE label (see Figure 6.2), with the entries for "Code Number," "Name of Material," "Lot Number," and "Date Received" completed.

Notes:
a. It is useful to have the QUARANTINE label, and the RELEASED and REJECTED labels (again, see Figure 6.2) printed in different colors, for example, for QUARANTINE, black print on a yellow background, for RELEASED green print on a white background, and for REJECTED red print on a white background.
b. In the examples shown, the intention is that, when the QC decision is made, the RELEASED (or REJECTED) label should be applied just over

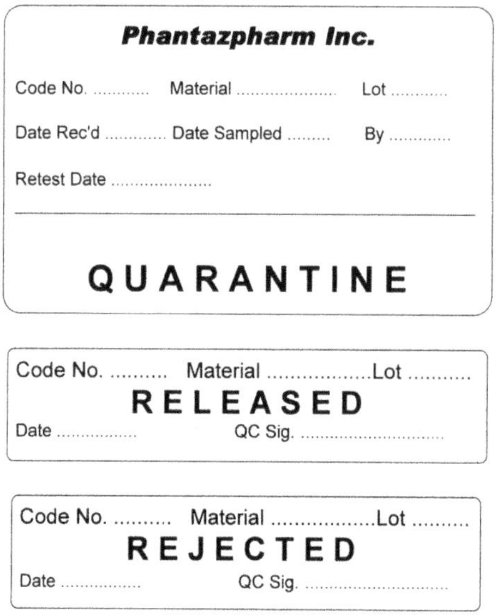

Figure 6.2   Quarantine, Released, and Rejected Labels

the lower QUARANTINE panel. This may seem an infringement of the golden rule about not applying new labels over old ones, but here (if the, say, RELEASED label falls off, or is removed) the labeled status of the material reverts to QUARANTINE, i.e., it is fail-safe. The benefit of the labeling system illustrated is the elimination of any possible error in transcribing the information originally entered on the QUARANTINE label.

c.   It is important that at least the QUARANTINE label is in a house style, with company name or logo, to avoid confusion with any other identity and status labels (e.g., those applied by vendors) that may already be on the container.

Goods Inwards then completes a Materials Receiving Report (Figure 6.3) in four copies, retaining one copy and sending the other three to Quality Control. They then make the appropriate entries (except for entries in the last two columns) in a departmental running record — a "Materials Delivery Record" (Figure 6.4). This can be a printed sheet or card, or manually drawn up in a record book (*or* a computer record).

Receipt of the copies of the Materials Receiving Report alerts the Quality Control Department that the material has been delivered, and is required to be sampled. Following sampling ("date sampled__" and "by__" on the QUAR-ANTINE label completed by the sampler) and testing against the agreed specification, the QC decision is entered on the copies of Materials Receiving Report, one copy being sent to the Purchasing Department (for information), one to Materials Inventory Control (so, if material is released, it may be allocated to manufacturing batches), and one retained on QC file, with the full analytical report. An authorized member of the QC Department then places a RELEASED (or REJECTED as appropriate) label, over the QUARANTINE portion, with the necessary details entered. He also enters a date at "retest

GOODS INWARDS - MATERIALS RECEIVING REPORT

Material... Code No. ....

INSTRUCTIONS: 1. Complete a separate Receiving Report for each delivery, and for each suppliers batch number within a delivery.
2. Retain one copy in Goods Inwards file, and send three copies to Quality Control.
3. **Quality Control**: On completion of testing, mark this report, where indicated, "RELEASED," "REJECTED," "HOLD" as appropriate, and send a copy to:
Purchasing Department
Materials Inventory Control

Retain one copy on Quality Control files

Date goods received ....

Supplier ....

Supplier's batch no. ....

Quantity received ....

Number of containers ....

Purchase order no. ....

Assigned lot no. ....

General condition/cleanliness of delivery

....
....
....
....

Delivery examined by (signed) .... Date ....

Remarks/Comments

QUALITY CONTROL DECISION

Figure 6.3   Materials Receiving Report

Date of Delivery	Material	Code No.	Lot No.	Quantity	No. of Containers	Supplier No.	Delivery Note	Supplier's Batch No. (s)	Supplier's Name for Material	Date Approved by QC	Location

Figure 6.4   Materials Delivery Report

date__" on the original label, to indicate when the material is due for reexamination.

(Note: only the Quality Control Department should be authorized to hold stocks of, and apply, RELEASED and REJECTED labels.)

On receipt of the QC decision, Goods Inwards either moves the released goods into the usable stock area of the stores, or the rejected material to a secure reject store. The two last columns of the Starting Material Delivery Record (Figure 6.4) are then completed ("date approved by QC" and "location").

PACKAGING MATERIALS

The purchase, receipt, sampling, release, and control of printed packaging materials and primary packaging materials (that is packaging materials that come into direct contact with the product, as compared with secondary packaging materials, which do not) need to be accorded the same level of attention as given to starting materials. Documents, records, and procedures analogous to those outlined above should be employed.

STANDARD OPERATING PROCEDURE

The procedure to be followed on receipt of materials will need to be defined in a Standard Operating Procedure (SOP), an example of which is shown in Figure 6.5. The content of this illustrative SOP defines the type of goods-receiving process outlined above. For the purposes of this book, it may also be taken as an example of general requirements for SOPs.

Standard features expected and required of SOPs in general, which this example illustrates, are as follows:

1. Each SOP should have a number, or an alphanumeric code, by which it can be specifically identified. To that number or code should be added a suffix (or an edition date displayed) so it is possible to check that the currently approved version of the SOP is in use.
2. The "date issued" and "supersedes" entries also serve to aid this assurance and to help make possible a document change control system.
3. The master copy should bear the (dated) signatures of the persons who wrote the document, who approved it (normally Production, or in this case the Warehousing, or Stores, manager) and who finally authorized it (normally a senior manager within the Quality function).
4. All pages should be numbered, as indicated.
5. An SOP should commence with a clear and unequivocal statement of purpose and scope.
6. Responsibility for ensuring the implementation of the SOP, and for its revision and updating, should be clearly stated.
7. The procedure to be followed should be stated in numbered steps in clear, simple, and direct language.

Phantazpharm Inc. **- Standard Operating Procedure**

S.O.P. No:	Date Issued:	Supersedes S.O.P. No: New Document	Review Date:	Page 1 of 4

STARTING MATERIALS - GOODS INWARDS PROCEDURE

Contents	Page
1. PURPOSE	2
2. SCOPE	2
3. RESPONSIBILITY	2
4. REVISION	2
5. PROCEDURE	3

Written by:	Approved by:	Authorised by:
Date:	Date:	Date:

Figure 6.5a   Standard Operating Procedure

1. Purpose

To define the procedure to be followed on receipt of a delivery of starting material (ingredients or raw materials).

2. Scope

This S.O.P. applies to all deliveries of starting materials whatever their nature or source.

3. Responsibility

Routine responsibility for ensuring that this procedure is implemented as and when necessary rests with the Head of the Goods Inwards Department.

Writing and approval/authorization of this procedure, and ensuring that it is revised and updated as necessary is the responsibility of _____

4. Revision

This procedure must be rewritten, approved and authorized whenever any change in method of operation, or any other circumstance, indicates the need. It must be reviewed every 12 months from the date of issue, in the light of current practice, to determine whether any revision is necessary .

All copies of S.O.P(s). which are superseded by any revision must be withdrawn from active use, and appropriate change-control records maintained to ensure effective implementation of this requirement.

Figure 6.5b

Edition: 3-7-99

S.O.P. No   **Page 3 of 4**

5. Procedure

5.1 On receipt, all deliveries must be carefully examined by a responsible member of Goods Inwards personnel for general condition, and to check for any signs of external damage, soiling, and dampness. At the same time, the labeled identity of the delivered material should be checked , and compared with the Goods Inwards copy of the Purchase Order and with any supplier's Delivery, or Advice, Note to confirm that the material delivered is, as far as its label(s) are concerned, the material that was ordered. If there is any doubt about the nature or the quality of the goods delivered, contact the Quality Control Department immediately.

5.2 At the same time, check the supplier's Batch (or Lot) numbers on the identity labels to confirm whether or not (in a delivery where there is more than one container) all containers contain material from the same supplier's batch. NOTE: Different suppliers' batches within one delivery should be segregated, and a different internal LOT NUMBER for each entered on the QUARANTINE label.

5.3 If the material (**all** containers) appears to be correct and in good condition, label each container with a standard QUARANTINE label, with the required entries on the label completed. That is, enter on the label(s):

> Code number
> Name of material
> Lot number
> Date received

5.4 Complete a "GOODS INWARDS - MATERIALS RECEIVING REPORT" (4-part set) , retaining one copy on Goods Inwards files and sending the other three copies to Quality Control.

5.5 At the same time, complete the columns of the STARTING MATERIALS DELIVERY RECORD as follows (NOTE : Make a separate series of entries for each internal Lot Number - see 5.2 above):

> 5.5.1 DATE - Enter date of delivery of the material

> 5.5.2 MATERIAL - Enter name of the material, as it appears on our Purchase Order.

> 5.5.3 CODE NUMBER - Enter our company code number for the material.

> 5.5.4 LOT NUMBER - Enter our internal Lot Number, which should be the same as the Lot Number that has been entered on the QUARANTINE label(s). See 5.2 above.

Figure 6.5c

Edition: 3-7-99

S.O.P. No   **Page 4 of 4**

5.5.5 QUANTITY - Enter the quantity received.

5.5.6 NUMBER OF CONTAINERS - Enter the number of containers received per each Lot Number.

5.5.7 SUPPLIER - Enter the name of the supplier of the material.

5.5.8 DELIVERY NOTE NUMBER - Enter the reference number on the supplier's Delivery Note (or Invoice - or other delivery documentation).

5.5.9 SUPPLIER'S BATCH NUMBER - Enter the supplier's batch number for the material, as it appears on the delivered container(s). See 5.2 Above.

5.5.10 SUPPLIER'S NAME FOR MATERIAL - Enter the name of the material as it appears on the supplier's label(s) and on the delivery note or invoice (etc.). If it is the same as the name already entered under "Material," enter "same."

5.6 That completes the action necessary immediately on receipt of material. The material should be held safely and securely, taking note of any special storage requirements (See SOP No.XXXX "Special Storage Requirements for Starting Materials"), until it has been sampled by Quality Control and a "RELEASED" or "REJECTED" decision has been made. An authorized member of the Quality Control Department will then apply "RELEASED" (or "REJECTED") labels to the Quarantined goods.

5.7 When the material is released for use by QC, enter the date in the "DATE APPROVED BY QC" column of the Delivery Record, and move the approved goods into usable stock. Make note of its location (by pallet position) under "LOCATION" in the Delivery Record.

5.8 If the material is rejected, enter "REJECTED," plus date, under "DATE APPROVED BY QC," and move the goods into the Reject Store to await a decision on the disposition of the material.

Figure 6.5d

FURTHER REGULATORY STATEMENTS

US cGMPs

Here we revert to Subpart E, Sec. 211.80, where, as already noted, there are four subsections; (a) and (d), which are concerned with written procedures and (b) and (c), which are concerned with physical aspects of storage.

Subpart E — Control of Components and Drug Product Containers and Closures

Sec. 211.80 (continued)

(b) Components and drug product containers and closures shall at all times be handled and stored in a manner to prevent contamination.
(c) Bagged or boxed components of drug product containers, or closures shall be stored off the floor and suitably spaced to permit cleaning and inspection.

EC GMP Guide

Chapter 5 Production

5.7 All materials and products should be stored under the appropriate conditions established by the manufacturer and in an orderly fashion to permit batch segregation and stock rotation.
5.10 At every stage of processing, products and materials should be protected from microbial and other contamination.

DISCUSSION

Both sets of regulatory requirements are in agreement that components or (starting materials) and containers, etc. should be stored and handled in a manner that will prevent contamination. The US cGMPs specifically require storage off the floor and suitable spacing "to permit cleaning and inspection." Such specific statements do not appear explicitly in the EC GMP Guide, but they could be considered implicit. The EC GMP Guide also adds that one of the objectives of storage "in an orderly fashion" is to "permit batch segregation and stock rotation" (FIFO, or "first in, first out").

All these requirements are more than just Good Manufacturing Practice. They represent Good Stores Management, Good Materials Handling, and, indeed, good sound common sense.

Storage off the floor guards against damage from flooding and liquid spillages. It also permits wet-cleaning of floors, without the risk of wetting the materials.

A well-laid-out, orderly store not only permits segregation of different types, lots, and batches of material (hence, aiding against contamination and mix-up) and rotation of stock, it also enables more (labor-, management-, and cost-) efficient running of the store.

Until somebody thinks of a better idea, and except in the smallest of operations, the most satisfactory practical solution is to stack materials (and products) segregated one from the other, on pallets, with the pallets held on steel pallet racking, with placing in and selection from stock being handled by fork trucks. Such a storage system satisfactorily meets the requirements. It permits relatively easy cleaning, good ease of selection, and facilitates stock rotation. A disadvantage is the initial capital cost of equipment. This, however, will be counter-balanced by the reduced labor content as compared with other storage and materials handling systems.

US cGMPs

Subpart E — Control of Components and Drug Product Containers and Closures

Sec. 211.82 Receipt and storage of untested components, drug product containers

(a) Upon receipt and before acceptance, each container or grouping of containers of components, drug product containers, and closures shall be examined visually for appropriate labelling as to contents, container damage or broken seals, and contamination.

(b) Components, drug product containers, and closures shall be stored under Quarantine until they have been tested or examined, as appropriate, and released. Storage within the area shall conform to the requirements of Sec. 211.80.

Sec. 211.84 Testing and approval or rejection of components, drug product containers

(a) Each lot of components, drug product containers, and closures shall be withheld from use until the lot has been sampled, tested, or examined, as appropriate, and released for use by the quality control unit.

(b) Representative samples of each shipment of each lot shall be collected for testing or examination. The number of containers to be sampled, and the amount of material to be taken from each container, shall be based upon appropriate criteria such as statistical criteria for component variability, confidence levels, and degree of precision desired, the past quality history of the supplier, and the quantity needed for analysis and reserve where required by Sec. 211.170.

(c) Samples shall be collected in accordance with the following procedures:
(1) The containers of components selected shall be cleaned where necessary, by appropriate means.
(2) The containers shall be opened, sampled, and resealed in a manner designed to prevent contamination of their contents and contamination of other components, drug product containers, or closures.
(3) Sterile equipment and aseptic sampling techniques shall be used when necessary.

(4) If it is necessary to sample a component from the top, middle, and bottom of its container, such sample subdivisions shall not be composited for testing.

(5) Sample containers shall be identified so that the following information can be determined: name of the material sampled, the lot number, the container from which the sample was taken, the date on which the sample was taken, and the name of the person who collected the sample.

(6) Containers from which samples have been taken shall be marked to show that samples have been removed from them.

(d) Samples shall be examined and tested as follows:

(1) At least one test shall be conducted to verify the identity of each component of a drug product. Specific identity tests, if they exist, shall be used.

(2) Each component shall be tested for conformity with all appropriate written specifications for purity, strength, and quality. In lieu of such testing by the manufacturer, a report of analysis may be accepted from the supplier of a component, provided that at least one specific identity test is conducted on such component by the manufacturer, and provided that the manufacturer establishes the reliability of the supplier's analyses through appropriate validation of the supplier's test results at appropriate intervals.

(3) Containers and closures shall be tested for conformance with all appropriate written procedures. In lieu of such testing by the manufacturer, a certificate of testing may be accepted from the supplier, provided that at least a visual identification is conducted on such containers/closures by the manufacturer and provided that the manufacturer establishes the reliability of the supplier's test results through appropriate validation of the supplier's test results at appropriate intervals.

(4) When appropriate, components shall be microscopically examined.

(5) Each lot of a component, drug product container, or closure that is liable to contamination with filth, insect infestation, or other extraneous adulterant shall be examined against established specifications for such contamination.

(6) Each lot of a component, drug product container, or closure that is liable to microbiological contamination that is objectionable in view of its intended use shall be subjected to microbiological tests before use.

(e) Any lot of components, drug product containers, or closures that meets the appropriate written specifications of identity, strength. Quality, and purity and related tests under paragraph (d) of this section may be approved and released for use. Any lot of such material that does not meet such specifications shall be rejected.

Sec. 211.86 Use of approved components, drug product containers, and closures

Components, drug product containers, and closures approved for use shall be rotated so that the oldest approved stock is used first. Deviation from this requirement is permitted if such deviation is temporary and appropriate.

EC GMP Guide

Chapter 5 Production

5.3 All incoming materials should be checked to ensure that the consignment corresponds to the order. Containers should be cleaned where necessary and labelled with the prescribed data.

5.4 Damage to containers and any other problem which might adversely affect the quality of a material should be investigated, recorded and reported to the Quality Control Department.

5.5 Incoming materials and finished products should be physically or administratively quarantined immediately after receipt or processing, until they have been released for use or distribution.

5.6 Intermediate and bulk products purchased as such should be handled on receipt as though they were starting materials.

5.25 The purchase of starting materials is an important operation which should involve staff who have a particular and thorough knowledge of the suppliers.

5.26 Starting materials should only be purchased from approved suppliers named in the relevant specification and, where possible, directly from the producer. It is recommended that the specifications established by the manufacturer for the starting materials be discussed with the suppliers. It is of benefit that all aspects of the production and control of the starting material in question, including handling, labelling and packaging requirements, as well as complaints and rejection procedures are discussed with the manufacturer and the supplier.

5.27 For each delivery, the containers should be checked for integrity of package and seal and for correspondence between the delivery note and the supplier's labels.

5.28 If one material delivery is made up of different batches, each batch must be considered as separate for sampling, testing and release.

5.30 There should be appropriate procedures or measures to assure the identity of the contents of each container of starting material. Bulk containers from which samples have been drawn should be identified (see Chapter 6, item 13).

DISCUSSION

As can be seen, both the US and the EU regulatory authorities agree on the importance of stock-rotation (FIFO), aimed at ensuring that the materials held in stock the longest are used before more recently received materials. This can be done by purely physical means, that is, by keeping different deliveries of the same material physically segregated one from the other and clearly identifying which should be used first. Alternatively, stock receipt and issue records can be adapted so as to ensure that older material is always allocated for manufacture before newer. Electronic stock control systems can be programmed so that newer material *cannot* be allocated for use while older material remains in stock.

Visual examination of materials on receipt is an important quality assurance measure. The first priority is to ensure that the correct material has been delivered from the correct authorized supplier and to check if any physical damage, or contamination, has occurred in transit. Are the containers of the material entire and undamaged? Are they securely sealed? Have they been exposed to adverse weather or other environmental conditions? Has there been any spillage on them? Is there any sign of tampering? (The possibility of deliberate sabotage should never be ignored.)

Any such instances should be reported to the Quality Control unit and carefully evaluated for potential impact on product quality. As a valuable additional check, all materials (components or starting materials and packaging materials) should also be similarly checked, before use, in the production area.

It is also important that the containers in which the materials are received should be cleaned before the goods are placed in quarantine. A record should be made of all goods received (See Figure 6.4 — Materials Delivery Record).

Both sets of regulatory GMP requirements agree that received materials (both components or starting materials and packaging materials) should be held in a quarantined state until they have been sampled, tested for compliance with specification, and formally released for use (or rejected and removed from stock). Quarantine status can be established and maintained by status labeling (see Figure 6.2), by secure physical segregation (for example in separate quarantine store, apart form the usable materials store) or by manual or electronic stock control systems. A combination of all three provides the greatest security against inadvertent use of material that has not been approved for use. All this is relatively simple. A much greater problem is that of how to take an appropriate sample.

SAMPLING — AND VENDOR EVALUATION

It will be recalled that the US cGMPs (Subpart E, Sec. 211.84, (b)) require that:

> Representative samples of each shipment of each lot shall be collected for testing or examination. The number of containers to be sampled, and the amount of material to be taken from each container, shall be based upon appropriate criteria such as statistical criteria for component variability, confidence levels and degree of precision desired, the past quality history of the supplier, and the quantity needed for analysis and reserve where required by Sec. 211.170.

The EC GMP Guide also states that:

> 4.22 There should be written procedures for sampling, which include the person(s) authorised to take samples, the methods and equipment to be used, the amounts to be taken and any precautions to be observed to avoid contamination of the material or any deterioration in its quality.

And that:

> 6.11 The sample taking should be done in accordance with approved written procedures that describe:
> - the method of sampling;
> - the equipment to be used;
> - the amount of the sample to be taken;
> - instructions for any required sub-division of the sample;
> - the type and condition of the sample container to be used;
> - the identification of containers sampled;

  – any special precautions to be observed, especially with regard to the
    sampling of sterile or noxious materials;
  – the storage conditions;
  – instructions for the cleaning and storage of sampling;
  – equipment.

Elsewhere (Annex 8), the EC GMP Guide refers to taking a "correct" sample.

Unfortunately, in neither of these two major regulatory documents is any information given on what, in fact, is a "representative" or a "correct" sample.

Samples and Sampling

Sampling is an activity that is of crucial significance to Quality Control in the pharmaceutical industry and is something that is required by the GMP regulations and guidelines. It is necessary to take samples, because (a) it would be a totally impractical proposition to attempt to carry out tests on the entire bulk of a delivery of starting material, or on a complete batch of product, and (b) because most of the laboratory tests carried out are destructive. Thus, while any attempt to test a complete delivery lot, or an entire production batch might, potentially, yield sound and valuable information, the enormity of the task would be overwhelming, and the effect on profit would be disastrous. So, we take samples. That being so, we need to always keep in mind that if those samples are not valid, that is if they do not adequately represent the batch, or lot, from which they were taken, then any conclusions drawn from them, and indeed the entire system of Quality Control, will in turn be invalidated.

Surprisingly, little of real value has been published on the theory and practice of taking samples for Quality Control purposes *in the pharmaceutical industry*. A number of papers and books have indeed been published on the statistical approach to sampling in the manufacturing industry in general. Although the approach expounded in such publications has considerable value and application in the fields, for example, of light engineering, the production of nuts and bolts, plastics bags, different grades of coal, and so on, they do not (on careful consideration) seem to have much relevance to what is done in the production of pharmaceuticals. The difference lies in the fact that statistical sampling relies on the use of sampling tables (for example the US Military Standard Tables and the British Standard 6000 series), which are based on assumptions that *some level of defective product is acceptable* (the Acceptable Quality Level or AQL), that the samples taken will be examined only for appearance or a few physical parameters, and that any defects are distributed uniformly throughout the batch or lot.

Inevitably, the approach in the context of the production of pharmaceutical products must be different. While acknowledging that, in this imperfect world, it is a philosophical impossibility to produce products that are 100% free of defect 100% of the time, it is difficult to accept *a priori* that there can be an agreed level of defective medicines. Certainly the patients who receive the defective products would find that difficult to accept. To illustrate the point: a common AQL in a number of manufacturing industries is 0.1. This denotes an agreement that 0.1% defective items is acceptable, i.e., one "wrong one" in every thousand is OK.

Applying this notion to various fields of human activity, this would mean, for example, two aircraft crashes per day at Chicago Airport, more than 100,000 microbially contaminated IV solutions infused worldwide per year, and 791 newborn babies sent home from hospitals with wrong mothers per year in the UK. Clearly, there are a number of important areas of activity where the acceptance of a 0.1% level of defectives is *not* acceptable.

The way samples are taken, the quantity to be taken, and what (if any) sampling plan is used will depend upon:

    a. What exactly is the purpose of taking the sample? That is, what is it that the QC system requires to know?

    b. To what extent can that knowledge be acquired, or inferred, from other sources?

The conclusions drawn from the examination of those samples will be affected by:

    a. The context in which they were taken

    b. The nature of the sampling techniques, instruments, and plans employed

    c. The devices, instruments, and methods used to examine and evaluate the sample

And the validity of those conclusions will depend upon:

    a. The quality of the sample (that is, its fitness for its intended purpose)

    b. The quality and the extent of the tests performed on it

    c. The quality (including reliability) of the powers of inference of those drawing the conclusions

One of the Annexes (8) to the EC GMP Guide is "Sampling of Starting and Packaging Materials." In this Annex, there is a subsection that deals specifically with the sampling of "starting materials" (starting material is defined as "any substance used in the production of a medicinal product but excluding packaging materials"). In effect, it means the same thing as ingredient or raw material and may be considered to be synonymous with the term "component," as used and defined in the US cGMPs. The first paragraph of this subsection of the EC annex reads:

> The identity of complete batch of starting material can normally only be ensured if individual samples are taken from all the containers and an identity test performed on each sample. It is permissible to sample only a proportion of the containers where a validated procedure has been established to ensure that no single container has been incorrectly labeled.

This is followed by a consideration of the factors that should be taken into account in this "validated procedure," and thus of the circumstances under which it may be considered permissible to forego the sampling and identity testing of the contents of each container of a multicontainer delivery of a starting material. The paragraph quoted above had its origins in Appendix 4 of the 1983 edition of the UK Orange Guide, where perhaps the point is made with greater clarity and indication of purpose, thus:

> The manufacturer of medicinal products must be aware of the possibility that containers of starting materials may be incorrectly labeled, and take steps to ensure that only the correct materials are used. Sampling and identity-testing the contents of each container can provide the necessary assurance...

In this same, original, passage it is, however, acknowledged that that "large deliveries in many containers can present practical and economic problems," and examples are given of circumstances under which "it may be possible to relax the policy of identity testing the contents of every container."

It is important to appreciate that, in both sets of guidelines (EC and original UK), the sampling of every container in a delivery refers only to sampling for *identification* purposes and not necessarily for the determination of quality and compliance with specification.

The background to this concern (in the UK at least) for the correct identity, as labeled, of starting materials was as follows. At some time in the mid-1970s, a highly reputable supplier of materials was discovered to have supplied a highly reputable pharmaceutical manufacturer with atropine mononitrate labeled as physostigmine sulphate, due to a labeling mix-up, and as a result of the same mix-up, to have supplied another equally reputable manufacturer with a similar quantity of physostigmine sulphate labeled as atropine mononitrate. Thus, both manufacturers were confronted with the potential danger of producing eyedrops that would have, most hazardously, the reverse of the intended physiological effect. In the event, both mislabeled lots were recovered in time and no patient harm was done. The event did, however, concentrate official and industrial minds on the potential hazards of mislabeled starting materials.

Around this same time, the Association of the British Pharmaceutical Industry (ABPI) asked its member companies to report, with details, the incidence of incorrectly labeled starting material deliveries over the preceding five years. Replies were received from 65 companies, of which 35 were nil returns. The other 30 reported a total of 66 occasions where chemical materials had been received incorrectly labeled. Of these 66 incidents, the entire delivery had been incorrectly labeled as to identity on 46 occasions. The remaining 20 were instances where only *some* of the containers were wrongly labeled, and which might thus have escaped detection in any sampling scheme that left some containers unsampled. A few examples of the errors reported were:

Containers Labeled	Contents	%Containers
Chloroform	Ether	10%
Isopropanol	Acetone	3%
Sodium hydroxide	Ammonium chloride	100%
Prednisone	Prednisolone	100%
Theophyline	Aminophyline	50%
Hydrochlorothiazide	Hydrochlorethiazole	100%
Sulphaguanidine	Theobromine	100%
Chromium trioxide	Sodium dichromate	5%
Iron oxide	Brown organic dye	15%

It is probably not necessary to state that the *partial* mislabeling of a delivery represents a greater potential hazard than a *complete* mislabeling of all the containers in a delivery. If all the contents of all the containers are not as labeled, then it should be spotted even if only one container is sampled.

It was considerations such as these that gave rise to the recommendations that, on receipt of a multicontainer delivery of a starting material, all containers in the delivery should be sampled and tested for identity, *unless adequate assurance against hazardous misidentification by the supplier can be obtained by other means.* The "other means" may be summarized as:

■ Where the source supplier, or plant, only produces and only supplies a single material, so there is no chance of mix-up.
■ Where the material comes directly from its producer or in that producer's own sealed and unbroken container, and where the purchaser has built up a history, over a period, of supplier reliability, and has been able to make a satisfactory assessment of the supplier's quality system through its own regular quality audits of that supplier. (The EC Guide, perhaps in anticipation of regulatory inspection of ingredient manufacturers, also accepts the possibility of such quality audits being performed by "an officially accredited body.")

The earlier UK GMP Guide (1983) also, very reasonably, allowed two other "other means," which are omitted (for no discernable reason) from the EC Guide. These were where:

> ... the pharmaceutical manufacturer's own manufacturing and quality control procedures, including assays of the end-product, would reveal the use of a wrong material (e.g., where a material is assayed in the finished product and the assay is specific).

And where:

> ... a (manufacturing) process would self-evidently fail if the wrong material was used.

The EC GMP Guide considers that, no matter the level of assurance of material identity that may be obtained by "other means," every container of materials to be used in the manufacture of injectable products should be sampled for identity test. This is based not on any suggestion that suppliers of materials to be used in injectables are more prone to labeling errors than other materials suppliers, but purely on a consideration of the relative patient hazards. The single example of potassium chloride will serve adequately to illustrate the point. Potassium chloride may be taken by mouth, without harm, in quantities that could prove fatal if injected by mistake for sodium chloride. The guidelines also consider that adequate assurance of identity (other than that obtained by sampling and identifying the contents of every container) *cannot* be obtained where the material is obtained via an intermediary (or series of intermediaries), particularly where the material is obtained from a broker who breaks bulk material and then repackages it in smaller quantities.

All this makes undeniably good sense. It does however, create a problem for the manufacturer who obtains relatively large quantities of material in many separate containers and is thus confronted with a by-no-means insignificant materials handling and cost implications. Many manufacturers will also look to purchase materials, at the best price (to themselves), in a world market. Indeed, some materials may only be obtainable from distant lands. In such circumstances, regular quality audits may not be an economic proposition. The healthcare manufacturer will need to carefully consider whether or not the immediate cost advantages of obtaining materials from far-off, not-too-well-known sources, outweigh the costs of 100% container sampling, or of obtaining assurance by those "other means," and also whether any patient hazard or commercial risk is justified. It is all a matter of balancing a number of different factors. The consideration of the safety, well-being, and protection of the ultimate consumer should always be paramount.

A notably worthwhile step in the direction of providing pharmaceutical manufacturers with a greater degree of assurance of the identity and quality of their purchased-in components or starting materials would be the rigorous implementation, worldwide, of regulatory inspection and licensing or certification of manufacturers of pharmaceutical ingredients (including nonactives — that is, Bulk Pharmaceutical Chemicals, or BPCs, not just Active Pharmaceutical Ingredients, or APIs).

The essential features of the regulatory statements on sampling (US and EC) may be summarized as:

a. Samples need to be taken, for test purposes, from deliveries of materials for use in manufacture (including packaging), and of the products produced by the process of manufacture.

b. Sampling is unquestionably an important activity, to be conducted with care and skill, and with a full awareness of all the implications of this activity.

c. Samples should be in some way representative of the total lot from which they are taken.

d. In some way, not clearly defined, sampling plans and statistical criteria have an important significance.

e. Precautions are necessary, when taking samples, to avoid contamination (chemical, microbial, or any other form) *of* the material being sampled, or of any other goods *by* the material being sampled.

f. Records need to be made that will ensure traceability of any given sample to not only the lot, but also to the individual container from which the sample was taken. Records also need to be made of the date of sampling, and of the name of the person who took the sample.

g. If the supplier of the material also supplies an analytical or test certificate of known reliability, it may not be necessary to take samples of delivered goods, other than those taken to confirm the identity of delivered lots of components (starting materials).

However, as far as the real-life approach to actually taking samples is concerned, some important things seem to be missing. This is not to criticize the regulators, who would generally declare that their job is to set-down *what* should be done, not to define *how* to do it; a view that many manufacturers, wishing to avoid increasingly prescriptive regulation, would applaud. So, having accepted these essential features (and who would wish to disagree with them?), what more does one need to know in order to establish and implement a rational and effective sampling policy? Well, in addition to information on the actual mechanics of sampling, the tools to be used, etc., what *are* the relevant "statistical criteria?" How may a suitable sampling plan be devised? In short, how many samples should be taken, from what number of (and which) containers of bulk material or product, and should or should not these individual samples be pooled (composited) for testing purposes — and if so, in what way?

If we turn again to Annex 8 (Sampling of Starting and Packaging Materials) of the EC GMP Guide, paragraph 4 initially seems to be offering some promise:

> "The quality of a batch of starting materials (sic) may be assessed by taking and testing a representative sample. The samples taken for identity testing could be used for this purpose. The number of samples taken for the preparation of a representative sample should be determined statistically and specified in a sampling plan. The number of individual samples which may be blended to form a composite sample should also be defined, taking into account the nature of the material, knowledge of the supplier and the homogeneity of the composite sample.

Unfortunately, as can be seen, this turns out to be just another example of adopting a position of (spurious) statistical authority, only to duck the crucial issue. This is a fairly common stance, with which many readers may well find sympathy — that is, mention statistics in a general way, to look or sound impressive, then run for cover before being cornered into offering any detailed comment.

A World Health Organization document[1] in the section (8.2) on "Sampling Plans for Consignments of Starting Materials Supplied in Several Sampling Units" (for "sampling units" read "bulk containers") at least attempts to offer some sampling plans. In fact, three of them may be summarized as follows:

WHO SAMPLING PLANS FOR SUPPLIES OF STARTING MATERIALS

The "n plan" (which "should be used with great caution and then only when the material is considered uniform and is supplied from a well-known source"):

$$n = \sqrt{N}$$

where N is the number of sampling units in the consignment. The value of n is rounded up to the next highest integer. Under this plan, original samples are taken from n sampling units selected at random, and each placed in a separate sample container. The control laboratory then inspects the appearance, and tests the identity, of the material in each original sample. "If the results are concordant" the original samples are pooled into a final sample from which an analytical sample is prepared, the remaining part being kept as a retention sample. "The n plan is not recommended for use by control laboratories of manufacturers who are required to analyze and release or reject each received consignment of the starting materials used to produce a drug product."

The "p plan" (which "may be used when the material is uniform [author's comment: how do you know?] *and is received from a source that is well known and where the main purpose is to check the identity"):*

$$p = \sqrt{0.4\ N}$$

where N is the number of sampling units. Under this plan, samples are taken from each of the N sampling units in the delivery, and each placed in a separate sample container. These original samples are each then visually inspected and tested for identity in the laboratory, and ("if the results are concordant") p final samples formed by "appropriate pooling of the original samples." (Note: for what specific purpose is not explicitly stated — presumably it is for a series of p complete analyses.)

The "r plan" (which "may be used when the material is suspected to be non-uniform and/or is received from a source that is not well known"):

$$r = \sqrt{1.5\ N}$$

where N is the number of sampling units. Under this plan samples are taken from each of the N sampling units in the delivery, and each placed in a separate sample container. These original samples are then tested for identity in the laboratory. "If the results are concordant," r samples are randomly selected and "individually subjected to testing." If the results are concordant the r samples are pooled for the retention sample.

Note:

1. *Under both the "p" and the "r" plans, samples taken from every container in a delivery are tested for identity.*
2. *No statistical basis is offered for any of these three plans — nor would there appear to be one.*

Sampling — Summary and Conclusions

Given that the totality of the advice that may be dredged from national and international regulations and guidelines is hardly complete or detailed (or even convincing), what advice can be given to the pharmaceutical manufacturer? The following comments, although not fully comprehensive, are offered.

General

1. Sampling of any sort of material or product is not something to be undertaken as a casual, menial job, by persons who have had no specific training in the task. It must be conducted in accordance with written procedures (prepared or approved by the Quality Control

Department) by persons who have been trained in the task and *who have a full understanding of the importance and also the potential hazards of what they are doing*, both to themselves and to the consumer of the company's products.

2. The training given to samplers should cover, at least:
   ■ Sampling plans
   ■ Application of the written sampling procedure(s)
   ■ The techniques and equipment for sampling
   ■ The risks, and prevention, of cross-contamination
   ■ Special precautions to be taken when sampling unstable, sterile, or hazardous substances
   ■ The importance of noting the visual appearance of materials, containers, and labels
   ■ The importance of recording any unexpected or unusual appearance, odor, etc.

3. The written sampling procedure should clearly define:
   ■ The method of sampling
   ■ The equipment to be used
   ■ The amount of sample to be taken
   ■ The number of samples to be taken
   ■ Instructions for any subdivision of the sample(s)
   ■ The type and condition of the sample container to be used
   ■ The identification of the containers sampled
   ■ Any special precautions to be observed, especially with regard to the sampling of sterile or hazardous materials
   ■ Storage conditions for the sample(s)
   ■ Instruction for the cleaning and storage of sampling equipment

4. Sampling should be performed (preferably) by QC personnel, or by other trained persons (e.g., production staff) who have been approved by QC and who are following the QC written procedures.

5. Great care needs to be taken to ensure that the act of sampling does not allow contamination, either *of* the material or product being sampled, or of other goods *by* that material or product. The written procedure should clearly define the precautions necessary to avoid any such contamination. Attention needs to be directed at the environment in which the sample is taken (for incoming supplies of starting materials, a separate dedicated sampling room is to be preferred), at the external cleanliness of the container holding the bulk material or product, of the sample containers, and of the sampling equipment.

6. Sampling devices and other equipment need to be scrupulously clean, both before and after use. They should be stored, pending further use, in a manner that maintains them in a clean condition. In many cases the best, and ultimately most cost-effective, solution will probably be to use single-use disposable sampling implements (scoops, spatulas, pipettes, dip-tubes, and the like).

7. The importance of the sampler carefully examining the general appearance of the bulk from which the sample is taken, and of

recording and drawing attention to anything unusual or untoward should not be underestimated.

8. Statistics-based sampling plans, and simple "rule of thumb" sampling formulae of the "n = √N +1" type must be considered as having little relevance, or application, to many aspects of sampling in the manufacture of pharmaceutical products. Indeed, it may be necessary to warn of the potential hazard of the sense of false security that a statistical sampling plan may engender and to emphasize that even the most sophisticated sampling plan, used to select a proportion containers to be sampled, will tell absolutely nothing about the quality (including identity) of the contents of the unsampled containers.

Components/Starting Materials

1. The importance of knowing your supplier, and of establishing a good working relationship in general with suppliers is, rightly, much stressed. It is vital to ensure that a supplier knows precisely what is needed, the form in which it is required and the containers in which it is to be delivered, and the purpose (in general terms, at least) for which it is required. That is, a detailed and agreed specification should form part of the order or contract. Knowing the supplier must be taken to mean much more than the establishment of a good drinking relationship, forged over lunch, between the manufacturer's freeloading buyers and the supplier's sales representatives. It means a thorough understanding of the supplier's practices and quality systems, gained through diligent quality audits.

2. Goods-Inwards personnel have an important role to play in the assurance of the quality of incoming materials. They should be required (by written procedures) and trained to examine carefully all incoming shipments for appearance, damage, spoilage, integrity of seals, labeling, and indeed for any and all signs that there is anything untoward — and to report anything that seems amiss.

3. The extent of sampling and the number of samples taken will depend upon the confidence gained through genuine knowledge of the supplier. The more slight the knowledge, the more remote the supplier, and the greater the number of hands through which the material has passed, the greater the need for extensive sampling and testing.

4. It is impossible to disagree with the statement that (with exceptions) "the identity of a complete batch of starting material can normally only be ensured if individual samples are taken from all the containers and an identity test performed on each sample" (EC GMP Guide). This must be particularly so in these days of fanatics whose inclination to hold others at ransom knows no bounds, and thus where the possibility of deliberate contamination or "switching" must, regrettably, be added to the purely inadvertent. However, the sheer handling problems that can be presented by large multicontainer deliveries cannot be denied. Reliable, rapid techniques that are easy to apply are clearly much to be desired. Near infrared

(NIR) techniques, which can be simply applied in the warehouse itself directly to material in the container (as delivered) without the need to take a sample and send it to the laboratory and without the need for highly specialized expertise (at the implementation stage) is now finding increasing application in the pharmaceutical and other industries, and may well prove to be the answer.[2-6] Smaller companies may be deterred by initial capital and setup costs, but they are, by the very nature of their business as small companies, unlikely to receive many large deliveries in many containers. Larger companies have found that savings accruing from reductions in the overall supply time (from placing of order to supply of dispensed material to production) more than compensate for the initial investment.

Retained (or Reference) Samples of Components (Starting Materials)

The EC GMP Guide requires that samples of starting materials shall be retained for at least two years after the release of the product in which they were used "if their stability allows." The EC Guide also states that "reference samples of materials and products should be of a size sufficient to permit at least a full re-examination." (For reasons that should be obvious, the prudent manufacturer will, where practicable, retain sufficient sample to permit *several* "full re-examinations").

The US cGMPs do not refer to reserve samples of components in Subpart E. This requirement appears in Subpart I, Laboratory Controls, (see Chapter 10). It does, however, apply only to *active* ingredients (APIs).

Testing

The persons responsible for laboratory management should ensure that suitable test-methods, validated in the context of available facilities and equipment (see Chapter 10 and Chapter 16), are adopted or developed.

Samples should be tested in accordance with the test methods referred to, or detailed, in the relevant specifications. The validity of the results obtained should be checked (and as necessary, any calculations checked) before the material is released or rejected.

Any in-process control work carried out by production staff should proceed in accordance with methods approved by the person responsible for Quality Control.

Contract Analysis

Although it is by no means an uncommon, and perfectly acceptable, practice for analysis and testing to be undertaken by an external Contract Analyst, the ultimate responsibility for Quality Control (still less for Quality Assurance) cannot be thus delegated to any external body, organization, or laboratory. The nature and extent of any such contract analysis should be formally agreed by both parties, clearly defined in writing, and procedures for taking samples should be established as set out above. The Contract Analyst should be supplied with full details of the test methods relevant to the material or product under examination. These will need to be validated as suitable for use by the contract laboratory. Formal arrangements will also need to be made for the retention of samples and of records of test results.

Certificates of Analysis

It is quite a common practice for deliveries of starting materials to be released for use in manufacture, substantially on the evidence of a certificate of analysis received from the supplier. This may be an acceptable procedure, provided that the document purporting to be a "certificate of analysis" is indeed genuinely just that, and not merely a copy of a standard specification, a statement of compliance with a specification, or a "typical batch analysis."

A true certificate of analysis should:

a. Clearly indicate the laboratory, or organization, issuing it
b. Be authorized (i.e., by signature, or by comparable electronic means) by a person demonstrably competent to do so
c. Clearly state the material, and the specific batch number, to which it refers
d. Clearly indicate by whom the material was tested and when
e. Clearly indicate the specification (e.g., USP, Pharm. Eur., purchaser's specification reference, etc.) and methods against which, and by which the tests were performed
f. State the test results obtained or assert that the results showed compliance with the stated specification

It is the responsibility of the Quality Control Department of the receiving company to satisfy itself that the person, or persons, issuing the certificate are competent to do so.

It needs to be firmly understood that the possession of a certificate of analysis does not (a) preclude the possibilities of a labeling mix-up or of damage to (or contamination of) the material in transit or (b) absolve the purchasing company from the ultimate responsibility for the quality (including identity) of the material to which the certificate refers, and as used in manufacture. Careful visual (at least) examination, and tests for identity are therefore (perhaps more than ever) necessary.

US cGMPs

Subpart E — Control of Components and Drug Product Containers and Closures

Sec. 211.87 Retesting of approved components, drug product containers, and closures

Components, drug product containers, and closures shall be retested or re-examined, as appropriate, for identity, strength, quality, and purity and approved or rejected by the quality control unit in accordance with Sec. 211.84 as necessary, e.g., after storage for long periods or after exposure to air, heat or other conditions that might adversely affect the component, drug product container, or closure.

EC GMP Guide

5.29 Starting materials in the storage area should be appropriately labelled. Labels should bear at least the following information:
– the designated name of the product and the internal code reference where applicable;
– a batch number given at receipt; - where appropriate, the status of the contents (e.g., in quarantine, on test, released, rejected);
– where appropriate, an expiry date or a date beyond which retesting is necessary.

When fully computerised storage systems are used, all the above information need not necessarily be in a legible form on the label.

The requirements of the US cGMPs are here direct, positive, unequivocal, and apply to containers and closures (packaging materials) as well as to components (starting materials). The corresponding EC requirement (indicated by the author's emphasis in the above) is rather more oblique and inferential, not to say ambiguous, stating only that an expiry or retest date should appear on the label applied to the material ("where appropriate," whatever that may mean). This clause also applies specifically to starting materials (components). That it applies also to packaging materials *may* perhaps be inferred from a later paragraph 5.40 in the EC GMP Guide, which reads:

5.40 The purchase, handling and control of primary and printed packaging materials shall be accorded attention similar to that given to starting materials.

Clearly, it is not sufficient merely to examine and test components and packaging materials only on receipt. It must be recognized that deterioration can occur over time and through exposure to adverse conditions. It therefore becomes necessary to establish and maintain systems that will ensure that materials that are at all liable to deteriorate, or that have been exposed to adverse conditions, are appropriately reexamined and tested. Proper storage in well-designed and maintained stores under controlled conditions should prevent damage due to adverse conditions, but it needs to be recognized that mistakes can happen and that action may be necessary. A first step toward guarding against deterioration over time is to label with an expiry or retest date. Written procedures should then indicate that no components or other materials should be issued from stores until the labels have been checked and the goods have been found to be "in date." Manual store records can be designed to flag when any lot has reached, or passed, its retest date. Perhaps best of all, computerized records of receipts, issues, and balances can be designed so that a lot of material that has passed its retest date cannot be issued for use in manufacture.

US cGMPS

Subpart E — Control of Components and Drug Product Containers and Closures

Sec. 211.89 Rejected components, drug product containers, and closures

Rejected components, drug product containers, and closures shall be identified and controlled under a quarantine system designed to prevent their use in manufacturing or processing operations for which they are unsuitable.

EC GMP Guide

Chapter 3 Premises and Equipment

(a)   Segregated areas should be provided for the storage of rejected, recalled or returned materials or products.

Chapter 5 Production

5.61 Rejected materials and products should be clearly marked as such and stored separately in restricted areas. They should either be returned to the suppliers, or, where appropriate reprocessed or destroyed. Whatever action is taken should be approved and recorded by authorised personnel.

Both of the two official publications, understandably, require that measures be taken to prevent the use of rejected materials (that is, rejected components or ingredients and packaging materials) in any manufacturing operation. The EC GMP Guide specifically requires that rejected materials should be stored, pending disposal, in a segregated, restricted area. The US cGMPs do not require this, presumably allowing that other systems of control may well be suitable. However, many companies (on both sides of the Atlantic) do have segregated reject storage areas, and it has to be said that this is a sensible, practical precaution against inadvertent use.

It is to be noted that the EC requirement here extends to *products* in addition to materials.

US cGMPs

Subpart E — Control of Components and Drug Product Containers and Closures

Sec. 211.94 Drug product containers and closures

(a)   Drug product containers and closures shall not be reactive, additive, or absorptive so as to alter the safety, identity, strength, quality, or purity of the drug beyond the official or established requirements.

(b)   Container closure systems shall provide adequate protection against foreseeable external factors in storage and use that can cause deterioration or contamination of the drug product.

(c)   Drug product containers and closures shall be clean and, where indicated by the nature of the drug, sterilized and processed to remove pyrogenic properties to assure that they are suitable for their intended use.

(d)   Standards or specifications, methods of testing, and, where indicated, methods of cleaning, sterilizing, and processing to remove pyrogenic properties shall be written and followed for drug product containers and closures.

The EC GMPs do not contain any clauses that directly correspond to paragraphs (a) and (b) above, in Sec. 211.94 of the US cGMPs. This is not to say that Europe is not concerned that "containers and closures shall not be reactive, additive" etc., nor that it is not expected that "container/closure systems shall provide adequate protection..." — far from it. The point is that in Europe, just as the establishment of the clinical efficacy and safety of a medicinal (or drug) product is not considered to be a part of Good *Manufacturing* Practice, nor is the establishment of the suitability of the container or closure to protect the product. (The assurance that, in routine manufacture, the correct as-specified container is used, is of course an important aspect of GMP.)

In Europe, the suitability of a container or closure system and the stability of a product or container unit, for any given product, is established on the basis of data submitted in support of an application for an authorization to market a product. (A "Marketing Authorisation," or in Britain a "Product License" — which corresponds roughly to the US NDA).

The requirement that containers should be clean appears in the EC GMP Guide in Chapter 5, Production, as follows:

5.48   Containers for filling should be clean before filling. Attention should be given to avoiding and removing any contamination such as glass fragments and metal particles.

The issues of sterilization and depyrogenation of containers are covered in the EC GMP Guide Annex 1 on the Manufacture of Sterile Products.

REFERENCES

1. WHO , Quality Assurance of Pharmaceuticals, a Compendium of Guidelines and Related Materials Volume 1, World Health Organization, Geneva, Switzerland, 1997.
2. Burns, D.A. and Ciuczak, E.W., *Handbook of Near Infrared Analysis*, Marcel Dekker Inc., New York, 1992.
3. Ciurczak, E.W., Uses of Near Infrared Spectroscopy in pharmaceutical analysis, *Applied Spectroscopy Review*, 23, 147–163,1987.
4. Ciurczakl, E.W., Application of NIR Spectroscopy in the Pharmaceutical Industry, *NIR News*, 2, 8–9, 1991.
5. Fox, L.E., Identification of Pharmaceutical Materials, Proc. Third International Conference on NIR, Brussels, Belgium, 429–433, 1990.
6. Higgins, M., Pinpointing production problems with NIR analysis, *Manufacturing Chemist*, 68/4, 38–39.

7

PRODUCTION AND PROCESS CONTROLS

REGULATORY STATEMENTS

In their coverage of this topic, the two GMPs differ significantly in their approach. In the US cGMPs, after an introductory statement (Section 211.100) of the basic principles that (a) "There shall be written procedures for production and process control ..." and that (b) "Written production and process control procedures shall be followed ...." a further series of seven sections (211.101 to 211.115) lay down requirements for the various stages of the overall activity of Production (which in this US context does not include packaging), from "Charge-in of components" to "Reprocessing," indicating at each stage the written procedures that should be followed.

The EC GMP Guide has a separate chapter (4) on Documentation, which does not have a direct parallel in the US cGMPs, and which (after statements of basic principles and of general documentation requirements) lists the documentation requirements (specifications, procedures, records, etc.) for each phase of manufacturing (*including* packaging). Chapter 4 of the EC GMP guide also gives an outline of the required content for each different type of document. The subsequent chapter in the EC GMP Guide (Chapter 5, Production) covers the various production (including packaging) *activities* that should be carried out and recorded in accordance with the documentation requirements established in the previous chapter (4) of the EC GMP Guide.

The EC GMP requirements regarding Specifications for starting materials (components) and packaging materials have already been covered in Chapter 6 of this book.

US cGMPs

Subpart F — Production and Process Controls

Sec. 211.100 Written procedures; deviations

(a) There shall be written procedures for production and process control designed to assure that the drug products have the identity, strength, quality, and purity they purport or are represented to possess. Such procedures shall include all requirements in this subpart. These written procedures, including any changes, shall be drafted, reviewed, and approved by the appropriate organizational units and reviewed and approved by the quality control unit.

(b) Written production and process control procedures shall be followed in the execution of the various production and process control functions and shall be documented at the time of performance. Any deviation from the written procedures shall be recorded and justified.

EC GMP Guide

Chapter 4 Documentation

Principle

Good documentation constitutes an essential part of the quality assurance system. Clearly written documentation prevents errors from spoken communication and permits tracing of batch history. Specifications, Manufacturing Formulae and instructions, procedures, and records must be free from errors and available in writing. The legibility of documents is of paramount importance.

General

4.1 *Specifications* describe in detail the requirements with which the products or materials used or obtained during manufacture have to conform. They serve as a basis for quality evaluation.
Manufacturing Formulae, Processing and Packaging Instructions state all the starting materials used and lay down all processing and packaging operations.
Procedures give directions for performing certain operations e.g., cleaning, clothing, environmental control, sampling, testing, equipment operation.
Records provide a history of each batch of product, including its distribution, and also of all other relevant circumstances pertinent to the quality of the final product.

4.2 Documents should be designed, prepared, reviewed and distributed with care. They should comply with the relevant parts of the manufacturing and marketing authorisation dossiers.

4.3 Documents should be approved, signed and dated by appropriate and authorised persons.

4.4 Documents should have unambiguous contents; title, nature and purpose should be clearly stated. They should be laid out in an orderly fashion and be easy to check. Reproduced documents should be clear and legible.

The reproduction of working documents from master documents must not allow any error to be introduced through the reproduction process. Documents should be regularly reviewed and kept up-to-date. When a document has been revised, systems should be operated to prevent inadvertent use of superseded documents.

4.6 Documents should not be handwritten; although, where documents require the entry of data, these entries may be made in clear, legible, indelible handwriting. Sufficient space should be provided for such entries.

4.7 Any alteration made to the entry on a document should be signed and dated; the alteration should permit the reading of the original information. Where appropriate, the reason for the alteration should be recorded.

4.8 The records should be made or completed at the time each action is taken and in such a way that all significant activities concerning the manufacture of medicinal products are traceable. They should be retained for at least one year after the expiry date of the finished product.

4.9 Data may be recorded by electronic data processing systems, photographic or other reliable means, but detailed procedures relating to the system in use should be available and the accuracy of the records should be checked. If documentation is handled by electronic data processing methods, only authorised persons should he able to enter or modify data in the computer and there should be a record of changes and deletions; access should be restricted by passwords or other means and the result of entry of critical data should be independently checked. Batch records electronically stored should be protected by back-up transfer on magnetic tape, microfilm, paper or other means. It is particularly important that the data are readily available throughout the period of retention.

Chapter 5 Production

Principle

Production Operations must follow clearly defined procedures; they must comply with the principles of Good Manufacturing Practice (GMP) in order to obtain products of the requisite quality and be in accordance with the relevant manufacturing and marketing authorisations. ...

5.2 All handling of materials and products, such as receipt and quarantine, sampling, storage, labelling, dispensing, processing, packaging and distribution should be done in accordance with written procedures or instructions and, where necessary, recorded.

US cGMPs

Sec. 211.101 Charge-in of components

Written production and control procedures shall include the following, which are designed to assure that the drug products produced have the identity, strength, quality, and purity they purport or are represented to possess:

(a) The batch shall be formulated with the intent to provide not less than 100 percent of the labeled or established amount of active ingredient.

(b) Components for drug product manufacturing shall be weighed, measured, or subdivided as appropriate. If a component is removed from the original container to another, the new container shall be identified with the following information:
(1) Component name or item code;
(2) Receiving or control number;
(3) Weight or measure in new container;
(4) Batch for which component was dispensed, including its product name, strength, and lot number.
(c) Weighing, measuring, or subdividing operations for components shall be adequately supervised. Each container of component dispensed to manufacturing shall be examined by a second person to assure that:
(1) The component was released by the quality control unit;
(2) The weight or measure is correct as stated in the batch production records;
(3) The containers are properly identified.
(d) Each component shall be added to the batch by one person and verified by a second person.

EC GMP Guide

Chapter 4 Documentation

Manufacturing Formula and Processing Instructions

Formally authorised Manufacturing Formula and Processing Instructions should exist for each product and batch size to be manufactured. They are often combined in one document.

4.14 The Manufacturing Formula should include:
(a) The name of the product, with a product reference code relating to its specification;
(b) A description of the pharmaceutical form, strength of the product and batch size;
(c) A list of all starting materials to be used, with the amount of each, described using the designated name and a reference which is unique to that material; mention should be made of any substance that may disappear in the course of processing;
(d) A statement of the expected final yield with the acceptable limits, and of relevant intermediate yields where applicable.
4.15 The Processing Instructions should include:
(a) A statement of the processing location and the principal equipment to be used;
(b) The methods, or reference to the methods, to be used for preparing the critical equipment (e.g., cleaning, assembling, calibrating, sterilising);
(c) Detailed stepwise processing instructions (e.g., checks on materials, pre-treatments, sequence for adding materials, mixing times, temperatures);
(d) The instructions for any in-process controls with their limits; Where necessary, the requirements for bulk storage of the products; includ-

ing the container, labelling and special storage conditions where applicable;

(f) Any special precautions to be observed.

Batch Processing Records

4.17 A Batch Processing Record should be kept for each batch processed. It should be based on the relevant parts of the currently approved Manufacturing Formula and Processing Instructions. The method of preparation of such records should be designed to avoid transcription errors. The record should carry the number of the batch being manufactured

Before any processing begins, there should be recorded checks that the equipment and work station are clear of previous products, documents or materials not required for the planned process, and that equipment is clean and suitable for use.

During processing, the following information should be recorded at the time each action is taken and, after completion, the record should be dated and signed in agreement by the person responsible for the processing operations:

(a) The name of the product;

(b) Dates and times of commencement, of significant intermediate stages and of completion of production;

(c) Name of the person responsible for each stage of production;

(d) Initials of the operator of different significant steps of production and, where appropriate, of the person who checked each of these operations (e.g., weighing);

(e) The batch number and/or analytical control number as well as the quantities of each starting material actually weighed (including the batch number and amount of any recovered or reprocessed material added);

(f) Any relevant processing operation or event and major equipment used;

(g) A record of the in-process controls and the initials of the person(s) carrying them out, and the results obtained;

(h) The product yield obtained at different and pertinent stages of manufacture;

(i) Notes on special problems including details, with signed authorisation for any deviation from the Manufacturing Formula and Processing Instructions.

Chapter 5 Production

5.12 At all times during processing, all materials, bulk containers, major items of equipment and where appropriate, rooms used should be labelled or otherwise identified with an indication of the product or material being processed, its strength (where applicable) and batch number. Where applicable, this indication should also mention the stage of production.

5.13 Labels applied to containers, equipment or premises should be clear, unambiguous and in the company's agreed format. It is often helpful in addition to the wording on the labels to use colours to indicate status (e.g., quarantined, accepted, rejected, clean, etc.).

5.30 ... Bulk containers (of starting materials) from which samples have been drawn should be identified.

5.31 Only materials which have been released by the QC Department and which are within their shelf life should be used.

5.32 Starting materials should only be dispensed by designated persons, following a written procedure, to ensure that the correct materials are accurately weighed or measured into clean and properly labelled containers.

5.33 Each dispensed material and its weight or volume should be independently checked and the check recorded.

5.34 Materials dispensed for each batch should be kept together and conspicuously labelled as such.

US cGMPs

Sec. 211.103 Calculation of yield

Actual yields and percentages of theoretical yield shall be determined at the conclusion of each appropriate phase of manufacturing, processing, packaging, or holding of the drug product. Such calculations shall be performed by one person and independently verified by a second person.

EC GMP Guide

Chapter 5 Production

Production/General

5.8 Checks on yields, and reconciliation of quantities, should be carried out as necessary to ensure that there are no discrepancies outside acceptable limits.

US cGMPs

Sec. 211.105 Equipment identification

(a) All compounding and storage containers, processing lines, and major equipment used during the production of a batch of a drug product shall be properly identified at all times to indicate their contents and, when necessary, the phase of processing of the batch.

(b) Major equipment shall be identified by a distinctive identification number or code that shall be recorded in the batch production record to show the specific equipment used in the manufacture of each batch of a drug product. In cases where only one of a particular type of equipment exists in a manufacturing facility, the name of the equipment may be used in lieu of a distinctive identification number or code.

EC GMP Guide

5. Production/General

    5.12 At all times during processing, all materials, bulk containers, major items of equipment and, where appropriate, rooms used should be labelled or otherwise identified with an indication of the product or material being processed, its strength (where applicable) and batch number. Where applicable, this indication should also mention the stage of production.

    5.14 Labels applied to containers, equipment or premises should be clear, unambiguous and in the company's agreed format. It is often helpful in addition to the wording on the labels to use colours to indicate status (for example, quarantined, accepted, rejected, clean, etc.).

US cGMPs

Sec. 211.110 Sampling and testing of in-process materials and drug products

(a) To assure batch uniformity and integrity of drug products, written procedures shall be established and followed that describe the in-process controls, and tests, or examinations to be conducted on appropriate samples of in-process materials of each batch. Such control procedures shall be established to monitor the output and to validate the performance of those manufacturing processes that may be responsible for causing variability in the characteristics of in-process material and the drug product. Such control procedures shall include, but are not limited to, the following, where appropriate:

    (1) Tablet or capsule weight variation;

    (2) Disintegration time;

    (3) Adequacy of mixing to assure uniformity and homogeneity;

    (4) Dissolution time and rate;

    (5) Clarity, completeness, or pH of solutions.

(b) Valid in-process specifications for such characteristics shall be consistent with drug product final specifications and shall be derived from previous acceptable process average and process variability estimates where possible and determined by the application of suitable statistical procedures where appropriate. Examination and testing of samples shall assure that the drug product and in-process material conform to specifications.

(c) In-process materials shall be tested for identity, strength, quality, and purity as appropriate, and approved or rejected by the quality control unit, during the production process, e.g., at commencement or completion of significant phases or after storage for long periods.

(d) Rejected in-process materials shall be identified and controlled under a quarantine system designed to prevent their use in manufacturing or processing operations for which they are unsuitable.

EC GMP Guide

Chapter 4 Documentation

    4.13 Specifications for finished products should include:

      (a) ...

      (b) ...

(c) ...
(d) directions for sampling and testing or a reference to procedures;
(e) the qualitative and quantitative requirements, with the acceptance limits;
(f) ...

4.15 The Processing Instructions should include:
(a) ...
(b) ...
(c) ...
(d) the instructions for any in-process controls with their limits;
(e) ...

Chapter 6 Quality Control

6.11 The sample taking should be done in accordance with approved written procedures that describe:
   - the method of sampling;
   - the equipment to be used;
   - the amount of the sample to be taken;
   - instructions for any required sub-division of the sample;
   - the type and condition of the sample container to be used;
   - the identification of containers sampled;
   - any special precautions to be observed, especially with regard to the sampling of sterile or noxious materials;
   - the storage conditions [of *what* not stated]
   - instructions for the cleaning and storage of sampling equipment

6.18 All the in-process controls, including those made in the production area by production personnel, should be performed according to methods approved by Quality Control and the results recorded.

Chapter 3 Premises and Equipment

3.23 Segregated areas should be provided for the storage of rejected, recalled or returned materials or products.

Chapter 5 Production

5.61 Rejected materials and products should be clearly marked as such and stored separately in restricted areas. They should either be returned to the suppliers or, where appropriate, reprocessed or destroyed. Whatever action is taken should be approved and recorded by authorised personnel.

US cGMPs

Sec. 211.111 Time limitations on production

When appropriate, time limits for the completion of each phase of production shall be established to assure the quality of the drug product. Deviation from established time limits may be acceptable if such deviation does not compromise the quality of the drug product. Such deviation shall be justified and documented.

EC GMP Guide

Chapter 4 Documentation

Batch Processing Records

> 4.17 A Batch Processing Record should be kept for each batch processed. It should be based on the relevant parts of the currently approved Manufacturing Formula and Processing Instructions. ...

During processing, the following information should be recorded at the time each action is taken ... :

> (a) the name of the product;
> (b) dates and times of commencement, of significant intermediate stages and of completion of production;
> (c) ...

Note for UK readers: Those referring to the MCA's "Rules and Guidance for Pharmaceutical Manufacturers ..." may notice a discrepancy here as compared with the quotation above. The "Rules and Guidance..." text reads here:

> (b) dates and times of commencement of significant intermediate stages and of completion of production;

In the transition from the EC original to their Orange "Rules and Guidance ..." the MCA has left out the comma after "commencement" and has thus significantly distorted the meaning of the sentence.

US cGMPs

Sec. 211.113 Control of microbiological contamination

> (a) Appropriate written procedures, designed to prevent objectionable microorganisms in drug products not required to be sterile, shall be established and followed.
> (b) Appropriate written procedures, designed to prevent microbiological contamination of drug products purporting to be sterile, shall be established and followed. Such procedures shall include validation of any sterilization process.

EC GMP Guide

Chapter 5 Production

General

> 5.10 At every stage of processing, products and materials should be protected from microbial and other contamination.

Note: In addition to this general statement, applicable to all types of manufacture, the EC GMP Guide has a detailed annex (Annex 1) on

Manufacture of Sterile Medicinal Products. The opening Principle to this annex reads, in part:

Annex 1 Manufacture of Sterile Medicinal Products
Principle

The manufacture of sterile products is subject to special requirements in order to minimise risks of microbiological contamination, and of particulate and pyrogen contamination ...

Chapter 5, Production, of the EC GMP Guide also has a three-paragraph section on Prevention of Cross-Contamination in Production. It refers to both chemical and microbial cross-contamination, and it reads as follows:

Chapter 5 Production
Prevention of cross-contamination in production

5.18 Contamination of a starting material or of a product by another material or product must be avoided. This risk of accidental cross-contamination arises from the uncontrolled release of dust, gases, vapours, sprays or organisms from materials and products in process, from residues on equipment, and from operators' clothing. The significance of this risk varies with the type of contaminant and of product being contaminated. Amongst the most hazardous contaminants are highly sensitising materials, biological preparations containing living organisms, certain hormones, cytotoxics, and other highly active materials. Products in which contamination is likely to be most significant are those administered by injection, those given in large doses and/or over a long time.

5.19 Cross-contamination should be avoided by appropriate technical or organisational measures, for example:
   (a) production in segregated areas (required for products such as penicillins, live vaccines, live bacterial preparations and some other biologicals), or by campaign (separation in time) followed by appropriate cleaning;
   (b) providing appropriate air-locks and air extraction;
   (c) minimising the risk of contamination caused by recirculation or re-entry of untreated or insufficiently treated air;
   (d) keeping protective clothing inside areas where products with special risk of cross contamination are processed;
   (e) using cleaning and decontamination procedures of known effectiveness, as ineffective cleaning of equipment is a common source of cross-contamination;
   (f) using "closed systems" of production;
   (g) testing for residues and use of cleaning status labels on equipment.

5.20 Measures to prevent cross-contamination and their effectiveness should be checked periodically according to set procedures.

US cGMPs

Sec. 211.115 Reprocessing

(a) Written procedures shall be established and followed prescribing a system for reprocessing batches that do not conform to standards or specifications and the steps to be taken to insure that the reprocessed batches will conform with all established standards, specifications, and characteristics.

(c) Reprocessing shall not be performed without the review and approval of the quality control unit.

EC GMP Guide

Chapter 5 Production

Rejected, recovered and returned materials

5.61 Rejected materials and products should be clearly marked as such and stored separately in restricted areas. They should either be returned to the suppliers or, where appropriate, reprocessed or destroyed. Whatever action is taken should be approved and recorded by authorised personnel.

5.62 The reprocessing of rejected products should be exceptional. It is only permitted if the quality of the final product is not affected, if the specifications are met and if it is done in accordance with a defined and authorised procedure after evaluation of the risks involved. Record should be kept of the reprocessing.

5.63 The recovery of all or part of earlier batches which conform to the required quality by incorporation into a batch of the same product at a defined stage of manufacture should be authorised beforehand. This recovery should be carried out in accordance with a defined procedure after evaluation of the risks involved, including any possible effect on shelf life. The recovery should be recorded.

5.64 The need for additional testing of any finished product which has been reprocessed, or into which a recovered product has been incorporated, should be considered by the Quality Control Department.

Note: The precise meaning and intention of 5.63, when considered in relation to 5.62, may be considered to be somewhat obscure. By checking back through earlier editions in order to determine their origins, it is possible to conclude that, whereas 5.62 refers to products that have failed to comply with their quality specification but which it is possible to reprocess with the object of producing a compliant product, 5.62 refers to such things as "tailings" or "leftovers" (for example, residual granules in the hopper of a tablet compression machine, or residual powder-mix left in the hopper of a capsule filling machine).

DISCUSSION

Although there are differences in approach, treatment, and emphasis, the UScGMPs and the EC GMP Guide are in close agreement on the essential requirements. These, in brief, are:

- There shall be written, and formally approved, procedures covering all stages of production and process control.
- These written procedures shall be followed.
- As production proceeds, check signatures or initials of those responsible for performing each step or stage in the process shall be entered on the batch record, to confirm that each step has been carried out in accordance with the written procedure, together with any results or readings as required.
- Any deviations from the written procedure, which must be justified and approved, shall be recorded.
- As components or materials are dispensed for manufacture the person(s) performing the dispensing operation shall make a recorded check that the correct amount of the correct component (material) of the correct receiving (or control) number has been dispensed, into a container which is correctly labeled with this same information, together with an indication of the product and batch for which the dispensed component is intended.
- The correctness of each of these details shall also be confirmed by check signatures or initials entered on the batch record by a second person.
- As each dispensed component or material is added to the batch in process of manufacture, there shall be a recorded check by the person making the addition, with verification by a second person.
- Only components or materials that have been released by the Quality Control Unit, and which are within their shelf life, shall be dispensed and used.
- Before any production operation is started, there shall be recorded checks to ensure that the equipment and work station are clear of previous products, documents or materials not required for the planned process, and that equipment is clean and suitable for use.
- As production proceeds, yields shall be determined at the end of each phase and at the completion of the manufacture of the batch, with the percentage of the theoretical calculated. Any discrepancies should be within previously established permissible limits.
- At all times during processing, all materials, bulk containers, major items of equipment, processing lines and, where appropriate, rooms used, should be labeled or otherwise identified with an indication of the product or material being processed, or held within a container, and its batch number. As necessary or applicable, the stage of production should be indicated.
- Major items of equipment shall be permanently identified by a distinctive identification number or code. This identification shall be recorded in the batch record, to show the specific item(s) of equipment that have been used in the manufacture of the batch of product. (Where the facility has only one of a particular item of equipment, its name alone will serve to identify it.)
- In-process products shall be sampled and tested in accordance with established written procedures in order to determine conformance with the relevant in-process material specification. The samples shall be taken in accordance with established sampling procedures.
- In-process materials and products that failed to conform to specification, and have thus been rejected, should be marked as such and quarantined in a segregated area until a decision is made on their disposition (for example, destruction or reprocessing).

- The date and time of commencement of a batch of product, and of significant intermediate stages, and of completion of the batch shall be recorded in the batch record. Any deviation from preestablished time limits shall be justified and documented.
- At every stage of processing, products and materials should be protected from microbial and other contamination, in accordance with previously established written procedures.
- Provided that the quality of the final product is not compromised, product batches that do not conform to specification may, in some circumstances, be reprocessed, but only with the approval of the Quality Control Unit, and only in accordance with previously established written procedures. (It may be necessary to obtain regulatory approval for reprocessing.)

Comparison

Significant differences between the two regulatory documents may be summarized as follows:

US cGMPs	EC GMP Guide
"Production," as in "Subpart F — Production and Process Controls" does not embrace Packaging and Labeling, which is covered in Subpart G — Packaging and Labelling Controls	"Production" (Chapter 5) is taken to include packaging and labeling
No great detail on the form and content of the various items of documentation	Considerable detail on the form and content of documents. (cf separate chapter on "Documentation"); specifically permits, with provisos, electronic, photographic, etc. data storage systems; whatever system is used, Batch Manufacturing Records to be retained for at least one year after the expiry date of the finished product
Any deviation from the written procedure shall be "recorded and justified"	Any deviation, which should be recorded, shall have "signed authorisation"
Written procedure is not specifically required to state processing location and principal items of equipment to be used, although the "major equipment" which has, in fact, been used is required to be recorded in the Batch Manufacturing Record	There shall be "a statement of the processing location and the principle equipment to be used" in the written processing instructions
A recorded check on equipment/line/work station clearance not specifically required	"Before processing begins, there should be recorded checks that the work area and equipment is clear of previous products, documents or materials not required … and that equipment is clean and suitable for use"

-- continued

US cGMPs	EC GMP Guide
A recorded check on equipment/line/work station clearance is not specifically required	"Before processing begins, there should be recorded checks that the equipment are clear of previous products, documents or materials not required ... and that equipment is clean and suitable for use"
Gives examples of five different types of in-process control procedures	Requires in-process controls, but does not cite examples
Makes no explicit or implicit statement on whether in-process controls may be performed by production personnel	Implicit (6.18) that some in-process controls may be performed by production personnel, provided they are "performed according to methods approved by Quality Control"
"When appropriate time limits for the completion of each phase of production shall be established" in advance	Does not specifically require the establishment in advance of such time limits, although "dates and times of commencement, of significant intermediate stages and of completion of production" are required to be recorded
Requires written procedures "designed to prevent objectionable microorganisms in products not required to be sterile," which must be followed	"At every stage of production, products and materials should be protected from microbial and other contamination" (5.10); There are also three paragraphs (5.18–5.20) on "Prevention of cross-contamination in production"
Written procedures designed to prevent microbiological contamination of sterile products shall be established and followed; "such procedures shall include validation of any sterilization process"	There is a detailed Annex (1) on "Manufacture of Sterile Products." This requires, inter alia, that "All sterilisation processes should be validated"; there is, also an Annex (15) on "Qualification and Validation"

RATIONALE

The rationale underlying documentation requirements (that is, the requirement to have authorized written procedures, to follow those procedures, and to make records as work proceeds) has been discussed in the previous chapter of this book (Chapter 6.). To restate, briefly, the objectives are:

1. To state clearly, in advance and in writing, what is to be done
2. To do it — in accordance with those instructions
3. To record what was done, and the results of doing it

Documentation or records help to build up a detailed picture of what a manufacturing function has done in the past and what it is doing now, and thus provides a basis for planning what it is going to do in the future.

The formal, and recorded determination of intermediate and final yields and their comparison with preestablished standard theoretical yields is good sense.

If yields are significantly in excess of, or below, theoretical then something has gone wrong, which at very least requires careful investigation. The same applies to actual processing times, which either exceed or fall short of established time limits.

Distinctive identification of processing equipment, storage containers, processing lines, and so on are all measures designed to minimize the possibility of mix-up and contamination, as is the requirement that rejected materials and products should be clearly identified as such, and segregated.

The nature, hazards, and control of environmental contamination (microbial, chemical, and physical) have already been discussed in Chapter 3 ("Premises/Buildings and Facilities") of this book.

IMPLEMENTATION

It is important to keep in mind that the main purpose of process documentation is to ensure, and record, correct *action* and not to be an end in itself.

The illustrative documents shown (Figure 7.1 and Figure 7.2) are not intended to be immutably definitive and are only to be taken as examples of possible layouts, with a form and content that are generally considered to be both suitable and acceptable. Individual manufacturers will have their own preferences, style, and approach. It is also to be noted that a number of manufacturers, particularly the larger ones, have turned, or are turning, to electronic documentation and record keeping. Nevertheless, worldwide, the majority is still largely reliant upon manual and paper systems. For those who are not, these examples may still serve as a basic indicator of what is required to be written in, and the data to be captured by, an electronic documentation system.

Issue of Materials for Manufacture

In some manufacturing companies, the material dispensing and checking operation is recorded on a separate dispensing record sheet. It is more common, convenient, and reliable for this operation to be covered by part of a copy of the master of a "Batch Manufacturing Formula, Method and Record" (or some such similar title). This is the approach that is illustrated by the example document in Figure 7.1. This example should be largely self-explanatory and in conformity with the general requirements for documentation as discussed earlier.

In addition, the Starting Materials Store will need to keep a separate record, for each batch delivery of material, of receipts issues and balances. A possible layout is suggested in Figure 7.2. This information could be recorded in a card index, in a book, or on a computer. The importance of such a record extends beyond materials inventory control and accounting. It enables comparisons to be made between "book" and actual physical stocks. This could be invaluable in investigating, and perhaps preventing, manufacturing errors.

Edition: 1-5-98

Phantazpharm Inc.

BATCH MANUFACTURING FORMULA, METHOD and RECORD

DEXTERIUM ADIPATE OINTMENT 2.5%	BATCH No.
Form: ANHYDROUS OINTMENT	Batch Size: 2000 kg

PROCESS START AND COMPLETION TIMES
START: TIME:
  DATE:
	COMPLETION: TIME:
	  DATE:

FORMULA and WEIGHING RECORD

MATERIAL	CODE	AMOUNT	LOT NO.	DISPENSED BY	CHECKED BY
WHITE SOFT PARAFFIN UP	123	1750.00 Kg			
HEAVY LIQUID PARAFFIN UP	130	200.00 Kg			
DEXTERIUM ADIPATE UP	1,376	50.00 Kg			

Name of dispenser * .... Initials .... Name of checker * .... Initials ....
*Enter names in block capitals

MANUFACTURING LOCATION: Processing room no. 5b, Block C

MAJOR EQUIPMENT: Dedicated ointment manufacturing equipment -
  OINIMILL KZ3 MILL.- E132OO1
  Steam-jacketed PZ /2 S/S MIXING VESSEL, fitted with HITECH 2 STIRRER
  Mobile S/S holding tank
  Dedicated transfer pipework and pump

Batch document issued by .... Date ....

MASTER DOCUMENT EFFECTIVE 1-5-98	SUPERSEDES: 10 - 7- 95	
Prepared by: Date:	Approved by: QC: Date:	Production: Date:

Figure 7.1A   Batch Manufacturing Formula, Method, and Record

Edition: 1-5-98

BMR: DEXTERIUM ADIPATE OINTMENT 2.5%	BATCH SIZE: 200 kg

MANUFACTURING INSTRUCTIONS & RECORD

Name of operator * .... Initials .... Name of checker * .... Initials ....

*Enter names in block capitals

	DATE	TIME	OP'TOR	CHECK

1. Check the Plant Log, and enter details of following:

    1.1 Previous equipment use: Product ....
                                  B. No. ....
                                    Date ....

    1.2 Last equipment cleandown (SOP 03.16)
                             Date ....

2. Obtain clearance from Production supervisor and QC that equipment is clean and ready for use:

                   PRODUCTION CLEARANCE ... .

                   QC CLEARANCE ....

3. Add the 50.00 kg Dexterium adipate to the 200.00 kg heavy liquid paraffin in a clean stainless steel container and stir by hand, using a stainless steel rod.

4. Immediately transfer the slurry to the KZ3 mill .

5. Mill at MEDIUM speed setting for 15 minutes.

                            Mill start time

                            Mill stop time

6. Open valve to supply steam to jacket of PZ/2 mixing vessel, and transfer the 1750.00 kg White soft paraffin to the mixing vessel.

7. When the temperature of the molten White soft paraffin reaches 70 to 75°C close the steam valve and commence stirring at speed setting 5. Record the temperature of the melt: Temperature ....

8. While stirring continues, add the slurry from the mill to the molten White soft paraffin in the mixing vessel.

9. As mixture in mixing vessel cools, continue to stir for a total of 2.5 hours.   Stirrer start time

                            Stirrer stop time

Etc. Etc. Etc.

Figure 7.1B Batch Manufacturing Formula, Method, and Record

Edition: 1-5-98

BMR: DEXTERIUM ADIPATE OINTMENT 2.5%	BATCH SIZE: 200 kg

MANUFACTURING INSTRUCTIONS & RECORD (Contd.)

Etc. Etc. Etc.

	DATE	TIME	OP'TOR	CHECK

13. Pump ointment to tared mobile stainless steel holding tank.

14. Weigh holding tank (and contents) on floor scales, record gross weight, and enter net yield of bulk ointment.
Gross weight (tank + ointment) ....
Tare ....
Yield of bulk ointment ....
STANDARD ACCEPTABLE BULK YIELD - 1950 to 2050 kg

14. Arrange for sample of bulk product to be taken by Q.C. sampler

AUTHORIZED DEVIATIONS FROM STANDARD PROCEDURE:

PRIOR AUTHORIZATION: Production .... Date............ QC .... Date.............

NON-STANDARD EVENT RECORD (Include explanation of any non-standard yield):

PRODUCTION CERTIFICATION: Apart from any authorized deviations, or non-standard events, as recorded above, this batch has been manufactured in accordance with this BATCH MANUFACTURING FORMULA, METHOD, and RECORD.

Signed .... Production supervisor
Date ....

QUALITY CONTROL APPROVAL FOR PACKAGING

This bulk batch of DEXTERIUM ADIPATE OINTMENT 2.5%, B. No. ....
has been sampled and tested, and is RELEASED FOR PACKAGING.
Signed .... Quality Control
Date ....

IF BATCH IS NOT RELEASED, FOR ANY REASON, BOLDLY DELETE THIS SECTION AND RECORD REASON:

Page 4 of 4

Figure 7.1C **Batch Manufacturing Formula, Method, and Record**

MATERIALS STOCK RECORD

Code No. ....(a)............... Material ....(b) .... Supplier..................

Date received ....   Lot No. ....

Date released ....   Amount to stock ....(c)..........

Date	Amount issued ....
	Balance ....
Date	Amount issued ....
	Balance ....
Date	Amount issued ....
	Balance ....
Date	Amount issued ....
	Balance ....
Date	Amount issued ....
	Balance ....
Date	Amount issued ....
	Balance ....
Date	Amount issued ....
	Balance ....

(a) i.e., Internal company code number for material
(b) i.e., Name of material as ordered
(c) i.e., Quantity of released material passed to usable stock

---

Figure 7.2   Materials Stock Record

It is worth noting a few special points regarding the Batch Manufacturing Formula, Method and Record:

1. On the first page (under the Formula and Weighing Record), provision is made for the entry of the names, in block capitals, of the dispenser, and of the checker, against their respective initials. The same provision appears at the top of page 2, in relation to the manufacturing operator and checker. At the time of writing, this is not a common practice, but it deserves to be widely adopted. It overcomes the problem of attempting to interpret indecipherable squiggles, some time after the event. Perhaps more importantly, it also serves to concentrate the minds, before the event, of dispensers, operators, and checkers on the importance of what they are doing, and their personal responsibility to do it properly.

2. Page 1 clearly indicates the manufacturing location, and the major items of equipment that are to be used.

3. Against each step in the manufacturing process, it may be necessary to record a date, time, and a check, in addition to the operator initials. This requirement could well be variable, and this is illustrated by, and catered for, by the way the various boxes are arranged under each column heading, and against each instruction.

4. Before manufacture commences, a record is required of the previous product or batch that was manufactured using the equipment, plus confirmation that this equipment has been cleaned, plus confirmation that it is, in fact, clean (Instruction numbers 1 and 2).

5. At completion of manufacture of the bulk product, provision is made for the recording of:
   a. Authorized deviations
   b. Nonstandard events
   c. Certification by production that, apart from a. or b., the batch has been manufactured in accordance with these instructions
   d. Quality control approval for packaging

Sampling of Products

Problems similar to those discussed (in relation to materials) in Chapter 6, arise in the context of product sampling.

1. Well-stirred bulk liquid products in the form of readily soluble materials dissolved in a solvent vehicle (e.g., water) may be considered to be homogeneous, given that the manufacturing process has been adequately validated to that end, and that the validated process has been followed. Thus, any sample of the bulk liquid may reasonably be taken to be representative. A manufacturer may wish to sample from a bulk manufactured liquid, as a guard against the cost of filling and labeling a product that is later rejected for inhomogeneity. Even so, it clearly makes good sense to take samples for assay from the filling line at the beginning, the middle, and the end of the filling run.

2. Sampling of liquid suspensions, emulsions, and the like may be more problematic. Process validation should have established that, at the completion of the bulk batch the active ingredients are uniformly and homogeneously distributed. However, against the possibility of separation during holding and transfer to filling, sampling from the filling line (filled containers taken at the beginning, middle, and end of the run) must surely be obligatory. Similar considerations also apply to the sampling of creams and ointments.

3. Unit dose solids (tablets and capsules): The commonly employed in-process controls on tablet weights and thickness, and on capsule fill-weights will contribute to the assurance of the uniformity of these products. For an impressive analysis of the problems of sampling bulk powder mixtures, with comprehensive guidance on sampling filled capsules and compressed tablets, see PDA Technical Report No. 25, 1997.[1] An original impetus for this report was the "Wolin Decision" (that is, the judgment of Judge Wolin in the US versus Barr Laboratories case). The report clearly shows that, despite the great store that the US FDA placed upon it, the judge's decision on this aspect of the case was somewhat flawed, both in practical and statistical terms.

Retained (or Reference) Samples

The EC GMP Guide requires that reference samples (in their normal final packaging) from each product batch shall be retained until one year after the expiry date of the batch. This is not a requirement of the US cGMPs, Subpart E, but it is covered under Subpart I (see Chapter 10).

REFERENCES

1. PDA Technical Report No. 25, Blend Uniformity Analysis: Validation and In-process Testing, issued as a supplement to *PDA Journal of Pharmaceutical Science and Technology*, 51/S3, Nov/Dec 1997.

8

PACKAGING AND LABELING CONTROL

The US cGMPs devote a separate Subpart, G, to the topic of packaging and labeling control. The EC GMP Guide considers packaging and labeling to be part of Production and, accordingly, treats this topic in its Chapter 5, Production. Furthermore, a number of the documentary requirements for packaging and labeling are given in EC Guide Chapter 4, Documentation.

US CGMPS

Subpart G — Packaging and Labeling Control

Sec. 211.122 Materials examination and usage criteria

(a) There shall be written procedures describing in sufficient detail the receipt, identification, storage, handling, sampling, examination, and/or testing of labeling and packaging materials; such written procedures shall be followed. Labeling and packaging materials shall be representatively sampled, and examined or tested upon receipt and before use in packaging or labeling of a drug product.

(b) Any labeling or packaging materials meeting appropriate written specifications may be approved and released for use. Any labeling or packaging materials that do not meet such specifications shall be rejected to prevent their use in operations for which they are unsuitable.

(c) Records shall be maintained for each shipment received of each different labeling and packaging material indicating receipt, examination or testing, and whether accepted or rejected.

(d) Labels and other labeling materials for each different drug product, strength, dosage form, or quantity of contents shall be stored separately with suitable identification. Access to the storage area shall be limited to authorized personnel.

(e) Obsolete and outdated labels, labeling, and other packaging materials shall be destroyed.

(f) Use of gang-printed labeling for different drug products, or different strengths or net contents of the same drug product, is prohibited unless the labeling from gang-printed sheets is adequately differentiated by size, shape, or color.

(g) If cut labeling is used, packaging and labeling operations shall include one of the following special control procedures:
(1) Dedication of labeling and packaging lines to each different strength of each different drug product;
(2) Use of appropriate electronic or electromechanical equipment to conduct a 100-percent examination for correct labeling during or after completion of finishing operations; or
(3) Use of visual inspection to conduct a 100-percent examination for correct labeling during or after completion of finishing operations for hand-applied labeling. Such examination shall be performed by one person and independently verified by a second person.

(h) Printing devices on, or associated with, manufacturing lines used to imprint labeling upon the drug product unit label or case shall be monitored to assure that all imprinting conforms to the print specified in the batch production record.

EC GMP GUIDE

Packaging materials and packaging operations are covered, along with starting materials (the US components) and bulk product manufacture in EC Chapter 5, Production. This EC Chapter 5 has two relevant subchapters, headed respectively "Packaging materials" and "Packaging operations." (Note: In EC terminology "Primary packaging materials" are those that come into direct contact with the product.)

Chapter 5 Production

Packaging materials

5.40 The purchase, handling and control of primary and printed packaging materials shall be accorded attention similar to that given to starting materials.

5.41 Particular attention should be paid to printed materials. They should be stored in adequately secure conditions such as to exclude unauthorised access. Cut labels and other loose printed materials should be stored and transported in separate closed containers so as to avoid mix-ups …

5.42 Each delivery or batch of printed or primary packaging material should he given a specific reference number or identification mark.

5.43 Outdated or obsolete primary packaging material or printed packaging material should be destroyed and this disposal recorded.

Chapter 3 Premises and Equipment

3.25 Printed packaging materials are considered critical to the conformity of the medicinal product and special attention should be paid to the safe, and secure storage of these materials.

Packaging operations

5.50 The correct performance of any printing operation (for example code numbers, expiry dates) to be done separately or in the course of the packaging should be checked and recorded. Attention should be paid to printing by hand which should be re-checked at regular intervals.

5.51 Special care should be taken when using cut-labels and when over-printing is carried out off-line. Roll-feed labels are normally preferable to cut-labels, in helping to avoid mix-ups.

The requirements here are broadly similar. As a comparison:

US cGMPs	EC GMP Guide
Requires written procedures to be followed for receipt, identification, storage, sampling, examination and testing	Similar requirement implicit in "the purchase, handling and control of primary and printed packaging materials shall be accorded attention similar to ...starting materials"
Specifically requires examination and testing of packaging materials on receipt *and before use*	Further examination and testing *before use* not specified here. *but*, see EC 5.47 (later in this chapter) — "check on delivery to Packaging Department ..."
Statement regarding use of gang-printed labels; "prohibited ...unless ... adequately differentiated ..." etc.	No corresponding statement regarding gang-printed labels
Specifies three different conditions, one of which must be in place if cut labels are used	Simply states: "roll-feed labels are normally preferable to cut labels ..."
Special security and storage requirements apply to labels and "labeling materials"	Security and storage requirements apply to *all* printed packaging materials, not just labels

US CGMPS

Subpart G — Packaging and Labeling Control

Sec. 211.125 Labeling issuance

(a) Strict control shall be exercised over labeling issued for use in drug product labeling operations.

   (b)   Labeling materials issued for a batch shall be carefully examined for identity and conformity to the labeling specified in the master or batch production records.

   c)   Procedures shall be used to reconcile the quantities of labeling issued, used, and returned, and shall require evaluation of discrepancies found between the quantity of drug product finished and the quantity of labeling issued when such discrepancies are outside narrow preset limits based on historical operating data. Such discrepancies shall be investigated in accordance with Sec. 211.192. Labeling reconciliation is waived for cut or roll labeling if a 100 percent examination for correct labeling is performed in accordance with Sec. 2 11. 12-2 (g) (2)).

   (d)   All excess labeling bearing lot or control numbers shall be destroyed.

   (e)   Returned labeling shall be maintained and stored in a manner to prevent mixups and provide proper identification.

   (f)   Procedures shall be written describing in sufficient detail the control procedures employed for the issuance of labeling; such written procedures shall be followed.

EC GMP GUIDE

Chapter 5 Production

Packaging materials

5.41 ... Packaging materials should only be issued for use only by authorised personnel following an approved and documented procedure.

Packaging operations

5.47 All products and packaging materials to be used should be checked on delivery to the packaging department for quantity, identity and conformity with the Packaging Instructions.

5.56 Any significant or unusual discrepancy observed during reconciliation of the amount of bulk product and printed packaging materials and the number of units produced should be investigated and satisfactorily accounted for before release.

5.57 Upon completion of a packaging operation, any unused batch-coded packaging materials should be destroyed and the destruction recorded. A documented procedure should be followed if uncoded printed materials are returned to stock.

Here, again, the two sets of regulatory requirements are closely similar, with the US cGMPs being somewhat more detailed and specific. It is not entirely clear whether the statement, under Sec. 211.125 that:

   (b)   Labeling materials issued for a batch shall be carefully examined for identity and conformity to the labeling specified ...

is merely a reiteration of the requirement set-out in Sec. 211.122 ("Labelling and packaging materials shall be ... examined or tested upon receipt *and before use..."* (author's emphasis), *or* an indication of a requirement to examine and test for a third time.

The US cGMPs state that there should be reconciliation of "the quantities of labelling issued, used and returned ..." (may be waived for cut or roll labeling "if a 100 percent examination for correct labeling is performed ..."). The corresponding statement in EC GMP Guide is somewhat less positive:

> 5.56 Any significant or unusual discrepancy observed during reconciliation of the amount of bulk product and printed packaging materials and the number of units produced should be investigated and satisfactorily accounted for before release.

This does not appear to unequivocally require that reconciliation should be carried out. Perhaps, however, this may be taken to be covered in the paragraph 5.8, under **General,** at the beginning of EC Chapter 5, Production:

> 5.8 Checks on yields, and reconciliation of quantities, should be carried out as necessary to ensure that there are no discrepancies outside acceptable limits.

US CGMPS

Subpart G — Packaging and Labeling Control

Sec. 211.130 Packaging and labeling operations

> There shall be written procedures designed to assure that correct labels, labeling, and packaging materials are used for drug products; such written procedures shall be followed. These procedures shall incorporate the following features:
>
> (a) Prevention of mixups and cross-contamination by physical or spatial separation from operations on other drug products.
> (b) Identification and handling of filled drug product containers that are set aside and held in unlabeled condition for future labeling operations to preclude mislabeling of individual containers, lots, or portions of lots. Identification need not be applied to each individual container but shall be sufficient to determine name, strength, quantity of contents, and lot or control number of each container.
> (c) Identification of the drug product with a lot or control number that permits determination of the history of the manufacture and control of the batch.
> (d) Examination of packaging and labeling materials for suitability and correctness before packaging operations, and documentation of such examination in the batch production record.
> (e) Inspection of the packaging and labeling facilities immediately before use to assure that all drug products have been removed from previous operations. Inspection shall also be made to assure that packaging and labeling materials not suitable for subsequent operations have been removed. Results of inspection shall be documented in the batch production records.

EC GMP GUIDE

Chapter 5 Production

Packaging operations

5.44 When setting up a programme for the packaging operations, particular attention should be given to minimising the risk of cross-contamination, mix-ups or substitutions. Different products should not be packaged in close proximity unless there is physical segregation.

5.45 Before packaging operations are begun, steps should be taken to ensure that the work area, packaging lines, printing machines and other equipment are clean and free from any products, materials or documents previously used, if these are not required for the current operation. The line-clearance should be performed according to an appropriate check-list.

5.46 The name and batch number of the product being handled should be displayed at each packaging station or line.

5.47 All products and packaging materials to be used should be checked on delivery to the packaging department for quantity, identity and conformity with the Packaging Instructions.

5.48 Containers for filling should be clean before filling. Attention should be given to avoiding and removing any contaminants such as glass fragments and metal particles.

5.49 Normally, filling and sealing should be followed as quickly as possible by labelling. If it is not the case, appropriate procedures should be applied to ensure that no mix-ups or mislabelling can occur.

5.50 The correct performance of any printing operation (for example code numbers, expiry dates) to be done separately or in the course of the packaging should be checked and recorded. Attention should be paid to printing by hand which should be re-checked at regular intervals.

5.51 Special care should be taken when using cut-labels and when over-printing is carried out off-line. Roll-feed labels are normally preferable to cut-labels, in helping to avoid mix-ups

5.52 Checks should be made to ensure that any electronic code readers, label counters or similar devices are operating correctly.

5.53 Printed and embossed information on packaging materials should be distinct and resistant to fading or erasing.

5.54 On-line control of the product during packaging should include at least checking the following:
(a) general appearance of the packages;
(b) whether the packages are complete;
(c) whether the correct products and packaging materials are used;
(d) whether any over-printing is correct;
(e) correct functioning of line monitors.

Samples taken away from the packaging line should not be returned.

5.55 Products which have been involved in an unusual event should only be reintroduced into the process after special inspection, investigation and approval by authorised personnel. Detailed record should be kept of this operation.

5.56 Any significant or unusual discrepancy observed during reconciliation of the amount of bulk product and printed packaging materials and the number of units produced should be investigated and satisfactorily accounted for before release.

5.57 Upon completion of a packaging operation, any unused batch-coded packaging materials should be destroyed and the destruction recorded. A documented procedure should be followed if uncoded printed materials are returned to stock.

Noteworthy differences and variations are:

US cGMPs	EC GMP Guide
"There shall be written procedures..."	A similar requirement not stated here, but covered by paragraph 5.2 (Production, General): "All handling of materials and products, such as ... labelling, packaging and distribution should be done in accordance with written procedures or instructions and, where necessary, recorded"
Prevention of mix-ups and cross-contamination by physical or spatial separation required	Cross-contamination, mix-ups, or *substitution* to be *minimized* by not packaging different lots "in close proximity unless there is physical separation"
Packaged, but unlabeled, product to be identified and handled (according to written procedures) to prevent mislabeling of individual containers	"Normally filling and sealing should be followed as quickly as possible by labelling"; if not, procedures should be applied to ensure no mix-up or mislabeling
Line clearance checks required before start of new run; to be documented	Line clearance checks required — "according to an appropriate check-list"
Packaged products to be identified by a lot or control number, which enables determination of batch history	This is a requirement of other EU legislation
No specific statement on checking functionality of printers, code readers, etc.	Requires checks on performance of printers, electronic code readers, label counters, etc.

US CGMPS

Subpart G — Packaging and Labeling Control

Sec. 211.132 Tamper-evident packaging requirements for over-the-counter (OTC) human drug products

This section of the US cGMPs was introduced into the regulations, following the deaths, in 1982, as a result of the malicious addition of cyanide to Tylenol®

capsules. Strictly, in this author's view, this is *not* a GMP issue, and there is no corresponding section in the EC GMP Guide. (The issue of tamper-evident packaging is covered in separate national and European legislation.) It will, therefore, not be considered further in this book.

US CGMPS

Subpart G — Packaging and Labeling Control

Sec. 211.134 Drug product inspection

   (a) Packaged and labeled products shall be examined during finishing operations to provide assurance that containers and packages in the lot have the correct label.

   (b) A representative sample of units shall be collected at the completion of finishing operations and shall be visually examined for correct labeling.

   (c) Results of these examinations shall be recorded in the batch production or control records.

EC GMP GUIDE

Chapter 5 Production

Packaging operations

5.54 On-line control of the product during packaging should include at least checking the following:
(a) general appearance of the packages;
(b) whether the packages are complete;
(c) whether the correct products and packaging materials are used;
(d) whether any over-printing is correct;
(e) correct functioning of line monitors.

Samples taken away from the packaging line should not be returned.

Both documents are in agreement in requiring in-process (or "on-line") checking. The EC GMP Guide is more detailed and specific in requiring examination for appearance, completeness of packages, correctness of product, packaging materials, and any overprinting and "correct functioning of line-monitors" (presumably this refers to code readers and label counters — see above.) The US cGMPs refer only to examination for correct labeling.

The EC GMP Guide makes the important point that "samples taken away from the packaging line should not be returned." The US cGMPs do not. It is an important point. Ill-controlled return of in-process samples to packaging lines has been a significant cause of product mix-up and contamination.

US CGMPS

Subpart G — Packaging and Labeling Control

Sec. 211.137 Expiration dating

(a) To assure that a drug product meets applicable standards of identity, strength, quality, and purity at the time of use, it shall bear an expiration date determined by appropriate stability testing described in Sec. 211.166.

(b) Expiration dates shall be related to any storage conditions stated on the labeling, as determined by stability studies described in Sec. 211.166.

(c) If the drug product is to be reconstituted at the time of dispensing, its labeling shall bear expiration information for reconstituted and unreconstituted drug products.

(d) Expiration dates shall appear on labeling in accordance with the requirements of Sec. 201.17 of this chapter.

(e) Homeopathic drug products shall be exempt from the requirements of this section.

(f) Allergenic extracts that are labeled "No U.S. Standard of Potency" are exempt from the requirements of this section.

(g) New drug products for investigational use are exempt from the requirements of this section, provided that they meet appropriate standards or specifications as demonstrated by stability studies during their use in clinical investigations. Where new drug products for investigational use are to be reconstituted at the time of dispensing, their labeling shall bear expiration information for the reconstituted drug product.

(h) Pending consideration of a proposed exemption, published in the Federal Register of September 29, 1978, the requirements in this section shall not be enforced for human OTC drug products if their labeling does not bear dosage limitations and they are stable for at least 3 years as supported by appropriate stability data.

There is no comparable coverage of expiration dating in the main body of the text of the EC GMP Guide, although there are a few references to the topic in the EC Guide Annex 18, Good Manufacturing Practice for Active Pharmaceutical Ingredients (closely similar to the FDA's Guidance for Industry — Q7A Good Manufacturing Practice for Pharmaceutical Ingredients). In the EU, this is an issue that must be addressed in an application for a Marketing Authorisation (in UK also called a Product License — Comparable to the US NDA).

DISCUSSION AND IMPLEMENTATION

It may be worth considering the simple question of what is the purpose of a packaging a drug (or pharmaceutical) product?

There are at least two very simple, very obvious, reasons for putting a product in a package:

To *contain* a convenient (or useful) quantity of the product. (We obviously would not want to supply our products to the ultimate consumer "loose," or by the pocketful or handful.)

and

To protect the product during storage, transport and distribution.

But there are a number of other good reasons. The main functions of a package, including the two obvious ones, can be summarized as follows:

1. To *hold*, or contain, a defined quantity of the product
2. To *protect* the product
3. To *identify* the product (what it is, its batch number, who manufactured it, and so on)
4. To indicate required storage conditions, expiry date or shelf life
5. To provide *other information*, for example:
   ■ Directions for use, dosage
   ■ Warning of any hazards in use
   ■ Information on side effects, etc.
6. To present the product in a form that is *easy to use*
7. To help in ensuring that patients take their medicines as, when, and how they should

One of the important things a package does is to *give information.* This is not just a good idea. Much of this (product identity, batch number, manufacturer's name, expiry date, storage conditions, and so on) is also required by law. The crucial point is that this information must be correct. A drug product that is in a package that gives false information could, quite literally, be a matter of life and death.

Protection of the product, as one of the important purposes of packaging, raises the question, protection from what? Products can be harmed (and as a consequence, lose activity, fail to work, or even become dangerous) through:

■ Mechanical damage — Shaking, jarring, dropping, or other impact can break up tablets or capsules or break or crack a container, allowing the product to leak out or contamination to enter.
■ Heat — A number of products will break down or deteriorate at raised temperatures. Although protection against heat is mainly a matter of storage, the package can help.
■ Light — A number of products are sensitive to light, which can cause breakdown and discoloration. Opaque, or colored-glass, containers (or containers placed in cartons) can help protect against this.
■ Humidity — Moisture can severely damage products, so the packaging must provide protection against the product getting damp.
■ The package itself can also harm the product inside it if it is a wrong or badly designed package. Obviously, the product must not be able to leak or escape from the package. Some materials (for example, some plastics) can absorb substances from the product, or allow them to pass through. There is also the possibility that some of the materials used to make

containers can release substances (and what is more, dangerous or poisonous substances) into the product. This applies particularly to some plastics, but some inferior grades of glass can release chemicals or glass flakes into liquid products. Also, some materials used to make containers can react chemically with some products.

So, any material that will:

- Allow the product to leak, seep or permeate through it
- Release substances into the product
- React with the product

must not be used to make containers that will be in direct contact with the product.

It is therefore vitally important that the materials (containers, closures, labels, cartons, outers, and so on) to be used for the packaging of a product are chosen and specified with extreme care, based on a thorough understanding of the product and its chemical and physical relation to its package. In other words, it is crucial that the package is properly designed for the job and to fulfill the purposes listed above.

Over the years, the biggest number of recalls of drug products has been due to errors in packaging. Every effort must be made to prevent packaging errors, especially labeling errors.

Having properly designed a package, it is then at least as important to ensure that in routine, day-to-day packaging operations, product is filled and labeled to form a package that is precisely as designed and specified — day after day, batch after batch, run after run. That is a major concern of GMP in this area of activity.

As with the manufacture of the bulk drug products (where effective control must commence not merely before production begins, but in fact before components or starting materials are ordered), so it must be with packaging. Effective control must commence with the development, design, and specification of the packaging materials (including labels and other printed materials) and then proceed through selection of supplier(s), ordering, receipt, examination and testing, release, holding, and issuance for use.

This point is made in both the two regulatory documents, in statements that we have already encountered. That is:

US CGMPS

Subpart G — Packaging and Labeling Control

Sec. 211.122 Materials examination and usage criteria

(a) There shall be written procedures describing in sufficient detail the receipt, identification, storage, handling, sampling, examination, and/or testing of labeling and packaging materials; such written procedures shall be followed. Labeling and packaging materials shall be representatively sampled, and examined or tested upon receipt and before use in packaging or labeling of a drug product.

And

EC GMP GUIDE

5.40 The purchase, handling and control of primary and printed packaging materials shall be accorded attention similar to that given to starting materials. (Author's emphasis)

Note: To appreciate the full significance of paragraph 5.40 of the EC GMP Guide, it is necessary to refer to the comments in that guide on "Starting materials." Those comments (EC GMP Guide paragraphs. 5.25–5.34) are quoted in full in Chapter 6 of this book (q.v.). In this context, the following paragraphs are especially relevant:

EC GMP GUIDE

Starting materials

5.25 The purchase of starting materials is an important operation which should involve staff who have a particular and thorough knowledge of the suppliers.

5.26 Starting materials should only be purchased from approved suppliers named in the relevant specification and, where possible, directly from the producer. It is recommended that the specifications established by the manufacturer for the starting materials be discussed with the suppliers. It is of benefit that all aspects of the production and control of the starting material in question, including handling, labelling and packaging requirements, as well as complaints and rejection procedures are discussed with the manufacturer and the supplier.

Thus it is crucial that suppliers of packaging materials are selected with care. It must be established, by vendor audits and discussions, that a supplier is competent to supply goods consistently in accordance with requirements, and as specified. They should also be able to supply, on time, as required.

These considerations apply particularly to labels and other printed packaging materials. Errors with labeling or other printed packaging materials (cartons, outers, enclosure leaflets, etc.) have, over the years, been the biggest single cause of product recall throughout the world. The trouble can begin in the print shop where the labels are printed. A low-level mix-up (a "trace cross-contamination") at the printer may be difficult to detect later. Hence, it is crucially important to select a printer who is fully aware of, or who can be made fully aware of, the special hazards of label mix-up in pharmaceutical packaging. There are some printers who specialize in labels for the pharmaceutical industry, and they are worth investigating.

Measures, relevant to the design and supply phases, that will help to minimize the possibility of mix-up and mislabeling include:

1. The use of roll-feed rather than "loose" cut labels. *However* — it is important not to let roll-feed labels engender a sense of false security. They *are* more secure than cut labels, but printers do need to splice

between part rolls, and here there is a potential for mix-up. Printers should be required to clearly mark where a splice occurs.

2. Avoidance of closely similar looking labels — particularly for different strengths of the same product — despite the desire of marketing men to project a consistent company image.

3. Avoidance of gang printing (the printing of different labels on the same sheet for later cutting), unless there is adequate precaution against mix-up — for example, by having the different labels on the sheet very obviously distinguished from one another by size, shape, or color.

4. Including, in the basic label artwork, a distinctive bar (or other) code that can be automatically scanned on line during the packaging operation.

5. Requiring printers to destroy, or return, any obsolete plates or masters before commencement of printing any new or modified version of a label or other printed material.

Having obtained supplies of labels, which, hopefully, are as required and specified, it now becomes crucial to ensure that no mix-ups or "switches" occur during storage, handling, transportation and usage.

There are a number of things that can be done to help prevent mistakes with labels and other printed materials. These include:

1. **Secure storage.** Printed packaging materials, after Quality Control approval, should be held in secure stores (preferably apart from other stores areas), which are locked when they are unsupervised, and where entrance is permitted to authorized staff only.

2. **Security on issue.** When required for use, printed packaging materials should be issued only to authorized persons, in specified quantities, and on presentation of formal requisition documents. *There must be no casual, uncontrolled issue of labels* and other printed materials.

3. **Security in transportation.** Special care needs to be taken to ensure security of printed materials during transport:
    ■ From printer to manufacturer
    ■ From Goods-Receiving to stores
    ■ From stores to packaging line
    It is usually a good idea to use sealed boxes or "cages" for transportation of printed materials around the manufacturing facility.

4. **Following instructions.** All packaging operations should be performed precisely in accordance with the batch packaging instructions. These will normally indicate the materials that are to be used. Before the operation starts, all the materials issued must be checked against the batch packaging instructions to ensure that the correct materials and labels will be used.

5. **Counting and reconciliation.** That is, issuing printed materials in known and specified quantities, and at the end of a batch packaging run comparing the number of packs produced with the amounts of the materials issued, making allowances for any spoilages.

6. **Line clearance checks.** At the end of every packaging run, and before a new run starts, the entire line or location and surrounding areas

should be checked to ensure that there are no products, containers, labels, cartons, or leaflets left over from a previous batch.

7. **Code reading.** Labels and other printed materials can have printed on them codes that uniquely identify them. The most common is, of course, a bar code. Patterns of perforations have also been used. These codes can be read by electronic scanners, or code readers. The best way of doing this is to install scanners or code readers on the packaging line itself. A rather less satisfactory alternative is to perform the code reading separately, off line, before the materials are taken to the packaging line for use. It is crucially important to ensure that the scanners or readers are functioning correctly.

8. **Constant vigilance.** In spite of all mechanical and electronic aids, a major factor in the prevention of potentially fatal packaging errors is the constant care and attention of all packaging operators and checkers. It is impossible to overemphasize the importance of packaging people to the protection of patients and other consumers. A sense of this importance must be instilled into packaging operators during training.

THE PACKAGING OPERATION

Both of the two regulatory documents require that all packaging and labeling operations should be conducted in accordance with written procedures, and that records should be made of significant steps, as the work proceeds.

Figure 8.1 is an example of a master "Batch Packaging Instruction and Record." It is intended to be neither more nor less than an *example,* and other styles, formats, and contents are possible, but it does display a number of the quality checks and controls that are required. The intention, of course, is that an exact copy of such a master should be used for each packaging run and used to record the various phases of the operation.

The first page (of the four pages displayed) clearly indicates at its head, the name of the product, the size of the pack (20 g), and the expected yield of finished product (in package units). A space is provided for the entry of the packaging batch number.

Proceeding further down the first page, noteworthy features are:

■ Spaces for entry of time and date of starting, and completion of this packaging run
■ Statement of the bulk product required, with the amount required and with spaces for entry of the batch number of the bulk product, and for the signature of the person who issued it to the packaging department, and who checked it was the correct bulk product, as required
■ Statement of the packaging materials required, per unit pack, with internal code reference and total quantity, with spaces for entry of material lot numbers, who issued them, and who checked them

Edition : 2-7-01

Phantazpharm Inc.

BATCH PACKAGING INSTRUCTIONS and RECORD

DEXTERIUM ADIPATE OINTMENT 2.5%	BATCH No.
20g Tubes	Batch Size: 2000 kg Bulk Expected Yield: 95,200 20-g tubes
PACKAGING START AND COMPLETION START: TIME: DATE:	COMPLETION: TIME: DATE:

BULK PRODUCT REQUIRED:

MATERIAL	AMOUNT	BATCH NO.	ISSUED BY	CHECKED BY
DEXTERIUM ADIPATE 2.5% OINT.	2000 kg			

UNIT PACKAGING MATERIALS REQUIRED:

MATERIAL	CODE	AMOUNT	LOT NO.	ISSUED BY	CHECKED BY
PRINTED TUBES - "DEXTERIUM ADIPATE 2.5%"	PT 035	95,200			
LEAFLET	OA 014	95,200			
PRINTED UNIT CARTON	PC 021	95,200			

MULTIPLE PACKAGING MATERIALS: UNITS per MULTIPLE - 10x CARTONED TUBES

MATERIAL	AMOUNT	LOT NO.	ISSUED BY	CHECKED BY
PRINTED 10X "DEXTERIUM ADIPATE" OUTER CARTON	9,520			

PACKAGING LOCATION: Room no. 4, Block B

PACKAGING LINE: SCHMIT-WORTLEBERRY SEMI-AUTOMATIC OINTMENT PACKAGING AND CARTONING LINE NUMBER 2.

Batch document issued by .... Date ....

MASTER DOCUMENT EFFECTIVE 2-7-01	SUPERSEDES: 10-8-99	
Prepared by: Date:	Approved by: QC: Date:	Production: Date:

Figure 8.1A Batch Packaging Instructions and Record

BATCH PACKAGING RECORD: DEXTERIUM ADIPATE 2.5% 20-g TUBES	BATCH SIZE: 200 kg/ ca. 95,200 x 20-g TUBES

PACKAGING INSTRUCTIONS

A. Line clearance and cleaning check

1. Check that the packaging line and equipment are clear of all previous, or any other product, packaging materials, labels, leaflets instructions, and the like, and that the line and all associated equipment have been cleaned in accordance with SOP XXX "INSTRUCTIONS FOR CLEANING SCHMIT-WORTLEBERRY AUTOMATIC OINTMENT PACKAGING AND CARTONING EQUIPMENT AND LINE"

LINE CLEARANCE and CLEANING CHECKED by .... CONFIRMED by ....
TIME/DATE .... TIME/DATE ....

2. Record details of previous product and batch packaged on this line:

Previous product .... Batch number .... Signed ....

B. On-line print data check

1. Check that the tube-crimper has been setup to print the correct batch code number on the tubes.

Batch code to be printed on tubes is ....
Checked by .... Time/Date ....

2. Check that the batch code and other on-line printed information to be printed on the unit cartons are as specified. Take a sample carton, printed with this information, and attach to the back of this record.

On-line carton printer setup to print required data . Checked by..............................
Time/Date ....

C. Packaging operation

1. Check that the general appearance/condition of the bulk ointment is satisfactory, and that its labeled identity and batch number are as entered under "Bulk Product Required" on page 1. Checked by ....

2. Set the ointment filler to fill 20.5 g per tube, and proceed to fill and package the ointment in accordance with SOP XXY "INSTRUCTIONS FOR FILLING AND PACKAGING OINTMENTS, USING THE SCHMIT-WORTLEBERRY AUTOMATIC LINE."

Etc. Etc.

Figure 8.1B   Batch Packaging Instructions and Record

Edition : 2-7-01

Page 3 of 4

BATCH PACKAGING RECORD: DEXTERIUM ADIPATE 2.5% 20-g TUBES	BATCH SIZE: 200 kg/ ca. 95,200 10-g tubes

IN-PROCESS CONTROL RECORD   **Batch Number** ....

Time -													
Checks for ID, Clean, Defects	a. Tubes b. Caps												
Fill Weight -													
Cap -	Sample 1												
	Sample 2												
Seal/Tightness	Sample 3												
	Sample 4												
	Tubes												
Checks for -	Leaflets												
- ID &	Cartons												
Defects	Outers												
Tube print	B N												
Carton	B N												
Overprint	EXP												
	MFG												
Shipping	B N												
Case	EXP												
Marking	MFG												
Checked by													
Comment													

Figure 8.1C   Batch Packaging Instructions and Record

Edition : 2-7-01

BATCH PACKAGING RECORD: DEXTERIUM ADIPATE 2.5% 20-g TUBES	BATCH SIZE: 200 kg/ ca. 95,200 10-g tubes

Etc. Etc.

13. On receipt of QC approval, send total yield of finished packaged goods to Finished Products Warehouse, obtaining a documented receipt for the batch and quantity supplied.

AUTHORIZED DEVIATIONS FROM STANDARD PROCEDURE:

PRIOR AUTHORIZATION: Production .... Date............ QC .... Date.............

NON-STANDARD EVENT RECORD (Include explanation of any non-standard yield):

PRODUCTION CERTIFICATION: Apart from any authorized deviations, or non-standard events, as recorded above, this batch has been manufactured in accordance with the relevant BATCH MANUFACTURING FORMULA, METHOD, and RECORD, and packaged in accordance with this BATCH PACKAGING RECORD.

Signed .... Production Manager
Date ....

QUALITY CONTROL RELEASE

This bulk batch of OPPROBRIUM APOSTATE OINTMENT 2.5%, B. No. ....
has been sampled and tested and found satisfactory, all documentation has been reviewed, and all relevant factors have been taken into consideration, and this batch is released for distribution.

Signed .... Quality Control
Date ....
IF BATCH IS NOT RELEASED, FOR ANY REASON, BOLDLY DELETE THIS SECTION AND RECORD REASON BELOW.

Figure 8.1D Batch Packaging Instructions and Record

- Statement of the outer packaging materials required (in this case "outers" of 10 tubes), with amount and spaces for entry of the outer lot number and who issued and who checked
- Statement of the packaging location, and the equipment to be used
- At the bottom of the page is a section that is a constant feature of the master. This, following a statement of the date from which it becomes effective, and of the date of the previous master that it supersedes, provides spaces for the signatures (with date) of:
  - The person who prepared the master
  - The person approving it on behalf of Quality Control
  - The person approving it on behalf of Production.

Page 2 commences with instructions for two important initial checks:

1. Check on line-clearance and cleaning, with record to be made of previous product and batch
2. Check on on-line print data, in this case the batch number to be impressed on the tubes by the crimping machine and the batch number to be printed on the unit carton

Following this are the instructions, only the first two of which are shown for illustrative purposes, for the packaging operation itself.

The third page illustrates an integral in-process control record and the final page (4) shows the last of the operational instructions, followed by sections for the recording of:

- Any deviations from standard procedure
- Any nonstandard event
- Certification by the Production Manager
- Release (or otherwise) by Quality Control

(In the European Union provision will be necessary for certification by a "Qualified Person.")

Keeping in mind the considerable potential for mix-up and contamination to occur in the packaging operation, the following measures will reduce that potential:

- Good spatial separation between packaging lines or stations
- Better yet, physical barriers between lines
- Adequate space for the orderly assembly of bulk product and packaging materials at the beginning of each line
- Adequate space for the orderly assembly of the finished packs at the end of the line, prior to transportation to the finished good warehouse
- Dust extraction over bulk tablet hoppers
- Coverage provided for other bulk containers, hoppers, open empty unit containers, and filled but unclosed containers
- Ensuring that during stoppages (refreshment breaks, shift changes, break-downs, alarm drills, etc.) all filled units are sealed, and that the line remains supervised

9

HOLDING AND DISTRIBUTION

US CGMPS

Subpart H — Holding and Distribution

Sec. 211.142 Warehousing procedures

Written procedures describing the warehousing of drug products shall be established and followed. They shall include:

(a)   Quarantine of drug products before release by the quality control unit.
(b)   Storage of drug products under appropriate conditions of temperature, humidity, and light so that the identity, strength, quality, and purity of the drug products are not affected.

Sec. 211.150 Distribution procedures

Written procedures shall be established, and followed, describing the distribution of drug products. They shall include:

(a)   A procedure whereby the oldest approved stock of a drug product is distributed first. Deviation from this requirement is permitted if such deviation is temporary and appropriate.
(b)   A system by which the distribution of each lot of drug product can be readily determined to facilitate its recall if necessary.

EC GMP GUIDE

Chapter 1 Quality Management

1.2 … The system of QA … should ensure that:
…

(viii) Satisfactory arrangements exist to ensure, as far as possible, that the medicinal products are stored, distributed and subsequently handled so that quality is maintained throughout their shelf life.

1.3 The basic requirements of GMP are that:

...

(vii) Records of manufacture, including distribution, which enable the complete history of a batch to be traced, are retained in a comprehensible and accessible form.

(viii) The distribution (wholesaling) of the products minimises any risk to their quality.

(ix) A system is available to recall any batch of product from sale or supply.

Chapter 5 Production

5.5   Incoming materials and *finished products should be physically or administratively quarantined immediately after receipt or processing, until they have been released for use or distribution.* (Author's emphases)

5.58   Finished products should be held in quarantine until their final release under conditions established by the manufacturer.

5.60   After release, finished products should be stored as usable stock under conditions established by the manufacturer.

Chapter 4 Documentation

Specifications for Finished Products

4.13 Specifications for finished products should include:
   (a)   the designated name of the product and the code reference where applicable;
   (f)   *the storage conditions and any special handling precautions,* where applicable; (Author's emphasis)
   (g)   the shelf-life.

Other

4.25   records should be maintained of the distribution of each batch of a product in order to facilitate the recall of the batch if necessary.

Chapter 3 Premises and Equipment

Storage Areas

3.18   Storage areas should be of sufficient capacity to allow orderly storage of the various categories of materials and products: starting and packaging materials, intermediate, bulk and finished products, products in quarantine, released, rejected, returned or recalled.

3.19   Storage areas should be designed or adapted to ensure good storage conditions. In particular, they should be clean and dry and maintained within acceptable temperature limits. Where special storage conditions are required (e.g., temperature, humidity) these should be provided, checked and monitored.

3.20 Receiving and dispatch bays should protect materials and products from the weather. Reception areas should be designed and equipped to allow containers of incoming materials to be cleaned where necessary before storage.

3.21 Where quarantine status is ensured by storage in separate areas, these areas must be clearly marked and their access restricted to authorised personnel. Any system replacing the physical quarantine should give equivalent security.

3.22 There should normally be a separate sampling area for starting materials. If sampling is performed in the storage area, it should be conducted in such a way as to prevent contamination or cross-contamination.

3.23 Segregated areas should be provided for the storage of rejected, recalled or returned materials or products.

3.24 Highly active materials or products should be stored in safe and secure areas.

The most explicit European statement on stock rotation is to be found not in the main body of the EC GMP Guide, but in a supplementary set of guidelines, Guidelines on Good Distribution Practice (GDP) of Medicinal Products for Human Use (94/C 63/03). (See Annex 1 to this chapter, where these guidelines are given in full.) These guidelines were issued in support of the European Directive 2001/83/EC, which introduced the requirement for wholesale dealers to obtain a formal authorization to engage in such activities. The relevant paragraph reads:

From EU GDP Guidelines –

15. There should be a system to ensure stock rotation ('first in first out') with regular and frequent checks that the system is operating correctly. Products beyond their expiry date or shelf life should be separated from usable stock and neither sold or supplied.

DISCUSSION

This subpart of the US cGMPs is relatively short, consisting of just two quite brief sections, concerned, respectively, with warehousing procedures and distribution procedures. The EC GMP Guide does not contain a separate chapter on warehousing and distribution (however, see Annex 1 to this chapter). While it is, perhaps, not quite so emphatic regarding the primacy of *written procedures,* the net *practical* effect of the EC GMP Guide is closely similar.

It cannot, indeed *must not,* be considered that concern for the quality of the products of the pharmaceutical industry may cease at the point where the product is filled, sealed, labeled, and approved or released by Quality Control. True *Quality Assurance* should extend right up to the point where the product is delivered to the ultimate consumer — the patient. Certainly, there will come a point where the influence that the manufacturer is able to exert will significantly decline. For example, the manufacturer can do little more than *advise* the dispensing pharmacist on the correct handling and storage of his drug products. Thereafter, the influence of the manufacturer becomes distinctly tenuous. Despite warnings and advice to patients given in enclosure leaflets, it does seem that many patients neither handle nor take their medicines properly. That said, it is incumbent upon pharmaceutical manufacturers to ensure that having manufactured, packaged and labeled their products, the quality (i.e., "fitness") of these product remains unimpaired for as far along the supply chain as they are able to exert influence. In the EU, legislation requires that any organization engaging in the activity of wholesaling medicinal products must be in possession of an Authorization to do so (see Annex 1 to this chapter). Manufacturers who distribute via external wholesale dealers should thus ensure that any such wholesale dealer is, indeed, in possession of an Authorization.

IMPLEMENTATION OF GOOD PRACTICE IN HOLDING AND DISTRIBUTION

Whatever their size and type, stores or warehouses all have a few things in common — they receive and take in goods or materials, they hold them (hopefully, safely and securely) for a while, and then they send them out again. Put very simply, even naively, its just a matter of Goods In, Goods Hold, Goods Out. It may all *seem* simple, but it is worth pausing to think of how important it really is. Pharmaceutical products can do a lot of good — if they are of the right quality and are used properly. If they are incorrect, damaged, soiled, contaminated, wrongly labeled, have the wrong instructions for use, or have deteriorated, they could fail to have their desired good effects, and could be a danger to the health (or even the life) of the ultimate consumer or patient.

The principal function of a store or warehouse is to provide a buffer defense against, probably unpredictable and random, fluctuations in supply and demand. A few years ago a "hot" management fashion statement was "Just in Time." The ultimate expression of this concept would be the complete elimination of all stores and warehouses in all industries. Suppliers of materials and components would produce them just in time to deliver them to manufacturers just in time for them to use them. The manufacturers, in turn, would manufacture their products just in time for them to be delivered to their customers, just as they needed them. To no great surprise, "Just in Time" does not seem to have been successfully implemented — certainly not in the pharmaceutical industry. For the foreseeable future, we may expect to see stores and warehouses providing that "buffer defense."

The Goods In phase provides an opportunity of checking that purchased materials or bought-in products, or finished products delivered from an internal packaging line, are correct and in good condition. If rubbish is taken in, it will remain rubbish.

The Goods Hold stage is where it is necessary to ensure that the goods *remain* in good condition, and do not become harmed or damaged through incorrect or unsuitable storage conditions or bad handling. That is, it is important to ensure that quality goods are not reduced to rubbish.

Goods Out might well be the last chance of checking and ensuring that everything is in order before the goods leave a manufacturer's hands, to the next step in the distribution chain, on their way to the consumer.

It cannot be overstressed that *people* in stores and warehouses play a vital part in the Quality Assurance of pharmaceutical products. They must be properly trained and fully aware of the significance of the job they are doing.

Goods In

There are at least three different categories of goods that are received into a manufacturer's warehouse:

1. Finished Products received from the manufacturer's own packaging lines.
2. Products and materials that come from another supplier or manufacturer. Manufacturers will usually receive components (starting materials) and packaging materials. They also may receive products, in either bulk or finished-pack form, from other manufacturers.
3. Returned or recalled products.

It might be thought that the receiving into a manufacturer's own warehouse of its own products from its own packaging line(s) is an inherently secure operation, certainly safer than the receipt of goods from outside the company. True it *should* be. But it is never a good idea to be overconfident, and it is better to remember that an essential feature of Quality Assurance is that there should be vigilant care and concern for quality, *all along the line*. The receiving of finished packaged products into the manufacturer's warehouse should be regarded as a further opportunity to check that everything is correct and in good condition. Details of these stock movements, like all others, must be recorded.

The receipt of goods or materials from other suppliers or manufacturers has already been considered in Chapter 6 (Materials Control). For the present, it is only necessary to reemphasize that whenever goods or materials of any sort are received by a manufacturer from a source outside the company, it is essential to ensure that the goods are:

■ The right goods — as ordered
■ The right quantity — as ordered
■ The right grade (or quality) — as ordered
■ Correctly labeled
■ Clean
■ Dry
■ Undamaged

Goods-In staff cannot, usually, be expected to carry out chemical tests to completely ensure the correct identity and quality of the goods received. What they can and must do is ensure that everything is in order as far as the external appearance and labeling of the goods concerned. The Goods Inwards (or Receiving) office will normally have a copy of the original purchase order, and the supplying company will usually send with the goods some form of delivery (or advice) note. The order, the delivery note, and the labeling on the goods should all be compared with each other to ensure that everything ties up. At the same time, the delivered goods should be checked for quantity, cleanliness, condition, and for any signs of damage or deterioration. If anything appears to be wrong, it should be reported by the Goods-Inwards staff immediately, so a decision can be made about accepting the delivery or sending the goods back.

There also needs also to be a check on the batch number(s) of the delivery, to see if they match up with the batch numbers on the supplier's delivery (or advice) note. When a delivery of a particular product or material consists of more than one supplier's batch number, the different suppliers' batches should be kept apart from each other, as far as recording, handling, and storing — and any sampling and testing that may be required — are concerned.

Recording the Delivery

A record must be made of the receipt of each delivery. The record should include:

- Date of receipt of goods
- Manufacturer's name and reference number for the goods
- Manufacturer's order number
- Name of supplier
- Supplier's batch number(s)
- Quantity received
- Number of containers received
- Internal lot number given to the delivery
- A check signature (or initials) to confirm that the delivery has been examined for general condition, cleanliness, lack of damage, etc. and was found to be satisfactory

All these activities should be performed, and records made, in accordance with the relevant SOP(s). If a period of quarantine is necessary, the goods should be labeled and stored appropriately, as required by SOP.

Care needs to be taken to ensure that goods that require special storage conditions (for example, low temperature, extra security) are placed in this special storage as soon as possible.

(Fuller details of the goods-receiving operation, with an example of a relevant SOP and suggested formats for records, etc., are given in Chapter 6, Materials Control, of this book.)

Returned or Recalled Products

It is usual to make a distinction between "returns" and "recalled products." Returns are products returned from the market to a manufacturer's warehouse, which are not specifically known to be seriously defective, but which have been sent back by a wholesale or retail customer because of overstocking, superficial damage, or some such similar reason. Recalled roducts are products that have been withdrawn from the market, at the request of the manufacturer, or the authorities, because of a known or suspected defect.

Every company will have its own policy and procedure for dealing with returns. Obviously, these goods must be kept securely apart from other stocks until a decision is made on what to do with them. The relevant subpart of the US cGMPs reads as follows:

US cGMPs

Subpart K — Returned and Salvaged Drug Products

Sec. 211-204 Returned drug products

> Returned drug products shall be identified as such and held. If the conditions under which returned drug products have been held, stored, or shipped before or during their return, or if the condition of the drug product, its container, carton, or labeling, as a result of storage or shipping, casts doubt on the safety, identity, strength, quality or purity of the drug product, the returned drug product shall be destroyed unless examination, testing, or other investigations prove the drug product meets appropriate standards of safety, identity, strength, quality, or purity. A drug product may be reprocessed provided the subsequent drug product meets appropriate standards, specifications, and characteristics. Records of returned drug products shall be maintained and shall include the name and label potency of the drug product dosage form, lot number (or control number or batch number), reason for the return, quantity returned, date of disposition, and ultimate disposition of the returned drug product. If the reason for a drug product being returned implicates associated batches, an appropriate investigation shall be conducted in accordance with the requirements of Sec. 211.192. Procedures for the holding, testing, and reprocessing of returned drug products shall be in writing and shall be followed.

The relevant paragraph of the EC GMP Guide states:

EC GMP Guide

> 5.65 Products returned from the market and which have left the control of the manufacturer should be destroyed unless without doubt their quality is satisfactory; they may be considered for re-sale, re-labelling or recovery in a subsequent batch only after they have been critically assessed by the Quality Control Department in accordance with a written procedure. The nature of the product, any special storage conditions it requires, its condition and history, and the time elapsed since it was issued should all be taken into account in this assessment. Where any doubt arises over the quality of the product, it should not be considered suitable for re-issue or re-use, although basic chemical reprocessing to recover active ingredient may be possible. Any action taken should be appropriately recorded.

Both the US and the European regulatory statements are clear and rational enough and need little further comment or justification, other than to suggest that, on the whole, it is preferable to *destroy* returns. Rarely can any economic justification be worth the risk of returning to the market products that may have been mishandled or tampered with in some undetectable way.

Recalled Products will also be received into the warehouse. These are products that are known or suspected to be defective and have been recalled from the market at the request of the regulatory authority or the manufacturer. It is required that a manufacturer should have a written recall procedure, and make and keep various records relating to complaints and recalls. This will be covered in chapter 11 of this book, Records and Reports. For now, it is only necessary to stress that on receipt of both returned and recalled products:

■ Full details of the receipt of the returned or recalled goods must be recorded.
■ Returned and recalled goods must be carefully marked (or labeled) as such and carefully and securely set apart from other goods until the final decision has been made about what to do with them.

Goods Holding

All goods must be stored in a clean, neat, orderly way, in conditions that will not affect their quality or cause them to deteriorate in any way. It is not just an issue of looking good. Untidy, scruffy stores are more difficult to run and control. They increase the possibility of mix-up and confusion — mix-up of different types of goods, mix-up of different batches (or lots), mix-up of goods of different status.

Neat, orderly stores are easier to keep clean. Dirt and dust cannot only spoil the appearance of products and materials, they could contaminate them.

It is very difficult to have effective Stock Rotation (FIFO) unless goods are stored in an orderly fashion. Also, in untidy, scruffy stores, goods are more difficult to handle and move and the chances of damage to products and their containers increases. It must be remembered that it is not only the damaged goods themselves that are spoiled — spillages from, say, broken bottles of a liquid can spoil or contaminate other goods. The environmental conditions in which products are stored can have a significant effect upon them. The impact varies from one sort of product to another. Some are tougher, and some are more sensitive than others. But a number of products can be badly affected by moisture (dampness), temperature (either too hot or too cold), and light.

Many things, not just pharmaceutical products, are spoiled by getting wet, so it will be no surprise that the goods we store must be protected from dampness, wetness, rain, and seepage. Hence, the importance of making sure that stores (as well as the Receiving and Dispatch bays) are protected from the weather.

As far as temperature is concerned, many of our goods remain stable (for at least as long as their shelf life, or expiry date) at normal room temperature. However, there are some goods that need to be stored in a cool place (generally below 15°C) and some that need to be stored in a refrigerator or even a freezer. But it needs to be noted that some pharmaceutical products can be spoiled if they are kept too cold. Some can be completely ruined if they are frozen.

Light, particularly bright sunlight, can also seriously affect some products, causing them to be completely ineffective — and even if this does not happen, light can cause fading, or even bleaching, of labels.

So, it is very important that products and materials are stored under the right conditions, that is, conditions that are right for them. However, where goods do require special temperature storage conditions, it is not good enough just to provide cool or cold storage cabinets, refrigerators, freezers, or cold rooms. It is very necessary to keep a regular or constant check that the correct temperature is being maintained. In a small unit, these checks may be made, at least daily, using a maximum/minimum thermometer, and recording the check in a log book. Larger cool rooms, or cold rooms, should be fitted with continuous recording device, such as a chart recorder or a digital printout. These need to be checked at least daily, and the checks recorded.

Goods Out — Distribution of Products

Some manufacturers may supply some of their products directly to retail outlets or to hospitals. Many products are supplied by manufacturers to wholesale dealers. In all cases, the same sort of Good Practices are necessary to ensure and maintain the quality of the products.

Particular care is necessary in the picking and assembly of orders for dispatch. It is vitally important to ensure that the items picked are as specified in the customer's order. But it goes further than that. This is perhaps the last chance to check that everything is OK. It is not only important to ensure that the right amounts of the right products, of the right strengths and sizes, are being picked for dispatch. It is also important that a watchful eye is kept open to check that the products being picked are in good condition, that they have been approved for distribution, and that they have not passed their expiry date (or shelf life).

It is necessary to make, and keep, a record of each order that is dispatched, which shows:

- Date of dispatch of goods
- Customer's name and address
- Quantity, name, batch number, and expiry date of each product dispatched

It would be a mistake to think that this is just another example of tiresome bureaucracy. If there is anything wrong with a product, it could well make a difference between life and death to have a chain of records that will enable a complete trace, connecting an individual pack of product with the manufacturer — right back to the details of the materials used in making the product. Without this sort of comprehensive record, it would be difficult, even impossible, to investigate the causes of any problem, and (more importantly) to prevent any further damage being done. The records of dispatch are a vital link in this chain. (The issue of recall procedures and other related documentation will be addressed in Chapter 11 of this book, Records and Reports.)

Cold Storage

A number of pharmaceutical products require storage at low temperatures. Higher temperatures can degrade some of them, and some are denatured by freezing. It is therefore very important that such products are held within a defined, lower temperature range.

The temperature in refrigerators used to store pharmaceutical products should be monitored continuously and records maintained of maximum and minimum daily temperatures. Sufficient space should be maintained within to permit adequate air circulation. Refrigerators used for vaccines and similar products should be capable of maintaining the internal temperature between 2°C and 8°C. Temperature monitoring devices should have an accuracy of +/- 0.5°. They should be readable from outside the refrigerator. Refrigerators should not be sited where extremes of temperature will affect their performance.

Large commercial refrigerators and walk-in cold rooms should be monitored with an electronic temperature-recording device that measures load temperature in one or more locations, depending on the size of the unit. Portable data loggers that can be downloaded onto a computer can be used instead of a fixed device. Records should be checked daily. Internal air temperature distribution should be mapped on installation of the facility, in the empty and full state, and thereafter annually under conditions of normal use.

Temperature alarms should be fitted to large and walk-in units and to those smaller units used to store products at risk from freezing.

Controlled Room-Temperature Storage

A maximum/minimum thermometer placed at selected locations within the room and read, recorded, and reset at least weekly (more frequently during periods of exceptionally hot or cold weather) will provide a simple means of temperature monitoring within a room-temperature store. Temperatures should be monitored at all levels, from floor level to top shelf, or pallet, level. Continuous temperature recording is desirable in large warehouses.

All warehouses should be temperature mapped to determine the temperature distribution under extremes of external temperature. Mapping should be repeated annually and after any significant modification to the premises, stock layout, or heating system. Pharmaceutical products should not be stored in areas shown by temperature mapping or other consideration to be unsuitable, e.g., at high level in poorly insulated stores, or next to heaters.

Transportation

Cold-Chain Distribution

The route and time of transportation, the local seasonal temperatures, and the nature of the load should all be considered when arranging distribution of goods that need to be transported under cold conditions (the "cold chain"). For small volumes of cold-chain goods, insulated containers may be satisfactory, but it is vital that products that can be damaged by freezing are prevented from coming into direct contact with ice packs at subzero temperatures.

Larger volumes of cold-chain goods should be shipped in refrigerated transport, particularly if transit times may be prolonged. Temperatures within loads of products at risk from freezing should be strictly controlled and monitored with recording probes or individual temperature monitoring devices. The temperature records for each consignment should be reviewed and there should be a procedure for implementing corrective action in the case of actual, or suspected, exceeding of temperature limits.

Distributors should ensure that consignments of cold-chain goods are clearly labeled with the required storage and transport conditions. Receivers should satisfy themselves that the goods have been transported under appropriate conditions and should place them in appropriate storage facilities as soon as possible after receipt.

Other Goods

Consideration should be given to the possible extremes of temperature inside uninsulated, unventilated delivery vehicle, and precautions should be taken to protect all products from heat damage. This should include representatives' samples kept in car boots and goods distributed using mail services.

Systems Checks and Calibration

The performance of temperature measuring and monitoring systems that are critical to ensuring the quality of the product should be tested and shown to be capable of achieving the desired result. Measuring and recording devices should be calibrated against a traceable reference device. Records should include pre- and post-calibration readings and details of any adjustments made or corrections to be applied. Alarms should be checked for correct functioning at the designated set temperatures.

Transit of Products

It is important that products are transported in a way that will ensure that:

- The identity of products is not lost
- They are not damaged, soiled, or spoiled
- They do not get contaminated and do not contaminate other products
- They are protected against breakage or spillage

Another point that needs to be remembered is that it is pointless to take great care to ensure that goods that need to be kept cool or refrigerated are stored under these conditions, and then to send them off in a hot truck. The required temperature conditions should also be maintained while the goods are being transported. (see above).

Facilities for Storage/Warehousing

The following is a kind of "summary digest," taken from the sections on facilities for stores or warehouses, found in various Good Practice regulations and guidelines:

1. Premises where products and materials are stored should (as far as possible) be sited where the risk of contamination from the local environment, or from other nearby activities, is low.
2. They should be soundly constructed, maintained in a state of good repair, and provide protection against the weather and the entrance of insects, pests, vermin, and birds.
3. Weather protection should be provided at receiving, unloading, loading, and dispatch bays.
4. Stores should be kept secure, with entrance allowed to authorized persons only.
5. Conditions inside stores (heating, lighting, ventilation, humidity) should be such that they do not affect the quality of the goods stored, and are comfortable and safe to work in.
6. The size (capacity) of the store should be sufficient to permit neat, orderly storage of goods of different types — for example, components (starting materials), packaging materials, bulk products, and finished-pack products) — and of goods of different status (for example approved, rejected, returned, recalled).
7. Stores should be laid out so as to allow effective rotation of stock (first in/first out, or FIFO).
8. Where goods are quarantined by storage in a separate area, then this area should be clearly marked, with access restricted to authorized persons only.
9. Stores should be maintained in clean and orderly condition, with provision made for the efficient disposal of waste. Care must be taken to avoid contaminating stored goods with any cleaning and pest-control materials used.
10. Where goods require special storage conditions (for example, low temperature, low humidity, or extra security for dangerous materials or controlled drugs), these should be provided, and any conditions, such as low temperature, should be monitored and recorded.
11. Particularly careful attention should be paid to the safe and secure storage of any printed packaging materials (labels, cartons, leaflets, etc.).

Concerning the location of premises for stores and warehouses, these (and any manufacturing facility of which they are a part) are often located where they are for logistical, economic, or for other (often long-forgotten) "historical" reasons. The local environment might well have changed for the worse since the facility was first established. There needs to be an awareness that if a store *is* sited in a particularly hostile environment (extra hot, extra wet, and so on) or next door to, say, a sewage treatment plant or a cement works, it requires extra care and effort from all those concerned with running and staffing the store to ensure that the goods do not become soiled or contaminated.

Store workers must be trained to do all they can to ensure that the store (or warehouse) remains clean, tidy, and in good repair. Careless maneuvering of forklift trucks, for example, can cause a lot of damage. Any damage to walls, roof, floor, or doors should be reported so that something can be done about it. It should be impressed on store workers that entrance and exit doors should be kept closed when not in use.

Rain and flood water can cause a lot of damage to stock, and adequate protection should be provided against it. Insects, vermin, birds, etc. can both damage and contaminate goods and materials. Appropriate pest-control measures should be in place.

Security of stores is another vital matter. It is not just a matter of loss to the company through stolen goods. If a store is broken into, and some of the goods stolen, there is also the danger that the remaining stock might have been damaged, contaminated, or mixed-up. All persons involved in stores work must be fully security conscious.

ANNEX 1 TO CHAPTER 9

EU Guidelines on Good Distribution Practice of Medicinal Products for Human Use (94/C 6Y03)

Introduction

These guidelines have been prepared in accordance with Article 10 of Council Directive 92/25/EEC of 31 March 1992 on the wholesale distribution of medicinal products for human use. They do not cover commercial relationships between parties involved In distribution of medicinal products nor questions of safety at work.

Principle

The Community pharmaceutical industry operates at a high level of quality assurance, achieving its pharmaceutical quality objectives by observing Good Manufacturing Practice to manufacture medicinal products which must then be authorised for marketing. This policy ensures that products released for distribution are of the appropriate quality.

This level of quality should be maintained throughout the distribution network so that authorised medicinal products are distributed to retail pharmacists and other persons entitled to sell medicinal products to the general public without any alteration of their properties. The concept of quality management in the pharmaceutical industry is described in Chapter 1 of the Community Guide to Good Manufacturing Practice for medicinal products and should be considered when relevant for the distribution of medicinal products. The general concepts of quality management and quality systems are described in the CEN standards (series 29 000).

In addition, to maintain the quality of the products and the quality of the service offered by wholesalers, Directive 92/25/EEC provides that wholesalers must comply with the principles and guidelines of good distribution practice published by the Commission of the European Communities.

The quality system operated by distributors (wholesalers) of medicinal products should ensure that medicinal products that they distribute are authorised in accordance with Community legislation, that storage conditions are observed at all times, including during transportation, that contamination from or of other products is avoided, that an adequate turnover of the stored medicinal products takes place and that products are stored in appropriately safe and secure areas. In addition to this, the quality system should ensure that the right products are delivered to the right addressee within a satisfactory time period. A tracing system should enable any faulty product to be found and there should be an effective recall procedure.

Personnel

1. A management representative should be appointed in each distribution point, who should have defined authority and responsibility for ensuring that a quality system is implemented and maintained. He should fulfill

his responsibilities personally. This person should be appropriately qualified: although a degree in Pharmacy is desirable, the qualification requirements may be established by the Member State on whose territory the wholesaler is located.

2. Key personnel involved in the warehousing of medicinal products should have the appropriate ability and experience to guarantee that the products or materials are properly stored and handled.

3. Personnel should be trained in relation to the duties assigned to them and the training sessions recorded.

Documentation

4. All documentation should be made available on request of competent authorities.

Orders

5. Orders from wholesalers should be addressed only to persons authorised to supply medicinal products as wholesalers in accordance with Article 3 of Directive 921251EEC or holders of a manufacturing or importing authorisation granted in accordance with Article 16 of Directive 75/319/EEC.

Procedures

6. Written procedures should describe the different operations which may affect the quality of the products or of the distribution activity: receipt and checking of deliveries, storage, cleaning and maintenance of the premises (including pest control), recording of the storage conditions, security of stocks on site and of consignments in transit, withdrawal from saleable stock, records, including records of clients orders, returned products, recall plans, etc. These procedures should be approved, signed and dated by the person responsible for the quality system.

Records

7. Records should be made at the time each operation is taken and in such a way that all significant activities or events are traceable. Records should be clear and readily available. They should be retained for a period of five years at least.

8. Records should be kept of each purchase and sale, showing the date of purchase or supply, name of the medicinal product and quantity received or supplied and name and address of the supplier or consignee. For transactions between manufacturers and wholesalers and between wholesalers (i.e. to the exclusion of deliveries to persons entitled to supply medicinal products to the public), records should ensure the traceability of the origin and destination of products, for example by use of batch numbers, so that all the suppliers of, or those supplied with, a medicinal product can be identified.

Premises and equipment

9. Premises and equipment should be suitable and adequate to ensure proper conservation and distribution of medicinal products. Monitoring devices should be calibrated.

Receipt

10. Receiving bays should protect deliveries from bad weather during unloading. The reception area should be separate from the storage area. Deliveries should be examined at receipt in order to check that containers are not damaged and that the consignment corresponds to the order.

11. Medicinal products subject to specific storage measures (e.g., narcotics, products requiring a specific storage temperature) should be immediately identified and stored in accordance with written instructions and with relevant legislative provisions.

Storage

12. Medicinal products should normally be stored apart from other goods and under the conditions specified by the manufacturer in order to avoid any deterioration by light, moisture or temperature. Temperature should be monitored and recorded periodically. Records of temperature should be reviewed regularly.

13. When specific temperature storage conditions are required, storage areas should be equipped with temperature recorders or other devices that will indicate when the specific temperature range has not been maintained. Control should be adequate to maintain all parts of the relevant storage area within the specified temperature range.

14. The storage facilities should be clean and free from litter, dust and pests. Adequate precautions should be taken against spillage or breakage, attack by micro-organisms and cross contamination.

15. There should be a system to ensure stock rotation ("first in first out") with regular and frequent checks that the system is operating correctly. Products beyond their expiry date or shelf life should be separated from usable stock and neither sold nor supplied.

16. Medicinal products with broken seals, damaged packaging, or suspected of possible contamination should be withdrawn from saleable stock, and if not immediately destroyed, they should be kept in a dearly separated area so that they cannot be sold in error or contaminate other goods.

Deliveries to customers

17. Deliveries should be made only to other authorised wholesalers or to persons authorised to supply medicinal products to the public in the Member State concerned.

18. For all supplies to a person authorised or entitled to supply medicinal products to the public, a document must be enclosed, making it possible to ascertain the date, the name and pharmaceutical form of the medicinal product, the quantity supplied, the name and address of the supplier and addressee.

19. In case of emergency, wholesalers should be in a position to supply immediately the medicinal products that they regularly supply to the persons entitled to supply the products to the public.

20. Medicinal products should be transported in such a way that
    (a) their identification is not lost;
    (b) they do not contaminate, and are not contaminated by, other products or materials;
    (c) adequate precautions are taken against spillage, breakage or theft;
    (d) they are secure and not subjected to unacceptable degrees of heat, cold, light, moisture or other adverse influence, nor to attack by microorganisms or pests.

21. Medicinal products requiring controlled temperature storage should also be transported by appropriately specialised means.

Returns

Returns of non-defective medicinal products

22. Non-defective medicinal products which have been returned should be kept apart from saleable stock to prevent redistribution until a decision has been reached regarding their disposal.

23. Products which have left the care of the wholesaler, should only be returned to saleable stock if:
    (a) the goods are in their original unopened containers and in good condition;
    (b) it is known that the goods have been stored and handled under proper conditions;
    (c) the remaining shelf life period is acceptable;
    (d) they have been examined and assessed by a person authorised to do so. This assessment should take into account the nature of the product, any special storage conditions it requires, and the time elapsed since it was issued. Special attention should be given to products requiring special storage conditions. As necessary, advice should be sought from the holder of the marketing authorisation or the Qualified Person of the manufacturer of the product.

24. Records of returns should be kept. The responsible person should formally release goods to be returned to stock. Products returned to saleable stock should be placed such that the "first in first out" system operates effectively.

Emergency plan and recalls

25. An emergency plan for urgent recalls and a non-urgent recall procedure should be described in writing. A person should be designated as responsible for execution and co-ordination of recalls.

26. Any recall operation should be recorded at the time it is carried out and records should be made available to the competent authorities of the Member States on whose territory the products were distributed.

27. In order to ensure the efficacy of the emergency plan, the system of recording of deliveries should enable all destinees of a medicinal product to be immediately identified and contacted. In case of recall, wholesalers

may decide to inform all their customers of the recall or only those having received the batch to be recalled.

28. The same system should apply without any difference to deliveries in the Member States having granted the authorisation for wholesaling and in other Member States.

29. In case of batch recall, all customers (other wholesalers, retail or hospital pharmacists and persons entitled to sell medicinal products to the public) to whom the batch was distributed should be informed with the appropriate degree of urgency. This includes customers in other Member States than the Member State having granted the wholesaling authorisation.

30. The recall message approved by the holder of the marketing authorisation, and, when appropriate, by the competent authorities, should indicate whether the recall should be carried out also at retail level. The message should request that the recalled products be removed immediately from the saleable stock and stored separately in a secure area until they are sent back according to the instructions of the holder of the marketing authorisation.

Counterfeit medicinal products

31. Counterfeit medicinal products found in the distribution network should be kept apart from other medicinal products to avoid any confusion. They should be clearly labelled as not for sale and competent authorities and the holder of marketing authorisation of the original product should be informed immediately.

Special provisions concerning products classified as not for sale

32. Any return, rejection, and recall operation and receipt of counterfeit products should be recorded at the time it is carried out and records should be made available to the competent authorities. In each case, a formal decision should be taken on the disposal of these products and the decision should be documented and recorded. The person responsible for the quality system of the wholesaler and, where relevant, the holder of the marketing authorisation should be involved in the decision making process.

Self inspections

33. Self-inspections should be conducted (and recorded) in order to monitor the implementation of and compliance with this guideline.

Provision of information to Member States in relation to wholesale activities

34. Wholesalers wishing to distribute or distributing medicinal products in Member State(s) other than the Member State in which the authorisation was granted should make available on request to the competent authorities of the other Member State(s) any information in relation to the authorisation granted in the Member State of origin, namely the nature of the wholesaling activity, the address of sites of storage and distribution point(s) and, if appropriate, the area covered. Where appropriate, the

competent authorities of this (these) other Member State(s) will inform the wholesaler of any public service obligation imposed on wholesalers operating on their territory.

10

LABORATORY CONTROLS

"Laboratory Controls" is the heading for one subpart of the US cGMPs. The corresponding issues are covered, in varying degrees of detail, in several different sections of the EC GMP Guide.

US CGMPS

Subpart I — Laboratory Controls

Sec. 211.160 General requirements

(a) The establishment of any specifications, standards, sampling plans, test procedures, or other laboratory control mechanisms required by this subpart, including any change in such specifications, standards, sampling plans, test procedures, or other laboratory control mechanisms, shall be drafted by the appropriate organizational unit and reviewed and approved by the quality control unit. The requirements in this subpart shall be followed and shall be documented at the time of performance. Any deviation from the written specifications, standards, sampling plans, test procedures, or other laboratory control mechanisms shall be recorded and justified.

(b) Laboratory controls shall include the establishment of scientifically sound and appropriate specifications, standards, sampling plans, and test procedures designed to assure that components, drug product containers, closures, in-process materials, labeling, and drug products conform to appropriate standards of identity, strength, quality, and purity. Laboratory controls shall include:

(1) Determination of conformance to appropriate written specifications for the acceptance of each lot within each shipment of components, drug product containers, closures, and labeling used in the manufacture, processing, packing, or holding of drug products. The specifications shall include a description of the sampling and testing procedures used. Samples shall be representative and adequately

identified. Such procedures shall also require appropriate retesting of any component, drug product container, or closure that is subject to deterioration.

(2) Determination of conformance to written specifications and a description of sampling and testing procedures for in process materials. Such samples shall be representative and properly identified.

(3) Determination of conformance to written descriptions of sampling procedures and appropriate specifications for drug products. Such samples shall be representative and properly identified.

(4) The calibration of instruments, apparatus, gauges, and recording devices at suitable intervals in accordance with an established written program containing specific directions, schedules, limits for accuracy and precision, and provisions for remedial action in the event accuracy and/or precision limits are not met. Instruments, apparatus, gauges, and recording devices not meeting established specifications shall not be used.

EC GMP GUIDE

Chapter 2 Personnel

2.6 The Head of the QC Department generally has the following responsibilities:
...
...

(iv) To approve specifications, sampling instructions, test methods and other QC procedures.

Chapter 6 Quality Control

Principle

Quality Control is concerned with sampling, specifications and testing as well as the organisation, documentation and release procedures which ensure that the necessary and relevant tests are carried out, and that materials are not released for use, nor products released for sale or supply, until their quality has been judged satisfactory. Quality Control is not confined to laboratory operations, but must be involved in all decisions which may concern the quality of the product. The independence of Quality Control from Production is considered fundamental to the satisfactory operation of Quality Control.

6.2 The principal duties of the head of Quality Control are summarised in Chapter 2. The Quality Control Department as a whole will also have other duties, such as to establish, validate and implement all quality control procedures, keep the reference samples of materials and products, ensure the correct labelling of containers of materials and products, ensure the monitoring of the stability of the products, participate in the investigation of complaints related to the quality of the product, etc. All these operations should be carried out in accordance with written procedures and, where necessary, recorded...

6.7 Laboratory documentation should follow the principles given in Chapter 4. An important part of this documentation deals with Quality Control

and the following details should be readily available to the Quality Control Department:
- specifications;
- sampling procedures;
- testing procedures and records (including analytical worksheets and/or laboratory notebooks);
- analytical reports and/or certificates:
- data from environmental monitoring, where required-,
- validation records of test methods, where applicable;
- procedures for and records of the calibration of instruments and maintenance of equipment.

Chapter 4 Documentation

4.10 There should be appropriately authorised and dated specifications for starting and packaging materials, and finished products; where appropriate, they should be also available for intermediate or bulk products.

Specifications for starting and packaging materials

4.11 Specifications for starting and primary or printed packaging materials should include, if applicable:
   (a) a description of the materials, including:
   - the designated name and the internal code reference;
   - the reference, if any, to a pharmacopoeial monograph;
   - the approved suppliers and, if possible, the original producer of the products;
   - a specimen of printed materials;
   (b) directions for sampling and testing or reference to procedures;
   (c) qualitative and quantitative requirements with acceptance limits;
   (d) storage conditions and precautions,
   (e) the maximum period of storage before re-examination.

Specifications for intermediate and bulk products

4.12 Specifications for intermediate and bulk products should be available if these are purchased or dispatched, or if data obtained from intermediate products are used for the evaluation of the finished product. The specifications should be similar to specifications for starting materials or for finished products, as appropriate.

Specifications for finished products

4.13 Specifications for finished products should include:
   (a) the designated name of the product and the code reference where applicable;
   (b) the formula or a reference to;
   (c) a description of the pharmaceutical form and package details;
   (d) directions for sampling and testing or a reference to procedures;

(e) the qualitative and quantitative requirements, with the acceptance limits;

(f) the storage conditions and any special handling precautions, where applicable;

(g) the shelf-life.

Chapter 6 Quality Control

6.17 The tests performed should be recorded and the records should include at least the following data:
(a) name of the material or product and, where applicable, dosage form;
(b) batch number and, where appropriate, the manufacturer and/or supplier;
(c) references to the relevant specifications and testing procedures;
(d) test results, including observations and calculations, and reference to any certificates of analysis;
(e) dates of testing;
(f) initials of the persons who performed the testing;
(g) initials of the persons who verified the testing and the calculations, where appropriate;
(h) a clear statement of release or rejection (or other status decision) and the dated signature of the designated responsible person.

6.18 All the in-process controls, including those made in the production area by production personnel, should be performed according to methods approved by Quality Control and the results recorded.

Chapter 3 Premises and Equipment

3.41 Measuring, weighing, recording and control equipment should be calibrated and checked at defined intervals by appropriate methods. Adequate records of such tests should be maintained.

DISCUSSION

What is perhaps, in this context, the key issue established in both the regulatory documents is that specifications for components (starting materials), packaging materials, in-process or intermediate products and materials, and finished pack products must be *approved* by the Quality Control (QC) function (specifically, US cGMPs refer to "the quality control unit" and the EC GMP Guide to "The Head of the QC Department").

The US cGMPs declare that the original drafting of specifications should be the task of "the appropriate organizational unit," but do not say who or what is "appropriate."

The EC GMP Guide is completely silent on this point — specifications are to be *approved* by QC, but who originates them? It is not specified.

In many instances, specifications for components or starting materials, and for products, will be based on the specifications and test methods set out in official compendia (USP, European Pharmacopoeia, BP). Test methods and equipment other than those indicated in the compendia are generally permitted provided that they are comparable to the official method, in terms of accuracy, precision, etc. Any dispute must be resolved by the application of official methods.

A mere reference to a compendium ("USP quality" or "BP quality") should not be regarded as a complete *manufacturer's* specification. Manufacturers should prepare their own written specifications for materials and products. These will usually include requirements not typically to be found in compendial monographs, for example bulk density of powders, bulk packaging units required, approved suppliers, sampling requirements, storage requirements, and so on.

The EC GMP Guide gives considerable specific detail on the required *content* of specifications; the US cGMPs do not.

The US cGMPs requirements for calibration of laboratory "instruments, apparatus, gauges and recording devices," although brief, are specific and precisely meaningful. The EC GMP Guide is even briefer, and is less explicit. Perhaps strangely, the EC requirement to calibrate appears under "Premises and Equipment," not in specific reference to QC laboratories. Nevertheless, it is normally taken to refer also to laboratory equipment.

Comparison

US cGMPs	EC GMP Guide
Specs to be approved by the quality control unit	Specs to be approved by the Head of the QC Department
Specs to be drafted by the "appropriate organizational unit"	Not stated who or what should draft or originate specs
Little detail on *content* of specs	Give detail on content of specs
More precise regarding calibration of lab instruments than EC	Somewhat indeterminate regarding calibration; requirement appears under "Premises and Equipment," but to be taken as embracing lab equipment

IMPLEMENTATION

An illustrative example of a Component/Starting Material Specification is shown in Figure 10.1. The same type of format will serve also for bulk, intermediate, and finished product specifications.

In the example shown, a (fictitious) compendial material is assumed, and the majority of the tests procedures to be employed are those of the relevant (fictitious) pharmacopoeia (the "UP," or "Universal Pharmacopoeia"). In the case of tests that are not pharmacopoeial, or cannot be defined by reference to other official standards, reference to an internal house procedure is necessary. For example, in Figure 10.1, for the bulk density test there is a reference to an in-house procedure (ASOP — Analytical Standard Operating Procedure — no. 142). Note, too, the listing of suppliers approved to supply this material.

Phantazpharm Inc. - STARTING MATERIAL SPECIFICATION

Dexterium adipate UP.   Code no. 1376

Molecular formula. $C_n H_y N_x P_z$, $2H_2O$

Relative molecular mass.   187.2

Pharmacopoeial Tests:

Description. Pale yellow fine crystalline powder

Solubility. Slightly soluble in water. Readily soluble in ethanol and acetone to give clear, faintly yellow solution.

Identification. Complies with the UP tests.

Melting point. 168° to 172°C

Loss on drying. Not more than 0.5%

Heavy metals. Not more than 15 ppm

sulfated ash. Not more than 0.1%

Related substances. Complies with the UP test

Assay. 99.9% to 101% as the dihydrate, calculated with reference to the dried material (UP method)

Additional tests:

Bulk density: 1.7 to 2.3 ml/g (ASOP no. 142)

Approved suppliers:
>   Chemolux Ltd.
>   Apimatic Inc.
>   Pharming Corp

Spec. no.	Supersedes	Prepared by:	Approved by:	Effective date:	Page 1 of 1

Figure 10.1   Starting Material Specification

CALIBRATION

The issue of calibration has already been discussed in Chapter 5, Equipment, of this book. There, it was considered largely in the context of measuring and recording devices used in *manufacture*. However, both the US cGMPs (explicitly) and the EC GMP Guide (implicitly) require calibration of laboratory instruments and measuring devices. The following is an extract from Chapter 5, to which the reader is referred, suitably adapted:

> The quality of any test procedure, and the validity of the results obtained, depends greatly on the quality (i.e., the fitness for purpose) of the instruments and measuring devices used. It is therefore important that the calibration of all measuring and testing equipment should not be conducted only when "it seems like a good idea," or when a device is clearly not functioning. It should be *managed* as a well-controlled operation, run according to preplanned programs and schedules, with written, approved procedures for the calibration of each type of instrument or device, and with records maintained of calibrations carried out. Whatever the format of the documentation system employed, it should clearly signal when an instrument or device is due for calibration. The main steps to Good Calibration Practice may be set out as follows:
>
> a. Carefully review all testing processes, to determine and record all the tests and measurements that need to be made, and to define the accuracy and precision required when making them. On this basis, select and obtain the necessary test and measuring equipment accordingly, or discard and replace any test equipment found to be unsuitable, or inadequate for the purpose.
>
> b. Mark, or by some other means (e.g., by reference in documents or records to plant or model numbers) identify all measuring equipment to ensure it is calibrated at defined intervals, against certified reference standards.
>
> c. Prepare and implement written calibration procedures, programs, schedules and records, that will ensure all instruments and measuring devices are indeed calibrated, as intended, at the prescribed time intervals. The careful determination of the intervals between routine calibrations of a given instrument or device is critical to the success, or otherwise, of a calibration program. It should not be a general, overall figure, applicable to all instruments. A specific time interval should be carefully selected for each device, after considering:
>
> - Type of device or instrument
> - How crucial is the accuracy and precision of the device in
> - Relation to quality and hence, consumer safety
> - Degree of accuracy and precision required
> - Device manufacturer's recommendations
> - Extent of use
> - Stress placed upon device in use
> - Any tendency of device to display drift
> - Previous history, and records, of device in use
> - Environmental conditions
>
> d. Keep calibration records, detailing what calibrations have been carried out, when, and by whom. Regularly review these records to ensure that the required calibrations are, in fact, being carried out at the specified intervals.

e. Ensure visibility of calibration status. That is, label the equipment with an indication of when it last was calibrated, and when it next is due, or record this information in an immediately accessible document or record book.

f. Ensure the calibration status of any measuring device or instrument before it is used.

g. Carry-out documented retrospective assessments of the validity of previous measurements/tests whenever a piece of measuring or test equipment is found to be out of calibration. (It is irresponsible to fail to consider the potential effects of potentially false earlier results when a measuring device is found to be reading incorrectly, and then fail to act accordingly, no matter the economic consequences.)

h. Ensure that measuring equipment is handled and stored so that its accuracy and general fitness for use is not hazarded.

i. Protect the equipment and any associated software against unauthorized adjustments that would invalidate its setting.

j. Ensure that calibrations, inspections, tests, and measurements are carried out under suitable environmental conditions, and that reference standards are very carefully protected against damage or deterioration.

The most important requirement is the need to ensure that calibration work is carried out by trained, experienced personnel who really know what they are doing and know the importance of what they are doing. It is also important that there is a formally assigned, accountable responsibility for calibration.

If calibration work is carried out under external contract, it should be subject to a formal written contract, clearly defining the nature and extent of the work required, and the content and format of the resultant test report(s).

US CGMPS

Subpart I — Laboratory Controls

Sec. 211.165 Testing and release for distribution

(a) For each batch of drug product, there shall be appropriate laboratory determination of satisfactory conformance to final specifications for the drug product, including the identity and strength of each active ingredient, prior to release. Where sterility and/or pyrogen testing are conducted on specific batches of short-lived radiopharmaceuticals, such batches may be released prior to completion of sterility and/or pyrogen testing, provided such testing is completed as soon as possible.

(b) There shall be appropriate laboratory testing, as necessary, of each batch of drug product required to be free of objectionable microorganisms.

(c) Any sampling and testing plans shall be described in written procedures that shall include the method of sampling and the number of units per batch to be tested; such written procedure shall be followed.

(d) Acceptance criteria for the sampling and testing conducted by the quality control unit shall be adequate to assure that batches of drug products meet each appropriate specification and appropriate statistical quality control criteria as a condition for their approval and release. The statistical quality control criteria shall include appropriate acceptance levels and/or appropriate rejection levels.

(e)   The accuracy, sensitivity, specificity, and reproducibility of test methods employed by the firm shall be established and documented. Such validation and documentation may be accomplished in accordance with Sec, 211.194(a)(2).

(f)   Drug products failing to meet established standards or specifications and any other relevant quality control criteria shall be rejected. Reprocessing may be performed. Prior to acceptance and use, reprocessed material must meet appropriate standards, specifications, and any other relevant criteria.

EC GMP GUIDE

Chapter 1 Quality Management

1.2 ... The system of QA appropriate for the manufacture of medicinal products should ensure that ...
   (vi) The finished product is correctly processed and checked according to the defined procedures ...
1.4 ... The basic requirements of QC are that ...
   (vi) The finished product correctly processed and checked according to the defined procedures ...

Chapter 4 Documentation

Specifications for finished products

4.13 Specifications for finished products should include:
   (a) the designated name of the product and the code reference where applicable;
   (b) the formula or a reference to;
   (c) description of the pharmaceutical form and package details;
   (d) directions for sampling and testing or a reference to procedures;
   (e) the qualitative and quantitative requirements, with the acceptance limits; ...

Chapter 6 Quality Control

Principle

Quality Control is concerned with sampling, specifications and testing as well as the organisation, documentation and release procedures which ensure that the necessary and relevant tests are carried out, and that materials are not released for use, nor products released for sale or supply, until their quality has been judged satisfactory.

6.7 Laboratory documentation should follow the principles given in Chapter 4. An important part of this documentation deals with Quality Control and the following details should be readily available to the Quality Control Department:
  – specifications;
  – sampling procedures;
  – testing procedures and records ...
  – analytical reports and/or certificates ...

6.11 The sample taking should be done in accordance with approved written procedures that describe:
  – the method of sampling;
  – the equipment to be used;
  – the amount of the sample to be taken;
  – instructions for any required sub-division of the sample;
  – the type and condition of the sample container to be used;
  – the identification of containers sampled;
  – any special precautions to be observed, especially with regard to the sampling of sterile or noxious materials;
  – the storage conditions
  – instructions for the cleaning and storage of sampling equipment.

6.15 Analytical methods should be validated. All testing operations described in the marketing authorisation should be carried out according to the approved methods.

6.16 The results obtained should be recorded and checked to make sure that they are consistent with each other. Any calculations should be critically examined.

6.17 The tests performed should be recorded and the records should include at least the following data:
  (a) name of the material or product and, where applicable, dosage form;
  (b) batch number and, where appropriate, the manufacturer and/or supplier;
  (c) references to the relevant specifications and testing procedures;
  (d) test results, including observations and calculations, and reference to any certificates of analysis;
  (e) dates of testing;
  (f) initials of the persons who performed the testing;
  (g) initials of the persons who verified the testing and the calculations, where appropriate;
  (h) a clear statement of release or rejection (or other status decision) and the dated signature of the designated Responsible Person.

Annex 3 (to EC GMP Guide) Manufacture of Radiopharmaceuticals

8. When products have to be dispatched before all tests are completed, this does not obviate the need for a formal recorded decision [to] be taken by the Qualified Person on the conformity of the batch ...

Chapter 5 Production

5.62 The reprocessing of rejected products should be exceptional. It is only permitted if the quality of the final product is not affected, if the specifications are met and if it is done in accordance with a defined and authorised procedure after evaluation of the risks involved. Records should be kept of the reprocessing.

DISCUSSION

Both the US cGMPs and the EC GMP Guide require that batches of drug products should be sampled and tested to ensure conformance to an approved specification before release for distribution. The EC GMP Guide specifically states that this testing "should be as described in the marketing authorisation ..." (the European equivalent of an NDA).

In the US it is required that any compendial product should comply with the official compendial specification (i.e., USP), unless it is stated on the label where and how the product is intentionally noncompliant. Similarly, in the European Union it is a requirement that any compendial product should comply with the relevant specification in the European Pharmacopoeia. For noncompendial products manufacturers will, during the research and development phase, have to establish their own product specification, to be submitted with their NDA, or application for a Marketing Authorization.

It would seem, at first sight, to be both obvious and rational that batches of finished products should be sampled and tested to ensure conformity to a finished product specification. There is, however, one specific exception to this general requirement. That exception is what is termed "parametric release" of terminally sterilized sterile products, *without* the performance of a sterility test. This has been allowed, given certain carefully formulated preconditions in both regulatory jurisdictions. (See later in this chapter, under the discussion arising from Sec. 211.167, Special testing requirements). For now, it should be noted that it can be (and indeed has been) argued that, *for any type of product,* given that there is assurance that it has been manufactured under controlled conditions, using the correct materials of assured quality, following a validated manufacturing procedure that incorporates appropriate in-process controls, surely (so the argument runs) the correct product, having all the right quality characteristics, *must* be produced.

So what is the point, the argument continues, of expensive and time-consuming end-product testing? It has to be conceded that there is some force in this argument. Modern analytical techniques become ever more accurate and precise, ever wider in their application. The weak link, however, lies in the problem of just how representative is the sample that has been submitted for test. As has already been suggested, more than once, this is indeed a very weak link. Does not the careful control and evaluation of production parameters provide a firmer basis for the assurance of product quality? This is a question that may well, with some justice, be asked. Indeed, the EC GMP Guide (in its Annex 17 on Parametric Release), although it focuses most specifically on parametric release for sterile products, seems clearly to accept that the principle may be extended to other types of product, thus:

EC GMP GUIDE

Annex 17 Parametric Release

> 2.1 It is recognised that a comprehensive set of in-process tests and controls may provide greater assurance of the finished product meeting specification than finished product testing.
>
> 2.2 Parametric Release may be authorised for certain specific parameters as an alternative to routine testing of finished products …

So, it could well be that, in the future, the concept of parametric release will be extended beyond the narrow confines of the application to the sterility of terminally sterilized products. It is, however, worth reflecting that there is a massive difference between a sterility test on the one hand, and a modern instrumental chemical analysis on the other. The latter will be a validatable technique of known accuracy, precision, and reliability. The former is a dubious procedure that, arguably, has little value or significance. It unquestionably makes sense to discard the sterility test in favor of other approaches that offer the possibility of higher levels of assurance. The case is by no means so clear-cut regarding end-product chemical (or physicochemical) analysis. For now, most manufacturers will wish to retain routine end-product testing in general, not only because it is a legal requirement, but also because it will seem to offer a better defense in the event of any possible action for damages arising from a purportedly defective product.

The confused and somewhat thorny issue of taking samples that are representative and statistically sound has already been discussed in Chapter 5, Materials Control of this book.

Both regulatory documents permit (US explicitly, EC implicitly) release of short-lived radiopharmaceuticals before all tests (specifically, sterility and pyrogen tests) have been completed, in recognition of the fact that any delay while such testing is completed could render a short-lived radiopharmaceutical unfit for its intended purpose.

The US cGMPs require "appropriate laboratory testing of … each batch of drug product required to be free of objectionable organisms." This is a somewhat strange comment, for it might well be thought that *all* batches of *all* drug products should be free of objectionable organisms. This requirement is presumably not intended to refer to products purported to be sterile (that these should be subject to sterility testing is covered in Sec. 211.167, Special testing requirements), but to nonsterile products that need to be tested to ensure the absence of specific organisms — for example, absence of pseudomonads and *S. aureus* from topicals. In any event, the EC GMP Guide does not state a similar, corresponding, requirement. It does however require (in Annex 1, Manufacture of sterile products) that batches of products required or purported to be sterile should be sterility tested, except where "parametric release has been authorised."

The US cGMPs explicitly permit reprocessing of rejected batches of product (given that the stated conditions are complied with). The EC GMP Guide, rather, frowns on such reprocessing, which "should be exceptional." Indeed, in some member states of the European Union it is required that before any such reprocessing is undertaken, a variation to the r elevant Marketing

Authorization must be obtained from the national health authority. (After all, reprocessing *is* a significant change in the manufacturing process.)

Both regulatory documents require the validation of analytical methods and procedures employed. The emphasis in US Sec. 211.165 is on the validation of the methods used in testing samples of manufactured drug products, although it may be taken to refer to the testing of components (starting materials) and all other test procedures. The EC GMP Guide simply makes the general statement:

> 6.15 Analytical methods should be validated ...

A summary and comparison can be made as follows:

US cGMPs	EC GMP Guide
Test products for conformance before release	Same as US cGMPs — test must accord with Marketing Authorization
Explicitly permits release of short-lived radiopharmaceuticals pending completion of sterility and pyrogen tests	"Dispatch (of radiopharmaceuticals) before all tests are completed" is implicitly permitted (Annex 3, parargaph 8)
"Appropriate testing ... of each batch of drug product required to be free of objectionable micro-organisms" required	Not stated as a specific requirement
	Sterility testing of sterile products required by Annex 1 — except where "parametric release has been authorised"
Requires written sampling procedures	Same as US cGMPs
Permits reprocessing of rejected batches of product — with conditions	"reprocessing should be exceptional"
Analytical test methods must be validated	Same as US cGMPs

Implementation

The content and format of a specification for a component (starting material) has already been discussed and illustrated earlier in this chapter. Suitably adjusted, a similar style and format will serve for a drug product specification.

ANALYTICAL VALIDATION

Any worthwhile analytical method must be scientifically sound and, when used by different operators with similar apparatus in different laboratories, be capable of giving reliable and (within limits) consistent results. In other words, it should have a rational basis, and it should "work," (i.e., it should be adequate for its stated purpose). The process of demonstrating that such a method works is called "Analytical Validation" or the "Validation of Analytical Methods." Both the US cGMPs and the EC GMP Guide require that analytical methods should be validated.

It is generally accepted that official pharmacopoeial methods (e.g., USP, BP, European Pharmacopoeia) *when applied to pharmacopoeial materials or prod-*

ucts, may be taken as validated. Other methods, or pharmacopoeial methods applied to nonpharmacopoeial materials, should be validated. In the US, the FDA require details of the validation of analytical methodology to be submitted with New Drug Applications (NDAs), and to a greater or lesser extent (depending on the quality and extent of the data submitted) will conduct method validation studies, on samples of the new drug, in their own laboratories. In Europe, details of analytical method validations are required by regulatory bodies when considering applications for Marketing Authorizations.

Regulatory bodies will require sufficient descriptive detail of the method to allow its repetition by the regulatory authorities themselves. They will normally require, for example, adequate information on the preparation of the sample, on any reference materials required, on the use of the apparatus and its calibration, on the number of replicates to be carried out, and on the methods of calculation of the results (together with details of any necessary statistical analysis).

Thus, as a preliminary to any analytical validation study, the test method and conditions will need to be precisely and formally defined. This formal definition should include:

- Sampling details (e.g., size and number of samples, method of sampling, sample container, any necessary pretreatment of sample)
- Any special sample-storage conditions
- Details of reagents and equipment to be used
- Description of the apparatus
- Any tests necessary to determine the satisfactory function of the apparatus
- System suitability tests (e.g., separating power of chromatographic columns)
- Exact test conditions, including reaction conditions and use of reagents for preparation of any derivatives
- Any precautions to be taken
- Method of calculation of results, and any necessary statistical analyses

The definition of reference materials may also need particular attention. In-house reference materials should be characterized and evaluated for their suitability for their intended uses and any working standards should be characterized against an authentic reference material.

Criteria for Analytical Validation

The following are the main criteria to be considered for validation studies. Their relative importance will depend on the use to which the method is to be put:

- Accuracy
- Precision (which embraces repeatability and reproducibility)
- Specificity
- Sensitivity
- Limit of detection
- Limit of quantitation
- Linearity
- Range

Note: The ICH guideline on the "Validation of Analytical Procedures" (1993)[1] introduces a further subclass (in addition to repeatability and reliability, of the more general class "precision"), which it terms "intermediate precision."

Not all of these criteria will need to be considered in all cases. Thus, for example, with an identity test, specificity will obviously be a key factor to be established. For an impurity control test not only the specificity, but also the limit of detection and the limit of quantitation will need to be confirmed. With quantitative assay procedures, the specificity, precision, accuracy, linearity, range, and sensitivity will all need to be considered.

The ICH text[1] contains a table illustrating "those validation characteristics regarded as the most important for the validation of different analytical procedures." For a version of that table, see Table 10.1.

SPECIFIC ASPECTS OF ANALYTICAL VALIDATION

Accuracy

Accuracy may be defined as the closeness of an experimental result to the true value. This raises the philosophical question of what *is* the true value, and in practical terms it is usually taken to represent the closeness of the mean value found using a number of repeat analyses to the "conventional true value" — e.g., that attributed to an in-house standard, or to an accepted reference value such as that attributed to a pharmacopoeial reference material.

Clearly, it is necessary to carefully evaluate the accuracy of most, if not all, assay methods. One approach is to compare the results obtained with the proposed new test procedure with those obtained using a previously validated or reference method (e.g., a pharmacopoeial method). This approach is often adopted when

Table 10.1
Analytical Validation (From ICH Guideline 1993)

Analytical Validation Criteria	Identification	Impurities tests Quantitation	Limit Tests	Assay (content/ potency/dissolution)
Accuracy	-	+	-	+
Precision				
- Repeatability	-	+	-	+
- Reproducibility	-	+[1]	-	+
Specificity	+	+	+	+[2]
Detection limit	-	+	+	-
Quantitation limit	-	+	-	-
Linearity	-	+	-	+
Range	-	+	-	+

Adapted from the table listing "those validation characteristic regarded as the most important for the validation of different types of analytical procedures" (ICH, 1993)

- signifies that this parameter is not normally evaluated; + signifies that this parameter is normally evaluated; (1) may be needed in some cases; (2) may not be needed in some cases.

validating analytical methods used for components (starting materials). With finished products, the test procedure can be evaluated by using samples or mixtures that have been "spiked" with known amounts of pure added analyte (that is, the substance, ion, functional group, etc. that is under test). Added confidence in the accuracy of a method can be obtained by taking the demonstrably pure analyte substance and spiking it with excipients or impurities and demonstrating that the assay result, in comparison with results on unspiked pure samples, is unaffected by the presence of the added materials.

The results of such tests can give a measure of the systematic errors associated with the method. Accuracy may be improved by studying and eliminating as many sources of systematic error as possible (for example, those due to interference, imprecise calibration, faulty equipment settings, and so on).

The determination of the accuracy of quantitative impurity tests using thin layer chromatography (TLC) — as in a test for related substances — may be approached by spiking a sample with the known, or suspected, impurities at the proposed specification limit, or at a series of levels up to that limit. After development, the plate should be examined by each of the proposed methods of detection. For a satisfactory validation, the impurity zones in the spiked sample should display similar responses to those generated in adjacent zones, by standard applications of the impurities.

A similar approach is suitable for high performance liquid chromatographic (HPLC) and gas chromatographic (GC) quantitative determination of impurities. Here, however, the test results will normally be calculated by electronic integration of detector responses, based on peak heights or areas. Validation is then based on comparison of integrator values for impurities in the spiked samples, with those generated by known levels of pure samples of the known or suspected impurities. For impurity levels in the range 0.1 to 1.0%, using modern chromatographic equipment, recoveries in the range 80 to 120% may be expected.

The accuracy of other methods for the quantitative determination of impurities (e.g., ion selective electrode potentiometric and atomic spectroscopic methods) can also be evaluated by recovery experiments, using samples spiked with known quantities of impurities.

A variety of analytical methods are used in the assay of bulk drug substances (or APIs), and thus the approach to evaluating accuracy must be selected accordingly, as appropriate to the analytical method concerned. For titrimetric methods, expected equivalence points can be calculated theoretically, taking into account the number of titratable functions and the molecular weight. It is, however, necessary to ensure that the expected stoichiometric relationships do indeed apply in practice, and this is best determined by performing the proposed titration procedure on a well-characterized reference standard. Recalling the relatively narrow assay tolerances common in specifications for bulk synthetic drug substances (often of the order of 99.5 to 100.5%), it is necessary to establish that this range is still valid in the presence of impurities at their proposed maximum limits, as these may significantly influence results due to additional titratable functional groups or large molecular weight differences.

In the determination of the accuracy of a UV light absorption assay of a bulk drug substance, based on a specific absorbance ($A_{1\%, 1cm}$) value, it is essential to ensure that the value selected is the appropriate one for the purpose. This may be verified by examination of the results achieved, using the selected $A_{1\%, 1cm}$ value in the assay of well-characterized reference samples. It is also important to consider the potential effects of the presence of impurities at their maximum limits, which could compromise the accuracy of the assay method.

The accuracy of assays of bulk substances by HPLC may be evaluated by methods analogous to those outlined above in relation to quantitative impurity determinations.

In the determination of the accuracy of assays of active content in a finished product, a prime consideration is that of the introduction of the further complicating factor of the common need to extract the analyte from the sample matrix. Thus, demonstration of the accuracy of the assay method *per se* may, on its own be insufficient. What is needed is a demonstration of the accuracy of the total package — extraction plus assay. A spectrophotometric assay, say, of a solution of an analyte, extracted from, say, a tablet, may well have a high level of accuracy. But it is to no avail if only a portion of the analyte has been extracted from the product, or if other interfering substances have been extracted with it.

A common approach to determination of the accuracy of assays for the active component of a finished product is to perform what are commonly termed "recovery experiments," in which the assay is performed on mixtures of excipients that have been spiked with accurately measured amounts of the pure active substance. Rational selection of the spiking range is important. For example, in the determination of the accuracy of an assay method used as a basis for product release, against a specification of 95 to 105%, then a spiking range of 80 to 120% should be employed, with the excipient mixture spiked with, say, five levels of the analyte equivalent to 80, 90, 100, 110, and 120% of the theoretical content. Errors, at each level of not more than ± 2% would normally be considered to be an adequate demonstration of the accuracy of the assay. For the assay used in a stability study, particularly when the product samples are subject to more-than-usual stress, and may thus be expected to degrade to below 80%, then accuracy validation over a wider spiking range is obviously appropriate.

In this type of accuracy validation, it may be that the relevant excipient mixtures are not available. Nevertheless recovery experiments are still possible, by spiking the actual product with pure active substance at carefully measured levels of, say, 5, 10, 15, and 20% above the theoretical level and assaying the samples as is (i.e., 100%) and at the 105, 110, 115, and 120% levels.

As an alternative approach, where recovery experiments as outlined above may not be possible, Carr and Wahlich (1990)[2] have suggested "recovery efficiency experiments." This is most appropriate for HPLC and GC methods, and requires an internal standard. In this approach to determination of the accuracy of a finished product assay, the sample is extracted with the solvent intended in the assay procedure, but with the addition of an internal standard. After centrifuging or filtering, about 75% of the supernatant (or the filtrate) is taken and subjected to the remainder of the assay procedure and the ratio of the analyte response to the internal standard response is noted. The sample residue (including the approximately 25% supernatant from the first extract) is then reextracted with a further volume of solvent, without internal standard. Following centrifuging (or filtration) this second extract is subjected to the remainder of the assay procedure, as before, and the ratio of analyte response to internal standard response is again noted. The variance between the two ratios should not be greater than $\pm 2\%$. If after the second extraction the peak response ratio remains unchanged, this indicates that no further analyte has been extracted, and the method may therefore be considered to be efficient. If the response ratio increases, this indicates that the extraction procedure was not efficient, as additional analyte has now been extracted. It does, however, need to be noted that if the active drug substances is so strongly absorbed onto the excipient(s) that it is not extracted by the second solvent treatment, the two response ratios will remain equal, and the problem will not be identified.

A further factor is the issue of sample aging. Analytical methods for products tend to be developed and validated using samples that have been freshly prepared, but it should be demonstrated that the method will remain accurate when applied to older samples. This is particularly relevant to the assay of samples of stability study materials that have been stored under stress conditions.

Precision

Precision is the closeness of the agreement, one with another, of a series of separate measurements or determinations made when applying a prescribed method to a series of samples all taken from the one homogeneous lot of material. It is a measure of the closeness of the "grouping" (or the wideness of the "scatter") of a series of results. It is thus possible for a method to be precise, without necessarily being very accurate. Tests for precision also reveal the random errors associated with the method.

The question arises of whether a method that is very precise, but notably inaccurate, is of any value — and the answer is that it can be if the inaccuracy is quantifiable and always in "the same direction." Then a systematic correction can be applied, to yield results that may be considered to be accurate.

Precision validation needs to be directed at the two subclasses of precision: determination of repeatability and of reproducibility.

Repeatability involves the evaluation of the results obtained by the same analyst, working under the same operating conditions, repeatedly using the same equipment, with identical reagents, over a relatively short time period. That is, it is about precision under the same, or very similar, conditions. The results may be expressed in terms of a repeatability standard deviation, repeatability coefficient of variation/relative standard deviation, and confidence interval of the mean value. The ICH text recommends that repeatability should be assessed by either (a) using a minimum of 9 determinations covering the specified range for the procedure (e.g., 3 replicates of each of 3 concentrations), or (b) a minimum of 6 determinations at 100% of the test concentration.

Reproducibility refers to variation between laboratories using different reagent sources and different analysts on different days and apparatus from different suppliers. That is, it is about precision under different conditions, and is assessed by a series of interlaboratory trials. The results may be expressed in terms of the reproducibility standard deviation, reproducibility coefficient of variation/relative standard deviation, and confidence interval of the mean value.

The precision of virtually all quantitative methods needs to be validated. Clearly, the validation of repeatability is crucial. However, it becomes something of a philosophical question as to when and how the reproducibility of an analytical method should be validated, and it has to be wondered if a method, intended only (and likely only) to be used in one laboratory, under standard and consistent conditions, really requires reproducibility. However, it does need to be noted that, in a "new product cycle" (from research and development, via scale up, to full-scale production) the new assay method may well be employed, even within the one company, in a series of different laboratories, from the analytical development lab, via pilot plant and (possibly) clinical trials manufacturing lab to routine production QC lab. So, even though the original intention is that a method is intended for use within the one organization, it may well be prudent, from a quality viewpoint, to validate reproducibility in the context of different laboratories, analysts, times, equipment, and sources of reagents.

When validating a finished product test method for precision, it is important that real samples, as distinct from spiked excipient mixes, are tested, as the latter can lead to apparently satisfactory results that cannot then be achieved in real life. Furthermore, the complete procedure should be applied to each replicate analysis, although it may be useful to examine separately the precision of the various stages of the analytical procedure, in order to reveal any steps that may be critical to the precision of the procedure.

The precision of a method also bears upon its value for use in stability studies. It might, for example, be expected that a product will degrade less than 1% during its anticipated (or simulated) shelf life. Even an HPLC method that is considered to be stability-indicating will not be able to discriminate, with 96% confidence, between samples at 99 and 100% of original potency, after a period of storage. It is thus necessary to monitor the test samples for individual degradation products, both for this reason as well as from the aspect of patient safety.

The determination of the aspect of precision termed *reproducibility* requires a number of different repeatability studies (as discussed above) to be performed under the various different conditions (i.e., different laboratories, different analysts, different equipment, different times, different batches — or suppliers — of reagents). It may not be necessary to involve all these variables, which should be selected as best to model the range of circumstances under which the method will be applied in routine use. From the results obtained in each set of circumstances, the mean, standard deviation and relative standard deviation should be calculated, from when it can be determined whether or not the different sets of values indicate acceptable reproducibility. A Student's *t*-test can be used to compare mean values, and an F-test to compare standard deviations. Reference to standard statistical tables for these values will indicate whether or not there are any significant differences when the method is performed under the various different circumstances.

Specificity (Selectivity)

The term "selectivity" often appears in the literature, used as if synonymous with "specificity." However, a distinction has been drawn between the two terms (e.g., in CPMP guidelines on analytical validation[3]). *Specificity* is a term to be applied to a method designed to make a quantitative determination of an analyte in a mixture with one or more other substances. *Selectivity* is a term to be applied to method designed to detect qualitatively the analyte in the presence of other substances, functional groups etc.

In more simple terms, the distinction has been expressed (Carr and Wahlich[4]) as: *specificity* is the ability of a nonseparative method to distinguish between different compounds, whereas *selectivity* is the ability of a separative method to resolve different compounds

The importance and significance of the validation of method specificity vary according to the use to which the test is put. For an identity test, the method should ensure the identity of the analyte. That is, the method should demonstrably be capable of specifically identifying the substance, compound, group, ion, etc. that it is intended to identify. With impurity control tests, it may be possible to consider the totality of the control methods and their *overall* adequacy in controlling factors, such as related substances, impurities, degradation products, heavy metals and catalyst residues, organic solvent residues, etc.

The assay method used for stability studies should be capable of detecting the signal from the analyte alone, without interference from excipients, degradation products, or other impurities. That is, it should provide a specific indication of the stability of the substance under study. On the other hand, for routine batch analysis, such a degree of specificity of the assay procedure may not be necessary, provided that additional tests adequately control such things as related substances and degradation products, which might otherwise interfere.

Validation of method specificity can be performed by spiking experiments, using known (or expected) impurities, degradation products, or excipients and then analyzing the spiked sample against an unspiked sample to demonstrate an adequate response from the specific analyte of interest, and a lack of interference from the other substances present.

Method selectivity may be conformed by spiking experiments. For example, when using a chromatographic method, if the spiked substance appears clearly in the chromatogram, the method may be considered to be selective. If, on the other hand, it does not appear, it may have been co-eluted with another substance in the sample mix, it may not be detectable by the method selected, or it may have been retained on the column. Applying the same chromatographic method to the pure substance alone should reveal which of these factors is operating. The selectivity of a method for finished product analysis can be established by comparing the results obtained when applying the method to the pure drug substance with those obtained on samples of the drug substance plus excipients.

Sensitivity/Linearity

Sensitivity may be defined as the capacity of an analytical method to record small variations in the concentration of the analyte. It may be determined by applying the method to samples containing increasingly small differences in the concentration of the analyte.

The linearity of a test procedure is a measure of its ability, within a given range, to yield results directly proportional to the amount (or concentration) in the sample of the substance under test (the analyte). Analytical procedures that are not strictly linear may be acceptable if some other mathematical relationship, or proportionality factor, is determined and applied.

For the establishment of linearity, a minimum of five different known concentrations of the analyte should be assayed, and the results obtained plotted against the known concentrations, to determine if the relationship between the known concentration and the assay response is linear, and if extrapolated back to zero concentration, the intercept passes through the origin. The range of concentrations over which linearity is determined should be selected with care, and in relation to the intended application of the analytical method. Appropriate ranges, as recommended by Carr and Wahlich,[4] are shown in Table 10.2.

Limit of Detection

Limit of Detection is the lowest amount of the analyte that can be detected, but not necessarily quantified, by the method. It is a parameter mainly of significance in the context of limit tests for impurities. It is determined by the analysis of samples with differing known concentrations of the pure analyte, and then establishing the minimum level at which the analyte can be detected.

Limit of Quantitation (Limit of Quantification)

Limit of Quantitation (or Quantification) is the lowest level at which a quantitative determination can be undertaken with a stated precision and accuracy under stated experimental conditions. That is, it is the lowest amount of an analyte in a sample that can be quantitatively determined (with suitable accuracy and precision). It is particularly relevant to the quantitative determination of low levels of impurities or degradation products in a sample. It is generally determined by the analysis of a series of samples with differing, and decreasing, low concentrations of the analyte, and thus establishing the minimum level at which the analyte can be quantified with acceptable accuracy and precision.

Range

Range is the interval between the upper and lower level of the amount of an analyte that can be demonstrated with suitable precision, accuracy, and linearity. It is normally derived from linearity studies.

Table 10.2   Linearity Validation

Intended Application	Typical Spec. Range (%)	Validation Range (%)
Release spec. assay	90–105	80–120
Assay of active in a stability study	90–105	80–120
Content uniformity test	75–125	60–140
Assay of preservative in stability study	50–110	40–120
Assay of degradation products in stability study	0–10	0–20

Robustness

A further term that appears in the literature (e.g., the ICH text on analytical validation methodology) is "robustness." The term "ruggedness" is also often used synonymously. The robustness of an analytical procedure is a measure of its capacity to remain unaffected by small but deliberate variations in method parameters, and thus provides an indication of the reliability of the method during normal usage, under various conditions.

If the results of an analytical procedure are susceptible to variations in the analytical conditions, such conditions must be properly controlled, and an appropriate cautionary statement included in the written specification or method.

Typical variations that may need to be evaluated include:

- Different makes of equipment
- Different analysts
- Instability of analytical reagents

In the case of liquid chromatography:

- Influence of variations of pH in a mobile phase
- Influence of variations in mobile phase composition
- Different columns (different lots and/or suppliers)
- Temperature
- Flow rate
- Temperature

In the case of gas chromatography:

- Different columns (lots and/or suppliers)
- Flow rate
- Temperature

US CGMPS

Subpart I — Laboratory Controls

See. 211.166 Stability testing

(a)   There shall be a written testing program designed to assess the stability characteristics of drug products. The results of such stability testing shall be used in determining appropriate storage conditions and expiration dates. The written program shall be followed and shall include:
   (1) Sample size and test intervals based on statistical criteria for each attribute examined to assure valid estimates of stability;
   (2) Storage conditions for samples retained for testing;
   (3) Reliable, meaningful, and specific test methods;
   (4) Testing of the drug product in the same container-closure system as that in which the drug product is marketed;

(5) Testing of drug products for reconstitution at the time of dispensing (as directed in the labeling) as well as after they are reconstituted.

(b) An adequate number of batches of each drug product shall be tested to determine an appropriate expiration date and a record of such data shall be maintained. Accelerated studies, combined with basic stability information on the components, drug products, and container-closure system, may be used to support tentative expiration dates provided full shelf life studies are not available and are being conducted. Where data from accelerated studies are used to project a tentative expiration date that is beyond a date supported by actual shelf life studies, there must be stability studies conducted, including drug product testing at appropriate intervals, until the tentative expiration date is verified or the appropriate expiration date determined.

(c) For homeopathic drug products, the requirements of this section are as follows:

(1) There shall be a written assessment of stability based at least on testing or examination of the drug product for compatibility of the ingredients, and based on marketing experience with the drug product to indicate that there is no degradation of the product for the normal or expected period of use.

(2) Evaluation of stability shall be based on the same container-closure system in which the drug product is being marketed.

(d) Allergenic extracts that are labeled "No U.S. Standard of Potency" are exempt from the requirements of this section.

EC GMP GUIDE

The main body of the EC text does not make reference to stability testing. There are, however, references to "Stability monitoring of APIs" in its Annex 18, "Good manufacturing practice for Active Pharmaceutical Ingredients," which is very closely similar to the guideline on APIs also adopted by the FDA. These references are as follows:

Annex 18 to EC GMP Guide

11.5 Stability monitoring of APIs

11.50 A documented, on-going testing program should be designed to monitor the stability characteristics of APIs, and the results should be used to confirm appropriate storage conditions and retest or expiry dates.

11.51 The test procedures used in stability testing should be validated and be stability indicating.

11.52 Stability samples should be stored in containers that simulate the market container. For example, if the API is marketed in bags within fiber drum, stability samples can be packaged in bags of the same material and in smaller-scale drums of similar or identical material composition to the market drums.

11.53 Normally the first three commercial production batches should be placed on the stability monitoring program to confirm the retest or retained expiry date. However, where data from previous studies show that the

API is expected to remain stable for at least two years, fewer than three batches can be used.

11.54 Thereafter, at least one batch per year of API manufactured (unless none is produced that year) should be added to the stability monitoring program and tested at least annually to confirm the stability.

11.55 For APIs with short shelf-lives, testing should be done more frequently. For example, for those biotechnological/biologic and other APIs with shelf-lives of one year or less, stability samples should be obtained and should be tested monthly for the first three months and at three month intervals after that. When data exist that confirm that the stability of the API is not compromised, elimination of specific test intervals (e.g., 9 month testing) can be considered. 11.56 Where appropriate, the stability storage conditions should be consistent with the ICH guidelines on stability.

And

17.5 Stability

17.50 Stability studies to justify assigned expiration or retest dates should be conducted if the API or intermediate is repackaged in a different type of container than that used by the API or intermediate manufacturer.

DISCUSSION

It must be stressed that the above quotations from the annex to the EC GMP Guide refer *only* to active pharmaceutical ingredients (APIs). It may seem strange and inconsistent that the compilers of the EC GMP Guide do not appear to consider, in the main body of their text, that stability studies form a part of Good Manufacturing Practice, yet nevertheless devote eight paragraphs to "stability monitoring" in an annex on Active Pharmaceutical Ingredients. The explanation probably lies in the way that the various annexes have been added to the main Guide from different sources at different times over the years. A note on the genesis of the EC GMP Guide has already been provided in Chapter 1 of this book. From this, it will be clear that, as far as the main body of the text is concerned, there is a single (albeit committee-driven) line of descent. In contrast, a number of the later-added annexes have been "bolted on" from a number of exterior sources. The Annex 18 on Good Manufacturing Practice for Active Pharmaceutical Ingredients is no exception. It is, in fact, the same as the document entitled "Guidance for Industry — QA7 Good Manufacturing Practice for Active Pharmaceutical Ingredients," which was prepared by the International Conference on Harmonisation (ICH), and which was published by the FDA in September 2001 with the statement that it "describes cGMPs for the manufacture of APIs" (*Federal Register*, September 25, 2001, www.fda.gov/cder/guidance/4286fnl.htm).

The European authorities issued this same document around the same time as Annex 18 to the EC GMP Guide. The original compilers of that Guide evidently did *not* consider that stability studies were a part of GMP (and it is reasonable to argue that they belong more in the general area of research and development), whereas ICH considered that they are, at least in regard to APIs.

As part of an NDA, or an application (in Europe) for a Marketing Authorization, convincing data derived from well-designed and conducted stability studies performed during the research and development phase of a new drug product will need to be submitted to the relevant regulatory authority. Those making such applications will need to obtain, and comply with, copies of the current regulatory requirements. Reference should be made to:

- Guideline for submitting documentation for the stability of human drugs and biologics. February 1987. Center for Drugs and Biologics. Food and Drug Administration. Department of Health and Human Services, Rockville, Maryland 20857.
- Stability tests for active ingredients and finished products (July 1988). Rules governing medicinal products in the European Community, Vol. III, Guidelines on the quality, safety and efficacy of medicinal products for human use (1989), ISBN 928596192. Office for Official Publications of the European Communities, L–29885, Luxembourg.

The following are some general observations.

The purpose of stability testing is to generate information that permits the establishment of a rational, science-based shelf life (or expiry date) for a drug product and to recommend appropriate storage conditions. Stability studies can also have an important role in the determination of which, of a number of possible pharmaceutical formulations, is to be preferred on the basis of stability.

The major factor is usually, and quite reasonably, considered to be the content of active drug substance(s) over a period of time. There are, however, other significant factors. For example:

- Limitation or absence of decomposition products of the active substance(s), with the consequent toxicity implications
- Physical appearance of the product and its container and labeling
- Content of other significant ingredients of the formulation, e.g., any microbial preservatives
- Physical properties
- Microbial properties
- Condition and integrity of the container or closure system
- Organoleptic properties — odour, taste, etc.

Thus the purpose of stability studies is to determine how the quality of a medicinal product, in relation to the various factors given above, varies as a function of time and under the influence of a variety of environmental factors.

It should be possible, on the basis of the information thus generated, to recommend storage conditions that will guarantee the maintenance of the quality of a product, in relation to its safety, efficacy, and acceptability, throughout its proposed shelf life, during storage, distribution, dispensing, and use. It is crucial that the studies are conducted on samples of the finished product *in the container, closure, or package system in which it is intended to be supplied to the market.*

The design of finished product stability studies to be carried out on a drug product should be based on the knowledge obtained from studies on the active drug substance and from development of pharmaceutics studies.

Information on the stability of the active drug substance should be obtained, for new active drug substance, by experimental studies. For already known and established active pharmaceutical ingredients, sufficient evidence will usually be available in the scientific literature. However, comparative accelerated stability studies may be necessary in some cases, for example, when there is a significant change in the route of synthesis from that approved by the regulatory authorities, or where there is a significant change in the production method. Where there are several possible manufacturers of the active pharmaceutical ingredient or different methods of obtaining this ingredient, consideration may need to be given to conducting studies on material from each of the sources.

A protocol should be prepared in advance, stating the general methodology of the study and giving details of proposed and any accelerated storage conditions, taking into account the physical and chemical properties of the active drug substance, temperature, humidity and light, and potential exposure to air, with details of planned duration of exposure under various conditions.

The analytical methods used must be validated (in addition to accuracy, precision, etc., particularly in relation to their specificity regarding their ability to separate the active drug substance from both its degradation products and the excipients.) That is, assay methods should be stability-indicating. Validated methods for the determination of degradation products should be stated in the study protocol.

Analysis of the results of the stability studies on the finished product should allow the determination of the shelf life, the recommendations for storage conditions (where relevant before and after opening the container), and the justification of any overage of the active ingredient added to guarantee potency at the end of the shelf life.

Specific "in-use" stability studies should be performed in cases where the product is labile once the container is opened or where the product is intended to be diluted or reconstituted before use.

STABILITY STUDY METHODS

Real-time studies should be carried out under a range of controlled test conditions, which will enable the shelf life and the product, container, closure, label, and package storage requirements to be defined. This should normally include studies that will allow the properties of the product at temperatures between 20° and 30°C to be evaluated.

For each study, the various test conditions, including the mean temperature, the temperature range, and the mean humidity, should be stated in both the study protocol and the report.

There remains a certain lack of clarity over the precise regulatory requirements regarding storage conditions (temperature, humidity, etc.) for products on stability testing. *Those performing such tests for the purposes of making regulatory applications should ensure that they are in possession of the current requirement of their local agency.*

Some have held that "ambient" conditions should be 30°C, whereas others have considered 25–30°C to be more appropriate. The issue has been clarified (somewhat) by the International Conference on Harmonization (ICH). [5]

Significant issues in this ICH Guideline include:

■ Stability storage conditions should normally involve long-term studies at 25° ± 2°C and 60% RH ± 5%, with at least 12 months of data and accelerated studies at 40° ± 2°C and 75% RH ± 5%, with at least 6 months of data.

■ Where "significant change" occurs during a 40°C accelerated study an additional intermediate condition should be studied, such as 30° ± 2°C/60% RH ± 5%. (Significant change is defined as a 5% loss of potency, any degradant exceeding its specification limit, exceeding pH limits, dissolution failures using 12 units, or failures of physical specifications — hardness, color, etc.)

■ For less stable products, the storage (and labeling) conditions may be reduced but the accelerated conditions should still be at least 15°C above those used for long-term evaluation.

■ For products where water loss may be important, such as liquids or semisolids in plastic containers, it may be more appropriate to replace the high RH conditions by lower RH such as 10 to 20%.

The ICH Guideline does not address the position of the samples during the study. This can have significance with regard to liquid products, where leakage and product closure interaction may need to be evaluated. One possible approach is to store the samples both upright and inverted, but only to test the inverted samples. The upright samples may be used as controls if problems are identified.

Depending on the nature and objectives of a stability study, it may also be necessary to consider the effect of *low* temperatures (for example freezer and refrigerator temperatures), of freeze-thaw recycling, and of light (either natural or artificial).

Normally, at least three different batches of the finished-pack product should be subject to stability study.

One curious feature of the US cGMPs, in this subpart (see Subpart I, Sec. 211.166 (c)) is the requirement to perform stability testing on homoeopathic drug products. Other than to marvel at how it could be considered possible to determine the stability of a product that, in effect, contains *no* active ingredient, this present author is unable to comment further.

US CGMPS

Subpart I—Laboratory Controls

Sec. 211.167 Special testing requirements

(a) For each batch of drug product purporting to be sterile and/or pyrogen-free, there shall be appropriate laboratory testing to determine conformance to such requirements. The test procedures shall be in writing and shall be followed.

(b) For each batch of ophthalmic ointment, there shall be appropriate testing to determine conformance to specifications regarding the presence of foreign particles and harsh or abrasive substances. The test procedures shall be in writing and shall be followed.

(c) For each batch of controlled-release dosage form, there shall be appropriate laboratory testing to determine conformance to the specifications for the rate of release of each active ingredient. The test procedures shall be in writing and shall be followed.

EC GMP Guide

Note : The EC GMP Guide does not carry any specific statement that testing for sterility, absence of pyrogens, absence of foreign, harsh, and abrasive substances (in ophthalmic ointments) or rate of release of active(s) from controlled-release products shall be performed. These requirement may be said to be covered by the requirements of the European and British Pharmacopoeias, and by the general statement that appears at 6.15, in Chapter 6, Quality Control, of the EC GMP Guide:

6.15 ... All testing operations described in the marketing authorisation should carried out according to approved methods.

With regard to sterility testing, although this is not stated explicitly as a requirement for sterile products, it is implicit in the subsection on Quality Control that concludes EC Annex 1 on Manufacture of Sterile Products:

Quality Control

91   The sterility test applied to the finished product should only be regarded as the last in a series of control measures by which sterility is assured. The test should be validated for the product(s) concerned.

92   In those cases where parametric release has been authorised, special attention should be paid to the validation and the monitoring of the entire manufacturing process.

93   Samples taken for sterility testing should be representative of the whole of the batch, but should in particular include samples taken from parts of the batch considered to be most at risk of contamination, e.g.,
   a. for products which have been filled aseptically, samples should include containers filled at the beginning and end of the batch and after any significant intervention,
   b. for products which have been heat sterilised in their final containers, consideration should be given to taking samples from (lie potentially coolest part of the load.

It should be further noted (a) that the principal thrust of the EC paragraph (91 of Annex 1) quoted above is not to emphasize the importance of the sterility test, but to stress that it is but one of a series of control measures, which by itself cannot be considered as an adequate criterion of batch sterility, and (b) that the possibility of parametric release is acknowledged.

DISCUSSION

The fact that the EC GMP Guide does not make explicit, specific, reference to these US "Special Testing Requirements" is a reflection of the clear distinction that is drawn, in European regulatory issues, between requirements arising from legislation relating to obtaining and holding a *Marketing* Authorization (or Product License, in common UK terminology) and that relating to obtaining and holding a *Manufacturing* Authorization (or Manufacturer's License, in UK terminology). In Europe, the issue of specific test requirements falls in the area of the Marketing Authorization. An essential prerequisite for obtaining and holding a Manufacturing Authorization is that a manufacturer "shall ensure that the manufacturing operations are carried out in accordance with good manufacturing practice." That, in turn, requires *inter alia* that each batch of each product is tested in accordance with the requirements of the Marketing Authorization. Thus, further explicit mention of specific testing requirements is not necessary in the EC GMP Guide.

The logic of the "special test requirements" of US Sec. 211.167 might perhaps be questioned. If it is felt necessary to make statements about these few testing procedures, important though they are, why omit others? Why, for example, omit mention of disintegration tests on tablets and capsules, content uniformity tests on low-dose tablets and capsules, assay of active content in all dosage forms, and so on? These are, perhaps, academic points, but it is worth reflecting further on the rationale of sterility testing.

STERILITY TEST

Table 10.3 (from Hugo and Russell[6]) shows the various probabilities of passing a standard pharmacopoeial sterility test (20 samples) at different levels of contamination of the batch. It is based solely on statistical probability considerations. This alone is hardly confidence-inspiring. There are also microbiological limitations, in particular the fact that there is no universal growth medium upon or in which all forms of microorganisms may be expected to grow. As generally practiced, sterility tests will not detect viruses, protozoa, exacting parasitic bacteria or many thermophillic and psychrophilic bacteria. Furthermore, organisms that have been damaged, but not killed, by exposure to sublethal levels of "sterilization," may not show up in the standard sterility test, as they may require conditions for growth, in terms of nutrients, temperature, and time, which the test does not provide.

Table 10.3   Probability of Passing a Standard Pharmacopoeial Sterility Test

No. of Units Contaminated (%)	Chance of Passing Test (%)
0.1	98
1.0	82
5.0	36
10.0	12

Despite these acknowledged limitations, the test continues to be performed, even by those who would accept that it has little real significance in terms of the quality of the product. This is probably due largely to regulatory requirements and to a nervous perception of potential legal implications. The EC GMP Guide appears to accept these limitations by declaring that "the sterility test applied to the finished product should only be regarded as the last in a series of control measures by which sterility is assured." The original UK GMP Guide[7] amplified this very point by adding that "compliance with the test does not guarantee sterility of the whole batch, since sampling may fail to select non-sterile containers, and the culture methods used have limits to their sensitivity…."

It is this fairly widespread acceptance of the dubious value of the sterility test in providing assurance of the sterility of a manufactured batch as a whole that has led to the stress placed on crucial importance of meticulous control of the entire sterile products manufacturing process. It has also given rise to the concept of parametric release.

Parametric Release

The concept of parametric release emerged in the early- to mid-1980s and originally was related solely to the sterility (or otherwise) of terminally heat-sterilized products. That is, it did not originally bear upon other release criteria, or on the release of any other products, sterile or otherwise.

One of the first (if not *the* first) regulatory publications on this subject is an FDA Compliance Policy Guide on "Parametric Release — Terminally Sterilized Drug Products" (CPG 7132a, issued October 21, 1987). It reads as follows:

Sec. 460.800 Parametric Release - Terminally Heat Sterilized Drug Products(CPG 7132a.13)

Background:

> In 1985, FDA approved supplemental new drug applications for certain large volume parenteral drug products, which substituted parametric release for routine lot by lot end-product sterility testing.
>
> Parametric release is defined as a sterility release procedure based upon effective control, monitoring, and documentation of a validated sterilization process cycle in lieu of release based upon end-product sterility testing (21 CFR 211.167). All parameters within the procedure must be met before the lot is released.

Policy:

> This policy applies only to parenteral drug products which are terminally heat sterilized. It does not apply to products sterilized by filtration or ethylene oxide. This policy does not pre-empt requirements of Section 505 of the FD&C Act. Approved supplements providing for parametric release are required for holders of new drug applications. (21 CFR 314.70(b))

Parametric release, in lieu of end product sterility testing, is acceptable when all of the following parameters are met and documented.

1.   The sterilization process cycle has been validated to achieve microbial bioburden reduction to 10^0 with a minimum safety factor of an additional six logarithm reduction. Cycle validation includes sterilizer heat distribution studies, heat distribution studies for each load configuration, heat penetration studies of the product, bioburden studies, and a lethality study referencing a test organism of known resistance to the sterilization process. All cycle parameters must be identified by the manufacturer as critical (e.g., time, temperature, pressure) or non-critical (e.g., cooling time, heat-up time). Under parametric release, failure of more than one critical parameter must result in automatic rejection of the sterilizer load (see paragraph D concerning biological indicators). (21 CFR 211.113(b))

2.   Integrity for each container/closure system has been validated to prevent in-process and post-process contamination over the product's intended shelf life. Validation should include chemical or microbial ingress tests utilizing units from typical products. (21 CFR 211.94)

3.   Bioburden testing (covering total aerobic and total spore counts) is conducted on each batch of presterilized drug product. Resistance of any spore-forming organism found must be compared to that of the organism used to validate the sterilization cycle. The batch is deemed non-sterile if the bioburden organism is more resistant than the one used in validation. (21 CFR 211.110)

4.   Chemical or biological indicators are included in each truck, tray, or pallet of each sterilizer load. For chemical indicators, time/temperature response characteristics and stability are documented and for each sterilization cycle minimum degradation values are established. Chemical indicators cannot be used to evaluate cycle lethality.

Documentation is required for biological indicators (BIs). Documentation for each BI lot shall include an organism's name, source and D-value, spore concentration per carrier, expiration date, and storage conditions. BIs can be used to evaluate cycle lethality where equipment malfunction prevents measurement of one critical cycle parameter. If more than one critical parameter is not met, the batch is considered non-sterile despite BI sterility. (21 CFR 2311.165(e) and 211.167)

Issued: 10/21/87

Regarding the definition of "parametric release" given in this guideline, it is perhaps worth the comment that if "sterility release" may be based on "effective control…" etc., in place (that is "in lieu") of a sterility test result, then the inverse corollary is surely implied that if a sterility test *has* been passed, then "effective control, monitoring and documentation of validated sterilization process" is not necessary — which is contrary to all the principles of quality assurance in the manufacture of sterile products, and is thus presumably not what the FDA really meant. It is also to be noted that the first sentence of the first paragraph under "… acceptable when all of the following parameters are met and documented" (that is "1. The sterilization process cycle has been validated to achieve microbial bioburden reduction to 10° with a minimum safety factor…") is given as it appeared originally, when the guideline was first issued. In the version currently available on the FDA Web site, it reads "1. The sterilization process cycle has been validated to achieve microbial bioburden reduction to 100 with a minimum safety factor…" It thus appears that there has been a change in the required "bioburden reduction" from 1 (= 10°) to 100. This may be no more than an error in cybertranscription, and in any event is of little microbiological significance.

Perhaps more importantly, it is reasonable to ask if (with the possible exception of the inclusion of biological indicators in *every* load) the list of parameters to be met are neither more nor less a list of things that should be done *anyway*, in the course of normal production, whether the lot is to be sterility tested or not. Care needs to be taken to avoid any thought that if a sterility test *is* performed, there is not a need to ensure that the listed parameters "are met and documented." That would be entirely wrong.

In this context, parametric release indicates that a notably unreliable test procedure (the sterility test) may be abandoned, with at least a theoretical possibility of regulatory approval, in favor of a rigorous concentration of effort on actions that will provide a significantly higher level of assurance of sterility. This is an excellent notion, in this context, and one that has been adopted (with official approval) by a few sterile products manufacturers. But they are relatively few. The reason for this probably lies in a not entirely irrational fear that, should legal action be taken for damages in the case of an alleged sterility failure, judges will probably consider "passed sterility test" a better defense than technical and statistical arguments they cannot understand.

It has already been noted above that, in the main body of its text, the EC GMP Guide allows for the possibility of parametric release, not only for terminally heat-sterilized products but (possibly) for other types of products as well. This wider application of the concept seems to be implicit in the quite recent Annex 17, Parametric Release, to the EC GMP Guide, which reads:

ANNEX 17 Parametric Release

> Note that this Annex should be read in conjunction with CPMP/QWP/3015/99 which was adopted by the Committee on Proprietary Medicinal Products (CPMP) in February 2001 (see http://www.emea.eu.int/htms/human/qwp/qwp-fin.htm).

1. Principle

> 1.1 The definition of Parametric Release used in this Annex is based on that proposed by the European Organization for Quality: 'A system of release that gives the assurance that the product is of the intended quality based on information collected during the manufacturing process and on the compliance with specific GMP requirements related to Parametric Release.'
>
> 1.2 Parametric Release should comply with the basic requirements of GMP with applicable annexes and the following guidelines.

2. Parametric Release

2.1 It is recognised that a comprehensive set of in-process tests and controls may provide greater assurance of the finished product meeting specification than finished product testing.

2.2 Parametric Release may be authorised for certain specific parameters as an alternative to routine testing of finished products. Authorisation for parametric release should be given, refused or withdrawn jointly by those responsible for assessing products together with the GMP inspectors.

3. Parametric Release for Sterile Products

3.1 This section is only concerned with that part of Parametric Release which deals with the routine release of finished products without carrying out a sterility test. Elimination of the sterility test is only valid on the basis of successful demonstration that predetermined, validated sterilising conditions have been achieved.

3.2 A sterility test only provides an opportunity to detect a major failure of the sterility assurance system due to statistical limitations of the method.

3.3 Parametric Release can be authorised if the data demonstrating correct processing of the batch provides sufficient assurance, on its own, that the process designed and validated to ensure the sterility of the product has been delivered.

3.4 At present Parametric Release can only be approved for products terminally sterilised in their final container.

3.5 Sterilisation methods according to European Pharmacopoeia requirements using steam, dry heat and ionising radiation may be considered for Parametric Release.

3.6 It is unlikely that a completely new product would be considered as suitable for Parametric Release because a period of satisfactory sterility test results will form part of the acceptance criteria. There may be cases when a new product is only a minor variation, from the sterility assurance point of view, and existing sterility test data from other products could be considered as relevant.

3.7 A risk analysis of the sterility assurance system focused on an evaluation of releasing non-sterilised products should be performed.

3.8 The manufacturer should have a history of good compliance with GMP.

3.9 The history of non sterility of products and of results of sterility tests carried out on the product in question together with products processed through the same or a similar sterility assurance system should be taken into consideration when evaluating GMP compliance.

3.10 A qualified experienced sterility assurance engineer and a qualified microbiologist should normally be present on the site of production and sterilisation.

3.11 The design and original validation of the product should ensure that integrity can be maintained under all relevant conditions.

3.12 The change control system should require review of change by sterility assurance personnel.

3.13 There should be a system to control microbiological contamination in the product before sterilisation.

3.14 There should be no possibility for mix ups between sterilised and non sterilised products. Physical barriers or validated electronic systems may, provide such assurance.

3.15 The sterilisation records should be checked for compliance to specification by at least two independent systems. These systems may consist of two people or a validated computer system plus a person.

3.16 The following information should be confirmed prior to release of each batch of product:
  – All planned maintenance and routine checks have been completed in the sterilizer used;
  – All repairs and modifications have been approved by the sterility assurance engineer and microbiologist;
  – All instrumentation was in calibration;
  – The sterilizer had a current validation for the product load processed.

Once Parametric Release has been granted, decisions for release or rejection of a batch should be based on the approved specifications. Non-compliance with the specification for Parametric Release cannot be overruled by a pass of a sterility test.

Note: Readers who find that this passage from the EC GMP Guide reads oddly, and in places, incomprehensibly, can take comfort in the thought that they are not alone. This is the sort of thing that tends to emerge from multinational, multilingual committees.

The next section of Subpart I of the US cGMP to be considered is Sec. 211.170:

US CGMPS

Subpart I — Laboratory Controls

Sec. 211.170 Reserve samples

(a) An appropriately identified reserve sample that is representative of each lot in each shipment of each active ingredient shall be retained. The reserve sample consists of at least twice the quantity necessary for all tests required to determine whether the active ingredient meets its established specifications, except for sterility and pyrogen testing. The retention time is as follows.

  (1) For an active ingredient in a drug product other than those described in paragraphs (a) (2) and (3) of this section, the reserve sample shall be retained for I year after the expiration date of the lot of the drug product containing the active ingredient.

  (2) For an active ingredient in a radioactive drug product, except for non-radioactive reagent kits, the reserve sample shall be retained for:
  (i) Three months after the expiration date of the last lot of the drug product containing the active ingredient if the expiration dating period of the drug product is 30 days or less; or
  (ii) Six months after the expiration date of the last lot of the drug product containing the active ingredient if the expiration dating period of the drug product is more than 30 days.

(3) For an active ingredient in an OTC drug product that is exempt from bearing an expiration date under Sec. 211.137, the reserve sample shall be retained for 3 years after distribution of the last lot of the drug product containing the active ingredient.

(b) An appropriately identified reserve sample that is representative of each lot or batch of drug product shall be retained and stored under conditions consistent with product labeling. The reserve sample shall be stored in the same immediate container-closure system in which the drug product is marketed or in one that has essentially the same characteristics. The reserve sample consists of at least twice the quantity necessary to perform all the required tests, except those for sterility and pyrogens. Except for those for drug products described in paragraph (b)(2) of this section, reserve samples from representative sample lots or batches selected by acceptable statistical procedures shall be examined visually at least once a year for evidence of deterioration unless visual examination would affect the integrity of the reserve sample. Any evidence of reserve sample deterioration shall be investigated in accordance with Sec. 211.192. The results of the examination shall be recorded and maintained with other stability data on the drug product. Reserve samples of compressed medical gases need not be retained. The retention time is as follows:

(1) For a drug product other than those described in paragraphs (b) (2) and (3) of this section, the reserve sample shall be retained for 1 year after the expiration date of the drug product.

(2) For a radioactive drug product, except for non-radioactive reagent kits reserve sample shall be retained for:
(i) Three months after the expiration date of the drug product if the expiration dating period of the drug product is 30 days or less; or
(ii) Six months after the expiration dating of the drug product is more than 30 days.

(3) For an OTC drug product that is exempt for bearing an expiration date under Sec. 211.137, the reserve sample must be retained for 3 years after the lot or batch of drug product is distributed.

In Europe, the requirements regarding reserve samples are covered in both the European Commission Directive 91/356/EEC, laying down the principles and guidelines of Good Manufacturing Practice, and in the EC GMP Guide. The relevant paragraphs of Directive 91/356/EEC and the EC GMP Guide are:

From EC Directive 91/356/EEC Article 11:

4. Samples of each batch of finished products shall be retained for at least one year after the expiry date. Unless in the Member State… a longer period is required, samples of starting materials (other than solvents, gases and water) used shall be retained for at least two years after the release of the product. This period may be shortened if their stability, as mentioned in the relevant specification, is shorter. All these samples shall be maintained at the disposal of the competent authorities.

EC GMP GUIDE

Chapter 6 Quality Control

6.12 Reference samples should be representative of the batch of materials or products from which they are taken. Other samples may also be taken to monitor the most stressed part of a process (e.g., beginning or end of a process).

6.13 Sample containers should bear a label indicating the contents, with the batch number, the date of sampling and the containers from which samples have been drawn.

6.14 Reference samples from each batch of finished products should be retained till one year after the expiry date. Finished products should usually be kept in their final packaging and stored under the recommended conditions. Samples of starting materials (other than solvents, gases and water) should be retained for at least two years (1) after the release of the product if their stability allows. This period may be shortened if their stability, as mentioned in the relevant specification, is shorter. Reference samples of materials and products should be of a size sufficient to permit at least a full re-examination.

ANNEX 18 GOOD MANUFACTURING PRACTICE FOR APIs

11.71 Appropriately identified reserve samples of each API batch should be retained for one year after the expiry date of the batch assigned by the manufacturer, or for three years after distribution of the batch, whichever is the longer. For APIs with retest dates, similar reserve samples should be retained for three years after the batch is completely distributed by the manufacturer.

11.72 The reserve sample should be stored in the same packaging system in which the API is stored or in one that is equivalent to or more protective than the marketed packaging system. Sufficient quantities should be retained to conduct at least two full compendial analyses or, when there is no pharmacopoeial monograph, two full specification analyses.

DISCUSSION

The requirements of the two regulatory documents are quite similar, but there are some significant differences. For the purposes of comparison it is convenient to consider the requirements relating to ingredients (or starting materials) separately from those relating to finished products:

Differences worth emphasizing are:

■ In the US cGMPs, the requirement to keep reserve samples applies only to *active* ingredients. In the EC GMP Guide and Directive, reserve samples of *all* ingredients (starting materials), with the exception of solvents, gasses, and water are required.

■ The US cGMPs demand that a quantity *twice* the amount needed to perform all specified tests should be retained. EC requires that only a quantity for "at least one full re-examination" should be retained.

US cGMPs — Reserve Samples of Ingredients	EC GMP Guide — Reserve Samples of Starting Materials
Apply to *active* ingredients	Apply to starting materials generally, not just actives (solvents, gases, and water excepted)
Samples shall be "representative"	Samples to be "representative," but "other samples *may* also be taken to monitor the most stressed part"
To be taken from "each lot in each shipment"	Not so explicit; refers only to "the batch"
Quantity shall be twice amount necessary for all specification tests, except sterility and pyrogens	"Should be of a size sufficient to permit at least a full re-examination"
Retain for one year after expiry date of drug product in which API is used *except* (a) if used in radioactive drug product and (b) if used in an OTC not bearing expiration date; (then retain for 3 years after distribution of last lot of OTC containing the API)	Retain for at least 2 years after the release of product; only exceptions — solvents, gases, water

■ In the US cGMPs, the retention period is until one year after the expiry date of the product in which the ingredient is used, with special requirements for ingredients used in radiopharmaceuticals and nonexpiration-dated OTCs. In EC, the retention period for reserve samples of all ingredients (except solvents, gasses, and water) is for at least two years after the release of the product(s) in which they are used.

■ The US cGMPs require reserve samples to be taken from each shipment of each lot of each active ingredient. Thus, if shipments of the same lot of an ingredient are received on more than one occasion, a sample from each delivery is to be retained. EC is not so precise or explicit on this point, although it is generally taken that a sample from each delivery should be retained.

The obvious rationale for retaining samples (of both ingredients and finished products) is to enable investigation in the event of a complaint, a report of a defect, or of untoward effects in patients, and to facilitate the taking of corrective action. Since a majority of complaints, defect reports, and the like, leveled at the pharmaceutical industry as a whole, are ultimately found to be without foundation, then retention samples and any subsequent examination thereof can be important elements in a manufacturer's defense against unwarranted complaints of faulty products, damage caused, and so on. Further, since incorrect or faulty nonactive ingredients can significantly affect the safety and efficacy of a product, the US cGMPs are less judicious than the EC in requiring that reserve samples of only *active* ingredients shall be retained. In any event, whatever the regulatory requirement, a prudent manufacturer will retain

US cGMPs — Reserve Samples of Finished Products	EC GMP Guide — Reserve Samples of Finished Products
Shall be "representative"	"Representative" — but (as for ingredients) "other samples *may* be taken …"
Samples shall be stored in conditions consistent with product labeling	Store "under the recommended conditions" (*as on the label* presumably intended)
" … shall be stored in the same immediate container-closure system" as the marketed drug product "or in one which has essentially the same characteristics."	" … should usually be kept in their final packaging"
Quantity — at least twice the quantity needed to perform all required tests (except sterility and pyrogens)	Quantity — at least sufficient to perform a "full re-examination"
Retain until 1 year after expiry date, except (a) radiopharmaceuticals and (b) nonexpiry-dated OTCs (until 3 years after distribution)	Same (i.e., 1 year after expiry date), but with no exceptions
Shall be examined visually at least once a year — unless examination would affect the integrity of the sample	No corresponding requirement

samples of *all* ingredients and (within the limits of storage space considerations) in quantities somewhat larger than the minimum specified.

As in relation to ingredients, the EC has the curiously indeterminate statement that "other samples *may* be taken …" The point of such a comment is obscure. Either it is a requirement that "other" samples *should* be taken, or it is not. And is it, or is it not, intended that any such "other" samples are to be held in reserve? There is probably no answer to such questions, other than that this is just another example of the sort of ambiguous indeterminism that comes from multinational, multilingual committees.

As with ingredients, the US requirement of the "at least twice" quantity is far more sensible than the EC's "sufficient for just one re-examination."

It is difficult to find a rationale for the US requirement for the visual examination at least once a year. It can hardly be of any great value, since in most cases any degradation would not be visually apparent.

US CGMPS

Subpart I — Laboratory Controls

Sec. 211.173 Laboratory animals

Animals used in testing components, in-process materials, or drug products for compliance with established specifications shall be maintained and controlled in a manner that assures their suitability for their intended use. They shall be identified, and adequate records shall be maintained showing the history of their use.

EC GMP GUIDE

Chapter 3 Premises and Equipment

3.33 Animal houses should be well isolated from other areas, with separate entrance (animal access) and air handling facilities.

Chapter 6 Quality Control

6.22 Animals used for testing components, materials or products, should, where appropriate, be quarantined before use. They should be maintained and controlled in a manner that assures their suitability for the intended use. They should be identified, and adequate records should be maintained, showing the history of their use.

DISCUSSION

There are two major issues relating to the housing and use of laboratory animals — their suitability for their intended purpose and the control of the product-contamination hazard they represent. The US cGMPs seem, in this context, to be concerned only with the former. The EC GMP Guide (6.22) shares this concern, but also (3.33) indicates the need to provide protection against the potential contamination hazard the animals could represent. As well as physical isolation and separate air handling systems, care needs to be taken over waste disposal and to the limitation of personnel access, both to and from animal houses.

Each animal should be identified by number, letter, or "name," and for each a record should be maintained of:

1. Identification
2. Species or variety
3. Source
4. Age
5. Date received
6. Procedure(s) used for
7. Date(s) used

US CGMPS

Subpart I — Laboratory Controls

Sec. 211.176 Penicillin contamination

If a reasonable possibility exists that a non-penicillin drug product has been exposed to cross-contamination with penicillin, the non-penicillin drug product shall be tested for the presence of penicillin. Such drug product shall not be marketed if detectable levels are found when tested according to procedures specified in 'Procedures for Detecting and Measuring Penicillin Contamination in Drugs,' which is incorporated by reference. Copies are available from the Division of Research and Testing (HFD-470), Center for Drug Evaluation and Research, Food and Drug Administration, 200 C St. SW., Washington, D.C. 20204, or available for inspection at the Office of the Federal Register, 800 North Capitol Street, NW., suite 700, Washington, D.C. 20408.

EC GMP GUIDE

Chapter 5 Production

Prevention of Cross-Contamination in Production

5.18 Contamination of a starting material or of a product by another material or product must be avoided. This risk of accidental cross-contamination arises from the uncontrolled release of dust, gases, vapours, sprays or organisms from materials and products in process, from residues on equipment, and from operators' clothing. The significance of this risk varies with the type of contaminant and of product being contaminated. Amongst the most hazardous contaminants are highly sensitising materials, biological preparations containing living organisms, certain hormones, cytotoxics, and other highly active materials. Products in which contamination is likely to be most significant are those administered by injection, those given in large doses and/or over a long time.

5.19 Cross-contamination should be avoided by appropriate technical or organisational measures, for example:

(a) Production in segregated areas (required for products such as penicillins, live vaccines, live bacterial preparations and some other biologicals), or by campaign (separation in time) followed by appropriate cleaning;

(b) Providing appropriate air-locks and air extraction;

(c) Minimising the risk of contamination caused by recirculation or re-entry of untreated or insufficiently treated air;

(d) Keeping protective clothing inside areas where products with special risk of cross contamination are processed.,

(e) Using cleaning and decontamination procedures of known effectiveness, as ineffective cleaning of equipment is a common source of cross-contamination.,

(f) Using "closed systems" of production,

(g) Testing for residues and use of cleaning status labels on equipment.

5.20 Measures to prevent cross-contamination and their effectiveness should be checked periodically according to set procedures.

DISCUSSION

Sec. 211.176 of the US cGMPs addresses solely the issue of contamination by *penicillin*. The reason for this concern arises, of course, from the well-known and potentially very dangerous hypersensitivity reactions that can occur in some persons when exposed to penicillin, penicillin derivatives, and other chemically related compounds. This section, however, refers only (and very specifically) to penicillin *per se* and, from the aspect of patient safety, could be considered to be rather too narrowly targeted. It is also to be noted that it does not, in fact, state categorically that other drug products shall not, or must not, be contaminated with penicillin, only that if "a reasonable possibility exists" that another product "has been exposed to contamination with penicillin," it shall be tested for penicillin and not marketed "if detectable levels are found ...."

In contrast, the EC GMP Guide takes a broader-brushed and rather more positive approach. It is noteworthy the first sentence of 5.18 displays a relatively rare (in the context of this Guide as a whole) instance of the use of "must," as distinct from the much more common, but somewhat less imperative "should." ("Contamination of a starting material or a product by another material or product *must* be avoided.")

This section of the EC GMP Guide thus recognizes the potential hazards of *all* forms of cross-contamination, but acknowledges that the patient risk varies both with the nature of the contaminant, and with the type of product contaminated. It also indicates some practical measures to avoid cross-contamination, including "production in segregated areas" for penicillin*s* (note the plural) and other products.

The control of contamination is a major GMP issue. Much depends on people and the way they behave and the protective clothing they wear. Other important control measures are the application of well-planned and proven cleaning and disinfection procedures. Crucial factors are also the design, structure, surface finishes, and layout of factories; the design, installation, and maintenance of equipment;, and the design, installation, efficiency, and maintenance of factory services, such as ventilation, heating, lighting, and water supply. Proper factory and equipment design and layout can also reduce the risk of what can perhaps be regarded as extreme cases of contamination — the complete mix-up of one product with another, of one ingredient with another, or of one packaging material with another.

REFERENCES

1. International Conference on Harmonization (ICH), Draft Consensus Text on Validation of Analytical Procedures, October 27, 1993.
2. Carr, G.P.R. and Wahlich, J.C., A practical approach to method validation in pharmaceutical analysis, *J. Pharm. Biomed. Anal.,* 8 (8-12), 613—618, 1990.
3. CPMP Notes for Guidance: *Analytical Validation,* Addendum to Volume III of the Rules Governing Medicinal Products in the European Community, Office for Official Publications of the European Community, Luxembourg, 1990.
4. Carr, G.P.R. and Wahlich, J.C., Analytical validation, in *International Pharmaceutical Product Registration — Aspects of Quality, Safety and Efficacy,* eds. Cartwright and Matthews, Ellis Horwood Ltd., New York, 1994.
5. ICH , Guideline on stability testing of new drug substances and products, *Federal Register,* 59 (183), 48754, 1994.
6. Hugo, W.D. and Russell, A.D., eds, *Pharmaceutical Microbiology,* 3rd ed., Blackwell Scientific, Oxford, 1983.
7. Sharp, J., ed., *Guide to Good Pharmaceutical Manufacturing Practice,* 3rd ed., Her Majesty's Stationery Office, London, 1983.

11

RECORDS AND REPORTS

In this chapter there will, inevitably, be a certain amount of repetition of, or back reference to, material that has already appeared in earlier chapters of this book. This is a consequence of US cGMPs Subpart J (upon which this chapter is based) being, in large part, a recapitulation from a different angle of material that has appeared in earlier subparts (for example, Subparts E, F, G, H, and I).

US CGMPS

Subpart J — Records and Reports

Sec. 211.180 General requirements

(a) Any production, control, or distribution record that is required to be maintained in compliance with this part and is specifically associated with a batch of a drug product shall be retained for at least 1 year after the expiration date of the batch or, in the case of certain OTC drug products lacking expiration dating because they meet the criteria for exemption under Sec. 211.137, 3 years after distribution of the batch.

(b) Records shall be maintained for all components, drug product containers, closures, and labeling for at least I year after the expiration date or, in the case of certain OTC drug products lacking expiration dating because they meet the criteria for exemption under Sec. 211.137, 3 years after distribution of the last lot of drug product incorporating the component or using the container, closure, or labeling.

(c) All records required under this part, or copies of such records, shall be readily available for authorized inspection during the retention period at the establishment where the activities described in such records occurred. These records or copies thereof shall be subject to photocopying or other means of reproduction as part of such inspection. Records that can be immediately retrieved from another location by computer or other electronic means shall be considered as meeting the requirements of this paragraph.

(d) Records required under this part may be retained either as original records or as true copies such as photocopies, microfilm, microfiche, or other accurate reproductions of the original records. Where reduction techniques, such as microfilming, are used, suitable reader and photocopying equipment shall be readily available.

(e) Written records required by this part shall be maintained so that data therein can be used for evaluating, at least annually, the quality standards of each drug product to determine the need for changes in drug product specifications or manufacturing or control procedures. Written procedures shall be established and followed for such evaluations and shall include provisions for:

   (1) A review of a representative number of batches, whether approved or rejected, and, where applicable, records associated with the batch.

   (2) A review of complaints, recalls, returned or salvaged drug products, and investigations conducted under Sec. 211.192 for each drug product.

(f) Procedures shall be established to assure that the responsible officials of the firm, if they are not personally involved in or immediately aware of such actions, are notified in writing of any investigations conducted under Secs. 211.198, 211.204 or 211.208 of these regulations, any recalls, reports of inspectional observations issued by the Food and Drug Administration, or any regulatory actions relating to good manufacturing practices brought by the Food and Drug Administration.

EC GMP GUIDE (ETC.)

A general statement, which most closely corresponds to the general requirements of US 211.180 (a) and (b) appears, not in the EC GMP Guide itself, but in Article 9.1 of European Directive 91/356/EEC. It reads as follows:

Article 9 of EC Directive 91/356/EEC

Documentation

9.1 The manufacturer shall have a system of documentation based upon specifications, manufacturing formulae and processing and packaging instructions, procedures and records covering the various manufacturing operations that they perform. ...Pre-established procedures for general manufacturing operations and conditions shall be available, together with specific documents for the manufacture of each batch. This set of documents shall make it possible to trace the history of the manufacture of each batch. The batch documentation shall be retained for at least one year after the expiry date of the batches to which it relates or at least five years after the certification ... (by the Qualified Person), and release whichever is the longer.

There is some ambiguity here. For example, it is not clear what precisely is meant by "this set of documents" and "the batch documentation," or whether or not these two expressions mean (more or less) the same thing. Thus, it is difficult to be sure if the stated retention period applies to batch manufacturing and packaging records only, or to documents relating to control, distribution, components, drug product containers, and labeling as well (as in US 211.180 (a) and (b)).

However, a specific requirement relating to the retention period for "Quality Control documentation" does appear in EC GMP Guide, thus:

Chapter 6 Quality Control

> 6.8 Any Quality Control documentation relating to a batch record should be retained for one year after the expiry date of the batch and at least 5 years after the certification (of the batch, by the Qualified Person).

Two further clauses in the EC GMP Guide also make it clear that distribution records are to be retained, but they do not *specifically* state a retention period. They are:

Chapter 4 Documentation

> 4.25 Records should be maintained of the distribution of each batch of a product in order to facilitate the recall of the batch if necessary (see Chapter 8).

And

Chapter 8 Complaints and Product Recall

> 8.12 The distribution records should be readily available to the person(s) responsible for recalls, and should contain sufficient information on wholesalers and directly supplied customers (with addresses, phone and/or fax numbers inside and outside working hours, batches and amounts delivered), including those for exported products and medical samples.

However, since a (*the?*) prime reason for retaining distribution records is so they can be used to facilitate a product recall, it is reasonable to assume that the intention is that they should be retained for a period of at least one year after the expiration date of the batch of product.

RATIONALE

The reasons for the requirements to make, maintain, and retain records have already been discussed in Chapter 6 and Chapter 7 of this book. The section (211.180, General Requirements), of Subpart J — Records and Reports of the US cGMPs, which we are discussing here, is largely concerned with the "after the event" *retention* aspects of documentation and record keeping. That is, having established formal, approved, written instructions for each job in order to ensure that there is no doubt about what has to be done and having defined standards for materials, equipment, premises, services, and, further, having confirmed, as work proceeds, that each step has been carried out (*and carried out correctly*) using the correct materials and equipment, there then remains the need to ensure the safe keeping of the records of those activities for a suitable period. This is to enable investigation of complaints, defect reports, and any other problems and to facilitate the taking of all necessary corrective action, including all action necessary to prevent reoccurrence. Retained records and reports also enable the performance of quality reviews and thus detection of any drifts away from defined quality standards. They also play a major role in self-inspections or quality audits.

An EC requirement corresponding to US Sec. 211.180 (c) ("All records... or copies of such records shall be readily available for authorized inspection ...") is not found in the EC GMP Guide. This requirement is, however, a requirement of primary European legislation, as exemplified for example by the UK Standard provisions for manufacturer's licenses Schedule 2, which reads, in part:

> 8. The licence holder shall keep readily available for inspection by a person authorised by the licencing authority, durable records of the details of manufacture and assembly of each batch of every medicinal product being manufactured or assembled [= packaging] under his licence and of the tests carried out thereon, including any register or other record ... in such a form that the records will be easily identifiable from the number of the batch as shown on each container in which the medicinal product is sold, supplied or exported, and he shall permit the person authorised to take copies or make extracts from such records. Such records shall not be destroyed without the consent of the licensing authority:
> (a) in relation to a medicinal product for human use, for the relevant period;
> (b) in any other case, for a period of five years from the date when the manufacture or assembly of the relevant batch occurred.

(The "relevant period" has, earlier in these regulations, been defined, in effect, as five years from release of the batch for distribution or one year after the expiry date, which ever is later.)

Thus, US and EC requirements, in this specific context, are closely similar.

It is to be noted that the retained copies do not have to be the originals. The US cGMPs allow the retention of records as the "original records or as true copies such as photocopies, microfilm, microfiche or other accurate reproductions ..."

The European position on records that are other than "paper originals" is set out is the EC GMP Guide thus:

Chapter 4 Documentation

> 4.9 Data may be recorded by electronic data processing systems, photographic or other reliable means, but detailed procedures relating to the system in use should be available and the accuracy of the records should be checked. If documentation is handled by electronic data processing methods, only authorised persons should be able to enter or modify data in the computer and there should be a record of changes and deletions; access should be restricted by passwords or other means and the result of entry of critical data should be independently checked. Batch records

electronically stored should be protected by back-up transfer on magnetic tape, microfilm, paper or other means. It is particularly important that the data are readily available throughout the period of retention.

And in the EC Directive (91/356/EEC), Laying down the principles and guidelines of Good Manufacturing Practice, thus:

EC Directive 91/356/EEC

9.2 When electronic, photographic or other data processing systems are used instead of written documents the manufacturer shall have validated the systems by proving that the data will be appropriately stored during the anticipated period of storage. Data stored by these systems shall be made readily available in legible form. The electronically stored data shall be protected against loss or damage of data. (e.g., by duplication or back-up and transfer onto another storage system).

Note: For "in legible form" in the above read "as hard copy."

A more detailed statement on electronic records and control systems appears in Annex 11, Computerised systems, to the EC GMP Guide. This is given, in full, in Annex 1 to this chapter.

It is noteworthy that the European Guide and Directive specifically allow the making and retention of *electronic* records. In March 1997, the FDA issued CFR Part 11 regulations "that provided criteria for acceptance by FDA, under certain circumstances, of electronic records and electronic signatures ... as equivalent to paper records and handwritten signatures executed on paper." However, in February 2003, the FDA issued a "Draft Guidance for Industry — Part 11, Electronic Records; Electronic Signatures — Scope and Application," which states *inter alia* that "we have determined that we will re-examine Part 11, and we may revise provisions of that regulation." That is the current "state of play" at the time of writing this book. The substance of the Draft Guidance for Industry is given in Annex 2 to this chapter. Manufacturers concerned with complying with FDA requirements in the area of electronic records and signatures should ensure that they have access to the latest regulations.

COMPUTER SYSTEMS VALIDATION

Although the expression "computer systems validation" is commonly used and heard, it needs to be realized that it is not only mainframe computer hard- and software that needs to validated, but all computer and microprocessor control systems. Thus, better expressions would perhaps be "validation of automated systems" or "validation of computer-related systems."

The validation process should establish documentary evidence that provides a high degree of assurance that an automated system will consistently function as specified and designed, and that any manufacturing process involving the automated system will consistently yield a product of the required and intended quality.

User specifications for both the hard- and software that comprise the overall system should be subject to design review and qualification, to ensure that the system will be, and remain, fit for the purpose intended. This design review and qualification should include a careful consideration of potential system failures, and of the possibility (and the consequences) of any undetected system failure that could adversely affect product quality.

Hardware must be:

a. Suitable, and of sufficient capacity, for the tasks required of it
b. Capable of operating, not merely under test conditions, but also under worst case production conditions (e.g., at top machine speed, high data input, high or continuous usage).

Hardware should be tested to confirm the above, with the tests repeated enough times to ensure an acceptable level of consistency and reproducibility. Hardware validation and revalidation studies should be documented, in accordance with basic documentation requirements.

Software should be validated to ensure that it consistently performs as intended. Test conditions should simulate worst case production conditions, e.g., of process speed, data volume, and frequency. Tests should be repeated a sufficient number of times to ensure consistent and reliable performance. Software validation and revalidation studies must be documented as for hardware validation.

Much of the necessary microprocessor or computer hardware and software validation may be performed by the machine, hardware, or software supplier. However, it must be stressed that the final responsibility for the suitability and reliability of any automated system used in pharmaceutical manufacture must rest with the pharmaceutical manufacturer.

Manufacturers should obtain (and retain) from the relevant third party sufficient data (specifications, programs, protocols, test data, conclusions, etc.) to satisfy themselves, and any enquiring regulatory body, that adequate validation work has been carried out to assure system suitability.

Those involved, or interested, in automated systems in pharmaceutical manufacturing should refer to the following:

■ The GAMP (Good Automated Manufacturing Practice) Supplier Guide for Validation of Automated Systems in Pharmaceutical Manufacture, Version 3.0, pub. GAMP Forum, 1998
■ The PDA Validation of Computer-Related Systems — Technical Report No. 18, PDA Journal of Pharmaceutical Science and Technology, Supplement to Vol. 49/1, January/February 1995.

A particular use for *retained* records, required by the US cGMPs but not, in general, required under European law and guidance, is the "at least annual" review. The nearest corresponding requirement is to be found in the short Chapter 9 of the EC GMP Guide, which reads:

Chapter 9 Self-Inspection

Principle

Self-inspections should be conducted in order to monitor the implementation and compliance with Good Manufacturing Practice principles and to propose necessary corrective measures.

9.1 Personnel matters, premises, equipment, documentation, production, quality control, distribution of the medicinal products, arrangements for dealing with complaints and recalls, and self inspection, should be examined at intervals following a pre-arranged programme in order to verify their conformity with the principles of Quality Assurance.

9.2 Self-inspections should be conducted in an independent and detailed way by designated competent person(s) from the company. Independent audits by external experts may also be useful.

9.3 All self-inspections should be recorded. Reports should contain all the observations made during the inspections and, where applicable, proposals for corrective measures. Statements on the actions subsequently taken should also be recorded.

In addition, a specific requirement to review complaints records "regularly" (no precise time interval is given) also appears in EC GMP Guide Chapter 8.

Subsection (f) of US cGMPs Sec. 211.180 addresses the issue of ensuring that top company management is made aware of adverse quality situations revealed by the company's own quality management and review system, or as a consequence of inspection by the FDA or others. This is an entirely laudable requirement and in line with the enlightened view that the attainment and maintenance of true Quality, requires the involvement and commitment of all concerned, at all levels, within an organization. Albeit not so specifically, this philosophy is more generally stated in the opening paragraph of the EC GMP Guide, thus:

Chapter 1 Quality Management

Principle

The holder of a manufacturing authorisation must manufacture medicinal products so as to ensure that they are fit for their intended use, comply with the requirements of the marketing authorisation and do not place patients at risk due to inadequate safety, quality or efficacy. The involvement of this quality objective is *the responsibility of senior management* (author's emphasis) and requires the participation and commitment by staff in many different departments and at all levels within the company ...

That said, it is difficult to believe that top management could remain unaware, or indeed *wish* to remain unaware, of incidents so potentially damaging to the company as recalls or adverse "inspectional observations."

As far as recalls are concerned, European requirements go beyond merely "informing the responsible officials of the firm." Article 33 of European Directive 75/319/EEC requires, *inter alia*, that the Committee on Proprietary Medicinal Products (CPMP, an EC body consisting of experts representative of the Member States) shall be informed of all recalls. Furthermore, the EC GMP Guide states:

Chapter 8 Complaints and Product Recall

> 8.11 All Competent Authorities of all countries to which products may have been distributed should be informed promptly if products are intended to be recalled because they are, or are suspected of being defective.

Significant points of comparison between US Subpart J, Sec. 211.180 and the corresponding EC statements may be summarized as follows:

US cGMPs	EC GMP Guide
Production, control, and distribution records to be retained until at least 1 year after expiry date …	"Batch documentation" to be retained until at least 1 year after expiry date, or 5 years after release, whichever is longer; same requirement for "QC documentation relating to a batch record"
… *except* OTCs with no expiry date — 3 years after distribution	No specific exception for OTCs — but covered by "5 years after release"
These records to be readily available for "authorized inspection"	Same requirement as US cGMPs, under EC Directive
Records may be retained as *copies* — photocopies, microfilm, or microfiche	Same as US cGMPs
US position regarding electronic records and signatures is, at the time of writing, fluid (to be "re-examined")	Electronic records are acceptable
Manufacturers shall use retained records for quality standards review ("at least annually")	Annual quality standards review not a specific requirement
Procedures required to ensure that "responsible officials of the firm" are informed, in writing, regarding recalls, returns, salvages, inspectional observations, regulatory actions etc.	Not a *specific* requirement

US CGMPS

Subpart J — Records and Reports

Sec. 211.182 Equipment cleaning and use log

A written record of major equipment cleaning, maintenance (except routine maintenance such as lubrication and adjustments), and use shall be included in individual equipment logs that show the date, time, product, and lot number of each batch processed. If equipment is dedicated to manufacture of one product, then individual equipment logs are not required, provided that lots or batches of such product follow in numerical order and are manufactured in numerical sequence. In cases where dedicated equipment is employed, the records of cleaning, maintenance, and use shall be part of the batch record. The persons performing and double-checking the cleaning and maintenance shall date and sign or initial the log indicating that the work was performed. Entries in the log shall be in chronological order.

EC GMP GUIDE

Chapter 4 Documentation

4.26 There should be written procedures and the associated records of actions taken or conclusions reached, where appropriate for:
  – Validation
  – Equipment assembly
  – *Maintenance, cleaning and sanitisation* (author's emphasis)
  – …

4.28 Log books should be kept for major or critical equipment recording, as appropriate, any validations, calibrations, maintenance, cleaning or repair operations, including the dates and identity of people who carried these operations out.

4.29 Log books should also record in chronological order the use of major or critical equipment and the areas where the products have been processed.

COMMENT

It should be noted that the requirement to maintain these logs applies only to *major* items of equipment. It is clearly not intended to apply to "minor" items such as spatulas, scoops, ladles, and buckets. This does not mean minor items are exempt from cleaning and from, in fact, being clean, only that they are exempt from being logged. Manufacturers should ensure that their written procedures clearly indicate what are major and what are minor items of manufacturing equipment.

The importance of maintaining a running log of the use of each major item of equipment lies in its potential use as a valuable aid to the investigation of any problems (for example, of product contamination) and in the taking of steps to prevent reoccurrence. It may well be possible, by reference to batch manufacturing records, to determine which batch (or batches) of which product (or products) were processed using a specific item of equipment before, or since, the batch in relation to which a problem has been highlighted. It will be quicker, simpler, and potentially more efficient to be able to check a usage log for evidence of potential contamination or for the effects of possible equipment malfunction.

The keeping of a record of equipment cleaning on the same (or probably less satisfactory, different) log serves the dual function of helping to ensure that equipment cleaning is, in fact, carried out as required and (after the event) confirming that it has indeed been performed.

The exclusion by the US cGMPs of "routine maintenance such as lubrication and adjustments" from the logging requirements is perhaps surprising. A prudent manufacturer should wish to maintain a record of *all* maintenance, adjustment and lubrication, "routine" or otherwise.

EQUIPMENT MAINTENANCE

As previously noted, all functioning machinery is subject to the deleterious effects of wear, dirt, stress, and corrosion, acting individually or in combination with another one. To minimize these adverse effects, and the inevitable consequent decline in machine performance, efficiency, and useful life, and (most importantly) in product quality, it is vital to take appropriate preventative measures. Thus, a comprehensive written maintenance program should be prepared, for each major item of mechanical production equipment, setting out each and every required maintenance activity in detail. It should include statements of the frequency with which each activity should be performed, in terms of real time (e.g., daily, weekly, monthly, yearly) or machine time (e.g., number of hours machine running time). The frequency and time base should be clearly defined in the written program(s) for each maintenance procedure to be carried out on each machine.

Machine maintenance should be carried out on a *planned preventative*, not on an emergency curative, basis. Formal maintenance records, which can be readily related to the overall maintenance program, should be compiled as each maintenance operation is performed, and held on file, in order to ensure, and to make it possible to demonstrate, that all required maintenance operations are indeed carried out as and when required by the program.

Unless maintenance programs, maintenance records, and change control procedures, are developed and implemented (to ensure that manufacturing equipment and attendant instruments and control devices remain in the same qualified and maintained state as they were during any validation studies) then any assurance hopefully derived from those validation studies could well be negated.

Requirements for equipment design, specification, qualification, calibration, and maintenance apply equally to equipment, installations, or services that are ancillary, subsidiary, or provide support to, manufacturing equipment — that is, to things such as electrical power supplies, HVAC systems, steam generators, air compressors, heat exchangers, chillers, water purification and supply systems, CIP and SIP systems, and to all measuring, indicating, controlling, monitoring, and recording instrumentation associated with these various items of equipment, systems, and services. All such items should be included in maintenance programs and maintenance records.

US CGMPS

Subpart J — Records and Reports

Sec. 211.184 Component, drug product container, closure, and labeling records

These records shall include the following:

(a) The identity and quantity of each shipment of each lot of components, drug product containers, closures, and labeling; the name of the supplier; the supplier's lot number(s) if known; the receiving code as specified in Sec. 211.80; and the date of receipt. The name and location of the prime manufacturer, if different from the supplier, shall be listed if known.

(b) The results of any test or examination performed (including those performed as required by Sec. 211.82(a), Sec. 211.84(d), or Sec. 211.122(a)) and the conclusions derived there from.

(c) An individual inventory record of each component, drug product container, and closure and, for each component, a reconciliation of the use of each lot of such component. The inventory record shall contain sufficient information to allow determination of any batch or lot of drug product associated with the use of each component, drug product container, and closure.

(d) Documentation of the examination and review of labels and labeling for conformity with established specifications in accord with Secs. 211.122(c) and 211.130(c).

(e) The disposition of rejected components, drug product containers, closure, and labeling.

EC GMP GUIDE

Chapter 4 Documentation

Procedures and records

Receipt

4.19 There should be written procedures and records for the receipt of each delivery of each starting and primary and printed packaging material.

4.20 The records of the receipts should include:
(a) the name of the material on the delivery note and the containers;
(b) the "in-house" name and/or code of material (if different from a);
(c) date of receipt;
(d) supplier's name and, if possible, manufacturer's name;
(e) manufacturer's batch or reference number;
(f) total quantity, and number of containers received;
(g) the batch number assigned after receipt;
(h) any relevant comment (e.g., state of the containers).

Chapter 5 Production

5.2 All handling of materials and products, such as receipt and quarantine, sampling, storage, labelling, dispensing, processing, packaging and distribution should be done in accordance with written procedures or instructions *and, where necessary, recorded.*

5.3 All incoming materials should be checked to ensure that the consignment corresponds to the order. Containers should be cleaned where necessary and labelled with the prescribed data.

5.4 Damage to containers and any other problem which might adversely affect the quality of a material should be investigated, *recorded and reported* to the Quality Control Department.

5.8 Checks on yields, *and reconciliation of quantities*, should be carried out as necessary to ensure that there are no discrepancies outside acceptable limits.

(Author's emphases)

Chapter 6 Quality Control

Testing

6.15 Analytical methods should be validated. All testing operations described in the marketing authorisation should be carried out according to the approved methods.

6.16 The results obtained should be recorded and checked to make sure that they are consistent with each other. Any calculations should be critically examined.

6.17 The tests performed should be recorded and the records should include at least the following data:
 (a) name of the material or product and, where applicable, dosage form;
 (b) batch number and, where appropriate, the manufacturer and/or supplier;
 (c) references to the relevant specifications and testing procedures;
 (d) test results, including observations and calculations, and reference to any certificates of analysis;
 (e) dates of testing;
 (f) initials of the persons who performed the testing;
 (g) initials of the persons who verified the testing and the calculations, where appropriate;
 (h) a clear statement of release or rejection (or other status decision) and the dated signature of the designated Responsible Person.

A *specific* requirement to record the "examination and review of labels and labelling," as in US cGMPs Sec. 211.1849 (d) does not appear in the EC GMP Guide but may be taken as implicit in Principle, which heads Chapter 6, Quality Control:

Principle

Quality Control is concerned with sampling, specifications and testing as well as the organisation, documentation and release procedures which ensure that the necessary and relevant tests are carried out, and that materials are not released for use, nor products released for sale or supply, until their quality has been judged satisfactory. ...

– when read in conjunction with paragraph 6.17 of the EC GMP Guide, as quoted above.
The statement in the EC GMP Guide that most closely corresponds to US cGMPs 211.184 (e) (recording of disposition of rejected components, etc.) is to be found in Chapter 5 Production, thus:

> 5.61 Rejected materials and products should be clearly marked as such and stored separately in restricted areas. They should either be returned to the suppliers or, where appropriate, reprocessed or destroyed. Whatever action is taken should be approved *and recorded* by authorised personnel. (author's emphasis)

Again, this should be read in conjunction with EC GMP Guide 6.17 (as quoted above), most specifically 6.17 (h), "a clear statement of release or rejection (or other status decision) and the dated signature of the designated Responsible Person."

Implementation

Examples of relevant procedures, records, and other documents have already been given in Chapter 6, Materials Control of this book.

US CGMPS

Subpart J — Records and Reports

Sec. 211.186 Master production and control records

(a) To assure uniformity from batch to batch, master production and control records for each drug product, including each batch size thereof, shall be prepared, dated, and signed (full signature, handwritten) by one person and independently checked, dated, and signed by a second person. The preparation of master production and control records shall be described in a written procedure and such written procedure shall be followed.

(b) Master production and control records shall include:
  (1) The name and strength of the product and a description of the dosage form;
  (2) The name and weight or measure of each active ingredient per dosage unit or per unit of weight or measure of the drug product, and a statement of the total weight or measure of any dosage unit;
  (3) A complete list of components designated by names or codes sufficiently specific to indicate any special quality characteristic;
  (4) An accurate statement of the weight or measure of each component, using the same weight system (metric, avoirdupois, or apothecary) for each component. Reasonable variations may be permitted, however, in the amount of components necessary for the preparation in

the dosage form, provided they are justified in the master production and control records;

(5) A statement concerning any calculated excess of component;

(6) A statement of theoretical weight or measure at appropriate phases of processing;

(7) A statement of theoretical yield, including the maximum and minimum percentages of theoretical yield beyond which investigation according to Sec. 211.192 is required;

(8) A description of the drug product containers, closures, and packaging materials, including a specimen or copy of each label and all other labeling signed and dated by the person or persons responsible for approval of such labeling;

(9) Complete manufacturing and control instructions, sampling and testing procedures, specifications, special notations, and precautions to be followed.

EC GMP GUIDE

Chapter 4 Documentation

Manufacturing Formula and Processing Instructions

Formally authorised Manufacturing Formula and Processing Instructions should exist for each product and batch size to be manufactured. They are often combined in one document.

4.14 The Manufacturing Formula should include:

(a) the name of the product, with a product reference code relating to its specification;

(b) a description of the pharmaceutical form, strength of the product and batch size;

(c) a list of all starting materials to be used, with the amount of each, described using the designated name and a reference which is unique to that material; mention should be made of any substance that may disappear in the course of processing;

(d) a statement of the expected final yield with the acceptable limits, and of relevant intermediate yields, where applicable.

4.15 The Processing Instructions should include:

(a) a statement of the processing location and the principal equipment to be used.,

(b) the methods, or reference to the methods, to be used for preparing the critical equipment (e.g., cleaning, assembling, calibrating, sterilising);

(c) detailed stepwise processing instructions (e.g., checks on materials, pre-treatments, sequence for adding materials, mixing times, temperatures);

(d) the instructions for any in-process controls with their limits;

(e) where necessary, the requirements for bulk storage of the products; including the container, labelling and special storage conditions where applicable;

(f) any special precautions to be observed.

Packaging Instructions

4.16 There should be formally authorised Packaging Instructions for each product, pack size and type. These should normally include, or have a reference to, the following:

(a) name of the product;

(b) description of its pharmaceutical form, and strength where applicable;

(c) the pack size expressed in terms of the number, weight or volume of the product in the final container;

(d) a complete list of all the packaging materials required for a standard batch size, including quantities, sizes and types, with the code or reference number relating to the specifications of each packaging material;

(e) where appropriate, an example or reproduction of the relevant printed packaging materials, and specimens indicating where to apply batch number references, and shelf life of the product;

(f) special precautions to be observed, including a careful examination of the area and equipment in order to ascertain the line clearance before operations begin;

(g) a description of the packaging operation, including any significant subsidiary operations, and equipment to be used;

(h) details of in-process controls with instructions for sampling and acceptance limits.

COMMENT

Here the two regulatory documents are closely similar. Most differences are terminological or presentational. They may be summarized as follows:

US cGMPs	EC GMP Guide
Refers to "*master* production and control records"	Refers to "manufacturing formula and processing instructions" ("master" understood)
In this context "production" includes packaging	Refers to separate "packaging instructions"
Accepts avoirdupois or apothecary units as well as metric	No specific comment, although metric units assumed in Europe
No statement of process location/equipment specifically required (but see Section 211.188)	Requires "master" statement of processing location and equipment
Requires a complete list of components.	Similar requirement, but quaintly adds: "mention should be made of any substance that may disappear (!) in the course of processing"
No specific requirement to include on the master a statement on methods of equipment cleaning, assembly, calibration, sterilization, etc.	Requires statement to be included in the processing instructions on "the methods, or reference to the methods, to be used for preparing the critical equipment (e.g., cleaning, assembly, calibration, sterilizing)"

-- continued

US cGMPs	*EC GMP Guide*
Requires specimen (or copy) of labeling	In addition , requires indication of where to apply batch number and expiry date on the specimen or copy
No specific requirement for an instruction on packaging line clearance check on master (but see Section 211.188)	Requires instruction on packaging line clearance check on master

US CGMPS

Subpart J — Records and Reports

Sec. 211.188 Batch production and control records

Batch production and control records shall be prepared for each batch of drug product produced and shall include complete information relating to the production and control of each batch. These records shall include:

(a)   An accurate reproduction of the appropriate master production or control record, checked for accuracy, dated, and signed;

(b)   Documentation that each significant step in the manufacture, processing, packing, or holding of the batch was accomplished, including:

(1) Dates;

(2) Identity of individual major equipment and lines used;

(3) Specific identification of each batch of component or in-process material used;

(4) Weights and measures of components used in the course of processing;

(5) In-process and laboratory control results;

(6) Inspection of the packaging and labeling area before and after use;

(7) A statement of the actual yield and a statement of the percentage of theoretical yield at appropriate phases of processing;

(8) Complete labeling control records, including specimens or copies of all labeling used;

(9) Description of drug product containers and closures;

(10) Any sampling performed;

(11) Identification of the persons performing and directly supervising or checking each significant step in the operation;

(12) Any investigation made according to Sec. 211.192.

(13) Results of examinations made in accordance with Sec. 211.134.

EC GMP GUIDE

Chapter 4 Documentation

Batch Processing Records

4.17   A Batch Processing Record should be kept for each batch processed. It should be based on the relevant parts of the currently approved Manufacturing Formula and Processing Instructions. The method of preparation of such records should be designed to avoid transcription errors. The record should carry the number of the batch being manufactured.

Before any processing begins, there should be recorded checks that the equipment and work station are clear of previous products, documents or materials not required for the planned process, and that equipment is clean and suitable for use.

During processing, the following information should be recorded at the time each action is taken and, after completion, the record should be dated and signed in agreement by the person responsible for the processing operations:

(a)   the name of the product;
(b)   dates and times of commencement, of significant intermediate stages and of completion of production;
(c)   name of the person responsible for each stage of production;
(d)   initials of the operator of different significant steps of production and, where appropriate, of the person who checked each of these operations (e.g., weighing);
(e)   the batch number and/or analytical control number as well as the quantities of each starting material actually weighed (including the batch number and amount of any recovered or reprocessed material added);
(f)   any relevant processing operation or event and major equipment used;
(g)   a record of the in-process controls and the initials of the person(s) carrying them out, and the results obtained;
(h)   the product yield obtained at different and pertinent stages of manufacture;
(i)   notes on special problems including details, with signed authorisation for any deviation from the manufacturing formula and processing instructions.

Batch Packaging Records

4.18   A Batch Packaging Record should be kept for each batch or part batch processed. It should be based on the relevant parts of the Packaging Instructions and the method of preparation of such records should be designed to avoid transcription errors. The record should carry the batch number and the quantity of bulk product to be packed, as well as the batch number and the planned quantity of finished product that will be obtained.

Before any packaging operation begins, there should be recorded checks that the equipment and work station are clear of previous products, documents or materials not required for the planned packaging operations, and that equipment is clean and suitable for use.

The following information should be entered at the time each action is taken and, after completion, the record should be dated and signed in agreement by the person(s) responsible for the packaging operations:

(a)   the name of the product;
(b)   the date(s) and times of the packaging operations;

(c) the name of the responsible person carrying out the packaging operation;

(d) the initials of the operators of the different significant steps;

(e) records of checks for identity and conformity with the packaging instructions including the results of in-process controls;

(f) details of the packaging operations carried out, including references to equipment and the packaging lines used;

(g) whenever possible, samples of printed packaging materials used, including specimens of the batch coding, expiry dating and any additional overprinting;

(h) notes on any special problems or unusual events including details, with signed authorisation for any deviation from the Manufacturing Formula and Processing Instructions;

(i) the quantities and reference number or identification of all printed packaging materials and bulk product issued, used, destroyed or returned to stock and the quantities of obtained product, in order to provide for an adequate reconciliation.

COMMENT

US cGMPs Sec. 211.186 (and the corresponding sections 4.14 to 4.16 of the EC GMP Guide) are concerned with the requirement for, and the content of, *Master* Production and Control documents. The US cGMPs, Section 211.188 (and the corresponding EC GMP Guide sections 4.17 and 4.18) are concerned with the issue of a copy of the relevant master for batch manufacturing (and packaging) and with the entries to be made on that copy during the ongoing course of production.

Both the US cGMPs and the EC GMP Guide stress the need for an *accurate* reproduction of the master to form the basis of a batch production record. The US cGMPs require that each reproduction shall be "checked for accuracy ...." It might well be questioned why it is necessary to check the accuracy of a reproduction made by photocopying, xerography, mimeographing, or computer printout. Nevertheless, it is sound sense to carefully and formally check all copies made by such methods — first to ensure that it is indeed a copy of the correct and current master, and then to see that it is complete, clearly legible and free from smudging, blurring, or any other defects that could compromise its accuracy and clarity. This specific requirement for a check for accuracy of reproduction might perhaps be taken as a tacit acceptance by the US FDA that copies of the master, to form a batch record, may be made by hand writing. Because of the high potential for error, and in view of the current ready availability of copying devices that largely eliminate transcription error, hand-written copies are best avoided. The opening clauses of EC GMP Guide sections 4.17 and 4.18 (see above) strongly imply this. Hand copying is positively barred in the opening General section of the EC GMP Guide, Chapter 4, Documentation, thus:

EC GMP GUIDE

Chapter 4 Documentation

4.4   … Reproduced documents should be clear and legible. The reproduction of working documents from master documents must not allow any error to be introduced through the reproduction process.

4.6   *Documents should not be handwritten;* although, where documents require the entry of data, these entries may be made in clear, legible, indelible handwriting. Sufficient space should be provided for such entries. (Author's emphasis)

Otherwise the two regulatory documents are closely similar in point of detail. An interesting difference is that, whereas the EC GMP Guide requires the "processing location and principal equipment to be used" to be dictated by the master document, the US cGMPs do not, requiring only that the "identity of individual major equipment and lines used" be entered in the batch production record. It would appear, therefore, that whereas an EC master processing instruction is prescriptive in declaring, in advance, the equipment to be used, the US cGMPs offer a degree of flexibility in this area.

Similarly, the EC GMP Guide requires an instruction to make "a careful examination … to ascertain the line clearance before operations begin" to be included on the master packaging instruction; US cGMPs require only that a record of "inspection of the packaging and labelling area before and after use" be made on the batch production and control record.

Since batch production records accompany materials, intermediates, and products through a (usually) multistage production cycle, it makes sense to ensure that they are printed (or otherwise copied) on substantial paper or other medium. To keep all documentary material, relating to a given batch, together, for the sake of neatness and to avoid loss, damage, or soiling, it is sound practice to keep all the records and instructions relating to a batch together in some form of plastic folder or bag.

MAKING MANUAL RECORDS

There are some important things (simple and obvious, but nevertheless very important things) that need to be remembered about making records, including entering a check signature or initials:

1. Records should always be made, or signatures, etc. entered, when (or immediately after) an action has been completed, a reading has been taken, or a check has been made. Records should always be made as things happen, not at the end of the shift, day, or week. They should be about current, real-time events, not history.

2. A person entering a second check signature or initials is confirming that he or she actually saw what was done (for example, a weighing) and has personally checked that *everything* was correct (product, material, batch, quantity, reading, or whatever). It is not good enough, and it could be very dangerous, for an operator to "trust their pal" to

have got it right, and write in second-check initials or signatures at, say, the end of the day or the shift.

3. Manuscript entries should always be neat and clear. They do not have to be beautiful, but operators need to be reminded that others may need to read these records in five or more years.

4. If a mistake is made when making a manuscript entry on a document, it should not be considered a crime. But if a mistake *is* made, it should not be obliterated or covered up. It should be crossed out neatly (so that it can still be read), the correction made and signed or initialed, and the date added, with any explanation that may be necessary.

DOCUMENT CONTROL AND REVISION

Issue and use of documents should be under formal control. They should be available to all who need them, and *not* available to those who do not. They should be kept up-to-date, but all revisions should be formal and authorized, not haphazard. The documentation system, overall, should be subject to review. It is vital that systems exist for the removal from active use of outdated or superseded documents.

Examples of format and content of master production and batch production records and their use have already been given in Chapter 7 and Chapter 8 of this book.

US CGMPS

Subpart J — Records and Reports

Sec. 211.192 Production record review

> All drug product production and control records, including those for packaging and labeling, shall be reviewed and approved by the quality control unit to determine compliance with all established, approved written procedures before a batch is released or distributed. Any unexplained discrepancy (including a percentage of theoretical yield exceeding the maximum or minimum percentages established in master production and control records) or the failure of a batch or any of its components to meet any of its specifications shall be thoroughly investigated, *whether or not the batch has already been distributed.* The investigation shall extend to other batches of the same drug product and other drug products that may have been associated with the specific failure or discrepancy. A written record of the investigation shall be made and shall include the conclusions and follow-up. (Author's emphasis)

EC GMP GUIDE

A requirement to critically examine all production and control records before a batch of drug product is released for sale or distribution is explicit or implicit in a number of sections of the EC GMP Guide. Perhaps the clearest and most explicit statement is to be found in an annex to the EC GMP Guide — Annex 16, Certification by a Qualified Person and Batch Release. (The nature and responsibilities of the European legal entity "Qualified Person" [QP] has been outlined in an annex to Chapter 2, Personnel, Organization, and Training, of this book). Section 8. of the EC GMP Guide Annex 16 reads:

8. Routine Duties of a Qualified Person

8.1 Before certifying a batch prior to release the QP doing so should ensure … that at least the following requirements have been met:

   (a) the batch and its manufacture comply with the provisions of the marketing authorisation (including the authorisation required for importation where relevant);

   (b) manufacture has been carried out in accordance with Good Manufacturing Practice or, in the case of a batch imported from a third country, in accordance with good manufacturing practice standards at least equivalent to EC GMP;

   (c) the principal manufacturing and testing processes have been validated; account has been taken of the actual production conditions and manufacturing records;

   (d) any deviations or planned changes in production or quality control have been authorised by the persons responsible in accordance with a defined system Any changes requiring variation to the marketing or manufacturing authorisation have been notified to and authorised by the relevant authority;

   (e) all the necessary checks and tests have been performed, including any additional sampling, inspection, tests or checks initiated because of deviations or planned changes;

   (f) *all necessary production and quality control documentation has been completed and endorsed by, the staff authorised to do so* (Author's emphasis);

   (g) all audits have been carried our as required by the quality assurance system;

   (h) the QP should in addition take into account any other factors of which he is aware — which are relevant to the quality of the batch.

These requirements for a formal review of *all* production and control records before a batch of product is released for sale or distribution could be regarded as a formalization of the view that, contrary to a "traditional" attitude that was all too widely held 30 or 40 years ago, the quality of a manufactured pharmaceutical product *cannot* be evaluated on the basis of end-product test results alone. Any such evaluation must be based on a critical consideration of *all* material, production, and control factors that bear upon the quality of the product.

The person making the release (or otherwise) must also be in a position to know that the condition of the environment and the equipment in which the product was made were appropriate for that type of product, that the equipment has been appropriately qualified and the manufacturing processes and analytical procedures validated, that correct (and approved) materials have been used, and that appropriate quality audits are regularly carried out.

To help ensure that the production record review has been properly and completely carried out, and as an aid in demonstrating compliance, it is useful to include, in the complete batch documentation package, a checklist, which includes:

1. Batch record current and approved as an accurate copy?
2. Correct, approved components used?
3. Correct quantities used?
4. Ingredients all within retest date?
5. All required data, readings, in-process control checks, and signatures or initials entered?
6. Correct product packaged in correct packaging components?
7. Correct batch number and expiry date printed on labels, cartons, etc.?
8. Yields entered and reconciled?
9. All deviations reported? Approved? Investigated?
10. All required testing and in-process controls performed and results within specification?
11. Samples taken for retention?

Care needs to be taken that the final review is not a mere mindless check on the absence of signatures or test data. It should be a meaningful, *informed* decision based on an intelligent awareness of all relevant factors. Emphasis should be on the actual *operations* and on ensuring that all operators are trained to understand the crucial importance of following procedures and instructions, and supervisory staff are competent to ensure that this is so.

US CGMPS

Subpart J — Records and Reports

Sec. 211.194 Laboratory records

(a) Laboratory records shall include complete data derived from all tests necessary to assure compliance with established specifications and standards, including examinations and assays, as follows:

(1) A description of the sample received for testing with identification of source (that is, location from where sample was obtained), quantity, lot number or other distinctive code, date sample was taken, and date sample was received for testing.

(2) A statement of each method used in the testing of the sample. The statement shall indicate the location of data that establish that the methods used in the testing of the sample meet proper standards of accuracy and reliability as applied to the product tested. (If the method employed is in the current revision of the United States Pharmacopoeia, National Formulary, Association of official Analytical Chemists, Book of Methods, or in other recognized standard references, or is detailed in an approved new drug application and the referenced method is not modified, a statement indicating the method and reference will suffice). The suitability of all testing methods used shall be verified under actual conditions of use.

(3) A statement of the weight or measure of sample used for each test, where appropriate.

(4) A complete record of all data secured in the course of each test, including all graphs, charts, and spectra from laboratory instrumentation, properly identified to show the specific component, drug product container, closure, in process material, or drug product, and lot tested.

(5) A record of all calculations performed in connection with the test, including units of measure, conversion factors, and equivalency factors.

(6) A statement of the results of tests and how the results compare with established standards of identity, strength, quality, and purity for the component, drug product container, closure, in-process material, or drug product tested.

(7) The initials or signature of the person who performs each test and the date(s) the tests were performed.

(8) The initials or signature of a second person showing that the original records have been reviewed for accuracy, completeness, and compliance with established standards.

(b) Complete records shall be maintained of any modification of an established method employed in testing. Such records shall include the reason for the modification and data to verify that the modification produced results that are at least as accurate and reliable for the material being tested as the established method.

(c) Complete records shall be maintained of any testing and standardization of laboratory reference standards, reagents, and standard solutions.

(d) Complete records shall be maintained of the periodic calibration of laboratory instruments, apparatus, gauges, and recording devices required by Sec. 211.160(b)(4).

(e) Complete records shall be maintained of all stability testing performed in accordance with Sec. 211.166.

EC GMP GUIDE

Chapter 4 Documentation

Testing

4.23 There should be written procedures for testing materials and products at different stages of manufacture, describing the methods and equipment to be used. The tests performed should be recorded (see Chapter 6, item 17).

Chapter 6 Quality Control

Documentation

6.7 Laboratory documentation should follow the principles given in Chapter 4. An important part of this documentation deals with Quality Control and the following details should be readily available to the Quality Control Department:

  – specifications;
  – sampling procedures;

    – *testing* procedures and *records (including analytical worksheets and/or laboratory notebooks);* (Author's emphasis)
    – analytical reports and/or certificates;
    – data from environmental monitoring, where required;
    – validation records of test methods, where applicable;
    – procedures for and records of the calibration of instruments and maintenance of equipment.

Sampling

6.13 Sample containers should bear a label indicating the contents, with the batch number, the date of sampling and the containers from which samples have been drawn.

Testing

6.15 Analytical methods should be validated. All testing operations described in the marketing authorisation should be carried out according to the approved methods.

6.16 The results obtained should be recorded and checked to make sure that they are consistent with each other. Any calculations should be critically examined.

6.17 The tests performed should be recorded and the records should include at least the following data:
    (a) name of the material or product and, where applicable, dosage form.
    (b) batch number and, where appropriate, the manufacturer and/or supplier.
    (c) references to the relevant specifications and testing procedures.
    (d) test results, including observations and calculations, and reference to any certificates of analysis.
    (e) dates of testing.
    (f) initials of the persons who performed the testing.
    (g) initials of the persons who verified the testing and the calculations, where appropriate.
    (h) a clear statement of release or rejection (or other status decision) and the dated signature of the designated responsible person.

6.18 All the in-process controls, including those made in the production area by production personnel, should be performed according to methods approved by Quality Control and the results recorded.

6.19 Special attention should be given to the quality of laboratory reagents, volumetric glassware and solutions, reference standards and culture media. They should be prepared in accordance with written procedures.

6.20 Laboratory reagents intended for prolonged use should be marked with the preparation date and the signature of the person who prepared them. The expiry date of unstable reagents and culture media should be indicated on the label, together with specific storage conditions. In addition, for volumetric solutions, the last date of standardisation and the last current factor should be indicated.

6.21 Where necessary, the date of receipt of any substance used for testing operations (e.g., reagents and reference standards) should be indicated on the container. Instructions for use and storage should he followed.

In certain cases it may be necessary to carry out an identification test and/or other testing of reagent materials upon receipt or before use.

The reasons for laboratory documentation, and its basic objectives, are the same as discussed earlier in this book (Chapter 7 and Chapter 8), in the context of manufacturing operations. Properly designed and maintained laboratory documentation will help to ensure the tests, assays, and other analytical procedures that *should* be carried-out are *in fact* carried out, and are in accordance with previously established and registered specifications and test methods. It should be designed, used, maintained, and retained in a manner that will facilitate rapid retrieval and review in the event of any later incident (complaint, adverse patient reaction, etc.) that demands rapid and efficient investigation. In particular, test results should be recorded in a manner that will facilitate comparative reviews of those results and thus the detection of trends. The details recorded in the formal records should include at least the following:

1. Name of product or material, and code reference
2. Batch or lot number
3. Date of receipt and sampling
4. Source of product or material
5. Date of testing
6. Indication of tests performed
7. Reference to the method used
8. Results obtained
9. Decision regarding release, rejection, or other status
10. Signature or initials of the analyst, and signature of person making the status decision

In addition, analysts' laboratory records (e.g., lab notebooks) should also be retained, with the raw data and calculations from which the test results were derived, together with all graphs, charts, spectra, and printouts obtained from laboratory printouts. These should be annotated so there is a clear, unequivocal, indication of the material, batch, or sample to which these data refer. (Note: To counter any potential suspicions that "out of specification" or other unexpected results may have been peremptorily discarded, and the relevant notes torn out and thrown away, it may be necessary to use notebooks with preprinted page numbers or to adopt some such similar device.)

Written specifications, prepared and authorized by Quality Control should be established and maintained for all components or starting materials, packaging materials, and bulk, intermediate, and finished products. An illustrative example of an analytical specification for a chemical component, or starting material ("Dexterium adipate"), has already been shown in Chapter 10 of this book.

Materials and products should be tested against the relevant specification and the results recorded. A system should be in place to ensure that a formal decision is made, on the basis of the test results, by an appropriately authorized person, and that this decision is securely and unequivocally conveyed to the persons (or organizational unit) who have to act upon it.

As the testing of the sample, against this written specification, proceeds following the raw-data, calculations etc. are commonly noted in the analyst's laboratory notebook. As the result of each test is obtained, it is entered on a formal analytical report form. Figure 11.1 shows an example of a useful format.

Phantazpharm Inc.   **ANALYTICAL REPORT**

Code no: 1376   Dexterium adipate   Lot no. 15927

Supplier *Apimatic*
Sample taken by *B Brown*   Location *W6 A*   Quantity *20 g*
Containers sampled *6*   Date *2/8/02*   Date sample rec'd *2/9/02*

TEST	SPECIFICATION	RESULT
Description	Pale yellow, fine cryst. pdr.	*pale yellow cryst. powder*
Solubility	Sl. soluble in water, v. sol in ethanol and acetone. Clear soln. in ethanol.	*complies*
Identification	Complies with UP tests	*complies*
Melting point	168° to 172° C	*169.5°C*
Loss on drying	Not more than 0.5%	*< 0.25%*
Heavy metals	Not more than 15 ppm	*5 ppm*
Sulphated ash	Not more than 0.1%	*0.06%*
Related substances	Complies UP test	*complies*
Assay	99.9% to 101%	*100.30%*
Bulk density	1.7 to 2.3 ml/g	*1.9 ml/g*

Sample tested according to Spec. no. *1376B* and report compiled by (Sign.) *JBlee* Date *2/11/02*
Lab. notebook reference/date *JS/5/15 2/10/02*
Report reviewed, and calculations checked, by (Sign.) *C Chemist* Date *2/11/02*

Release/Reject decision *Released*   Signed *RJintl* Date *2/12/02*

Comments

Form Prepared by	Approved by*	Supersedes	Effective from	Page 1 of 1
RWhyte	*C Chemist*	*First issue*	*12/8/98*	

* To include confirmation that this Report form is consistent with the current Specification ....

Figure 11.1   Analytical Report

This type of analytical report requires the preparation and printing of a series of report forms, which are specific to a given material or product. The great advantage is that the printing of a summary of the specification, against which each result is entered, permits an easy, rapid, and relatively secure evaluation of the compliance of the test results with the specification. In the example shown, the form has been completed manually and the decision (to release) taken, and formally entered.

At the same time, the test results should be entered on a summary report sheet, or card. An example is shown in Figure 11.2, with a manual entry made for the same lot of "Dexterium adipate," 15927. A record in this format facilitates checking for quality trends, and comparison of the standards attained by different suppliers. (Note that as this is not a primary record, and space is probably limited, a little shorthand — ticks, "complies," "OK," etc. — is usually considered acceptable.)

The message that the material is released (or rejected) is conveyed to those who need to know, by means of the Materials Receiving Report, previously shown in Chapter 6, but now reshown in Figure 11.3 as completed by Goods Inwards (GI), and now stamped "RELEASED" and signed by QC. The concept illustrated presupposes that the allocation of starting materials to manufacturing batches is the responsibility of a Materials Inventory Control unit, who are only able to allocate materials that have been released by QC. It is also at this stage that an authorized QC person applies the completed "RELEASED" labels to the quarantined goods (see Chapter 6). For completeness, he or she can, at the same time, "RELEASED" stamp and sign the GI copy of the Receiving Report, and GI can now complete the last two columns of the Materials Receiving Report (see Chapter 6).

A good insight into FDA concerns when reviewing laboratory records is provided by their Guide to Inspections of Pharmaceutical Quality Control Laboratories. This guide, addressed to field investigators, was issued by the FDA Division of Field Investigations in July 1993. Section 13 of this guide reads (see page 278):

FROM FDA "GUIDE TO INSPECTIONS OF PHARMACEUTICAL QUALITY CONTROL LABORATORIES"

13 Laboratory Records and Documentation

Review personal analytical notebooks kept by the analysts in the laboratory and compare them with the worksheets and general lab notebooks and records. Be prepared to examine all records and worksheets for accuracy and authenticity and to verify that raw data are retained to support the conclusions found in laboratory results.

Review laboratory logs for the sequence of analysis versus the sequence of manufacturing dates. Test dates should correspond to the dates when the sample should have been in the laboratory. If there is a computer data base, determine the protocols for making changes to the data. There should be an audit trail for changes to data.

ANALYTICAL REPORT SUMMARY: 1367 Dexterium adipate

Lot No.	Supplier	Description	Solubility	Solution	Ident.	M.Pt.	Loss on drying	Heavy metals	Sulphated ash	Related subs.	Bulk density	Assay	Decision	Date
15927	Apimatic	✓	Complies	Complies	✓	169.5°	<0.25%	5ppm	0.06%	Complies	1.9 m/g	100.3%	Release	2/12/02

ANREPSUM.SAM

Figure 11.2   Analytical Report Summary

GOODS INWARDS - MATERIALS RECEIVING REPORT

Material *DEXTERIUM ADIPATE* .... Code No. *1376* ......

INSTRUCTIONS: 1. Complete a separate Receiving Report for each delivery, and for each suppliers batch number within a delivery.
2. Retain one copy in Goods Inwards file, and send three copies to Quality Control.
3. **Quality Control**: On completion of testing, mark this report, where indicated, "RELEASED", "REJECTED", "HOLD" as appropriate, and send a copy to:
   Purchasing Department
   Materials Inventory Control

   Retain one copy on Quality Control files

Date goods received ...... *Feb. 1 '02*
Supplier ...... *Apimatic*
Supplier's batch no. *1375/1*
Quantity received ...... *100 Kg.*
Number of containers *10*
Purchase Order No. *02956*
Assigned Lot No. *15927*

General condition/cleanliness of delivery *Satisfactory*

Delivery examined by (Signed) *P. Rolands* Date *2/1/02*

Remarks/Comments

QUALITY CONTROL DECISION

QC RELEASED

Signed *Bruntl.*
Date *2/12/02*

Figure 11.3   Goods Inwards – Materials Receiving Report

We expect raw laboratory data to be maintained in bound (not loose or scrap sheets of paper) books or on analytical sheets for which there is accountability, such as pre-numbered sheets. For most of those manufacturers which had duplicate sets of records or "raw data," non-numbered loose sheets of paper were employed. Some companies use discs or tapes as raw data and for the storage of data. Such systems have also been accepted provided they have been defined (with raw data identified) and validated.

Carefully examine and evaluate laboratory logs, worksheets and other records containing the raw data such as weighings, dilutions, the condition of instruments, and calculations. Note whether raw data are missing, if records have been rewritten, or if correction fluid has been used to conceal errors. Results should not be changed without explanation. Cross reference the data that has been corrected to authenticate it. Products cannot be "tested into compliance" by arbitrarily labeling out-of-specification lab results as "laboratory errors" without an investigation resulting in scientifically valid criteria.

Test results should not have been transcribed without retention of the original records, nor should test results be recorded selectively. For example, investigations have uncovered the use of loose sheets of paper with subsequent selective transcriptions of good data to analyst worksheets and/or workbooks. Absorbance values and calculations have even been found on desk calendars.

Cut charts with injections missing, deletion of files in direct data entry systems, indirect data entry without verification, and changes to computerized programs to override program features should be carefully examined. These practices raise questions about the overall quality of data.

The firm should have a written explanation when injections, particularly from a series are missing from the official work-sheets or from files and are included among the raw data. Multiple injections recorded should be in consecutive files with consecutive injection times recorded. Expect to see written justification for the deletion of all files.

Determine the adequacy of the firm's procedures to ensure that all valid laboratory data are considered by the firm in their determination of acceptability of components, in-process, finished product, and retained stability samples. Laboratory logs and documents when cross referenced may show that data has been discarded by company officials who decided to release the product without a satisfactory explanation of the results showing the product fails to meet the specifications. Evaluate the justification for disregarding test results that show the product failed to meet specifications.

One thing that becomes apparent on reading this section of this FDA Guide is that the investigators are, among other things, investigating the possibility that a laboratory is concealing or destroying out-of-specification (OOS) or other "undesirable" results, without investigation and then, perhaps, repeat testing until the "right" result is obtained ("testing into compliance"). Note particularly that raw data is to be maintained "in bound (not loose or scrap sheets of paper) books, or on analytical sheets for which there is accountability, such as prenumbered sheets."

Much of the FDA concern over OOS results arises from the celebrated Judge Wolin verdict in the Barr L aboratories case. An understanding of the FDA view on this issue can be gained from a review of the FDA *Draft Guidance* for Industry on Investigating Out of Specification (OOS) Test Results for Pharmaceutical Production. It must be noted that this is *Draft Guidance,* distributed in September 1998 for comment purposes only. At the time of writing, a final definitive guidance document has yet to be issued. The substance of the draft is given in Annex 3 of this chapter.

Written sampling procedures stating amounts of sample to be taken, from when, and how will be required in relation to both starting materials and products. These should be laid down as SOPs. The same applies to testing procedures that are not as given in a pharmacopoeia or other official compendium, and to procedures and programs for the calibration and maintenance of laboratory instruments, including balances. Records of such calibration and maintenance need to be maintained. This is probably most conveniently done by means of hand-ruled, hardback notebooks held with, or close to, the equipment in question. Whatever system is employed, it should clearly flag when the next calibration, or maintenance service, is due.

USCGMPS

Subpart J — Records and Reports

Sec. 211-196 Distribution records

> Distribution records shall contain the name and strength of the product and description of the dosage form, name and address of the consignee, date and quantity shipped, and lot or control number of the drug product. For compressed medical gas products, distribution records are not required to contain lot or control numbers.

EC GMP GUIDE

Chapter 4 Documentation

> 4.25 Records should be maintained of the distribution of each batch of a product in order to facilitate the recall of the batch if necessary (see Chapter 8).

Chapter 8 Complaints and Recall

> 8.12 The distribution records should be readily available to the person(s) responsible for recalls, and should contain sufficient information on wholesalers and directly supplied customers (with addresses, phone and/or fax numbers inside and outside working hours, batches and amounts delivered), including those for exported products and medical samples.

The rationale for making and retaining records of the distribution of each batch of each product hardly needs to be stated. It is to facilitate a recall (or perhaps "freeze" stock, or "examine before dispatch"). This can be accomplished by recording product batch numbers on invoices or order-picking lists (and retaining copies), or by recording of the dates on which a specific product batch commenced and ceased distribution. The latter, however, is perhaps the least satisfactory in terms of speed and efficiency if and when a recall becomes necessary.

US CGMPS

Subpart J — Records and Reports

Sec. 211.198 Complaint files

(a) Written procedures describing the handling of all written and oral complaints regarding a drug product shall be established and followed. Such procedures shall include provisions for review by the quality control unit, of any complaint involving the possible failure of a drug product to meet any of its specifications and, for such drug products, a determination as to the need for an investigation in accordance with Sec. 211.192. Such procedures shall include provisions for review to determine whether the complaint represents a serious and unexpected adverse drug experience which is required to be reported to the Food and Drug Administration in accordance with Sec. 310.305 of this chapter.

(b) A written record of each complaint shall be maintained in a file designated for drug product complaints. The file regarding such drug product complaints shall be maintained at the establishment where the drug product involved was manufactured, processed, or packed, or such file may be maintained at another facility if the written records in such files are readily available for inspection at that other facility. Written records involving a drug product shall be maintained until at least I year after the expiration date of the drug product, or I year after the date that the complaint was received, whichever is longer. In the case of certain OTC drug products lacking expiration dating because they meet the criteria for exemption under Sec. 211.137 records shall be maintained for 3 years after distribution of the drug product.

　　(1) The written record shall include the following information, where known: the name and strength of the drug product, lot number, name of complainant, nature of complaint, and reply to complainant.

　　(2) Where an investigation under Sec. 211.192 is conducted, the written record shall include the findings of the investigation and follow-up. The record or copy of the record of the investigation shall be maintained at the establishment where the investigation occurred in accordance with Sec. 211.180(c).

　　(3) Where an investigation under sec. 211.192 is not conducted, the written record shall include the reason that an investigation was found not to be necessary and the name of the responsible person making such a determination.

EC GMP GUIDE

The EC GMP Guide devotes a whole (brief) chapter (8) to Complaints and Product Recall, thus:

Chapter 8 Complaints and Product Recall

Principle

All complaints and other information concerning potentially defective products must be reviewed carefully according to written procedures. In order to provide for all contingencies, and in accordance with Article 28 of Directive 75/319/EEC, a system should be designed to recall, if necessary, promptly and effectively products known or suspected to be defective from the market.

Complaints

8.1 A person should be designated responsible for handling the complaints and deciding the measures to be taken together with sufficient supporting staff to assist him. If this person is not the Qualified Person, the latter should be made aware of any complaint, investigation or recall.

8.2 There should be written procedures describing the action to be taken, including the need to consider a recall, in the case of a complaint concerning a possible product defect.

8.3 Any complaint concerning a product defect should be recorded with all the original details and thoroughly investigated. The person responsible for Quality Control should normally be involved in the study of such problems.

8.4 If a product defect is discovered or suspected in a batch, consideration should be given to checking other batches in order to determine whether they are also affected. In particular, other batches which may contain reworks of the defective batch should be investigated.

8.5 All the decisions and measures taken as a result of a complaint should be recorded and referenced to the corresponding batch records.

8.6 Complaints records should be reviewed regularly for any indication of specific or recurring problems requiring attention and possibly the recall of marketed products.

8.7 The Competent Authorities should be informed if a manufacturer is considering action following possibly faulty manufacture, product deterioration, or any other serious quality problems with a product.

Recalls

8.8 A person should be designated as responsible for execution and co-ordination of recalls and should be supported by sufficient staff to handle all the aspects of the recalls with the appropriate degree of urgency. This responsible person should normally be independent of the sales and marketing organisation. If this person is not the Qualified Person, the latter should be made aware of any recall operation.

8.9 There should be established written procedures, regularly checked and updated when necessary, in order to organise any recall activity.

8.10 Recall operations should be capable of being initiated promptly and at any time.

8.11 All Competent Authorities of all countries to which products may have been distributed should be informed promptly if products are intended to be recalled because they are, or are suspected of being defective.

8.12 The distribution records should be readily available to the person(s) responsible for recalls, and should contain sufficient information on wholesalers and directly supplied customers (with addresses, phone and/or fax numbers inside and outside working hours, batches and amounts delivered), including those for exported products and medical samples.

8.13 Recalled products should be identified and stored separately in a secure area while awaiting a decision on their fate.

8.14 The progress of the recall process should be recorded and a final report issued, including a reconciliation between the delivered and recovered quantities of the products.

8.15 The effectiveness of the arrangements for recalls should be evaluated from time to time.

The requirements of the two regulatory documents regarding the necessity of documenting and reviewing complaints and reports of purportedly defective products are closely similar. The US cGMPs require that "written records involving a drug product" (and in this context we must presume that this refers to written records of complaints, etc.) shall be retained until at least one year after the expiration date of the product concerned or one year after the complaint was received, whichever is longer (or for three years after distribution for nonexpiry dated OTC products). In this context, the EC GMP Guide does not state any specific retention times. Otherwise, the EC GMP Guide is rather more detailed and explicit in expressing, for example:

■ The need to designate a "responsible person" for handling complaints
■ The need to consider the possibility that batches, and products, other than the one complained of may also be suspect
■ The need to regularly review complaints records to determine if there are recurring problems, indicative of a hitherto unsuspected problem
■ The need to designate a person as responsible for managing and coordinating product recalls
■ The need to ensure that recall procedures are operable and will, indeed, operate *at any time*, not merely during normal working hours, during the normal working week
■ The need to ensure that recalled products are identified as such and segregated so as to avoid redistribution. (It has happened!)

Quality Assurance and GMP are about preventing errors. However, in this imperfect universe there is no such thing as an infallibly perfect system, and an essential feature of any QA system is a plan for dealing with complaints, or reports of faulty products, if they do occur. A requirement to cover this occurs in all notable GMP Guidelines. The ISO 9000 series also require the establishment of documented procedures for analyzing customer complaints and taking all necessary corrective action. It is important, therefore, that all complaints and defect reports should be examined and evaluated by competent, responsible personnel, no matter how trivial they may at first sight seem.

It should be noted that in both the US and Europe it is a regulatory requirement that the regulatory authorities be informed of any significant product defect. In the EU, in addition to the relevant national authority, the "competent authorities" of all countries to which the products may have been distributed must also be informed of any recall.

Manufacturers should establish an organizational and recording system for dealing with complaints or reports of defective products. Often, such complaints may seem, in isolation, to be trivial and unfounded and, on further investigation, may prove to be so. However, a full understanding of the nature and cause of a complaint, and a consideration of other possible reports of a similar nature, might well lead to a conclusion that there really is something amiss and that corrective action must be taken. That is why all complaints should be taken very seriously. They should be recorded and thoroughly investigated, with a decision on any necessary action being taken at an appropriately senior level. That decision *could* be to recall the product or batch.

All manufacturers should have a written recall procedure, with nominated persons responsible for implementing it as necessary, within, or outside of, normal working hours. Distribution records should be maintained, which will facilitate effective recall, and the written procedure should include emergency and off-hours contacts and telephone numbers.

The written procedure should state:

a. How the distribution or use of the product or batch should be halted
b. How contact should be made, as necessary, with the relevant regulatory authority
c. How, and how widely, the recall should be notified and implemented

Any notification of a recall should include:

a. Name of the product
b. Batch number(s)
c. Nature of defects
d. Action to be taken
e. Urgency of the action
f. Statement of reasons for the recall, and of potential risks

Account needs to be taken of any goods that may be in transit, and consideration must be given to the possibility that the fault may extend to other batches or products, e.g., due to a fault in equipment or machinery used in the manufacture of a number of different products or batches, or through the use of incorrect or faulty material in a number of different batches.

Figure 11.4 to Figure 11.6 are offered as example of how to implement the regulatory requirements in relation to complaints and recall. Figure 11.4 is an illustrative SOP for a Complaints and Defect Report procedure. Implicit in this procedure is the notion that investigation of a complaint or defect report may well lead to a conclusion that a product recall (or freeze) is necessary. The Complaints and Defect Report procedure is intended to be operated in conjunction with a Complaint and Defect Report Form (Figure 11.5), a copy of which, as is indicated, should form part of the SOP. Copies of this Report Form should be provided to *all* persons in the organization who may possibly be

Phantazpharm Inc. - Standard Operating Procedure

S.O.P. No:	Date Issued:	Supersedes S.O.P. No: New Document	Review Date:	Page 1 of 5

COMPLAINTS and DEFECT REPORT PROCEDURE

Contents   Page

Written by:	Approved by:	Authorized by:
Date:	Date:	Date:

Figure 11.4A   Complaints and Defect Report Procedure

S.O.P. No   **Page 2 of 5**

1. Purpose

To define the procedure to be followed on receipt of any product complaint or defect report.

2. Scope

This S.O.P. applies to all complaints and/or defect reports that are received or notified regarding products supplied/distributed by our Company, whatever their nature or source.

3. Responsibility

Routine responsibility for ensuring that this procedure is implemented as and when necessary rests with ....

Writing and approval/authorization of this procedure, and ensuring that it is revised and updated as necessary are the responsibility of ....

4. Revision

This procedure must be re-written, approved, and authorized whenever any change in method of operation, or any other circumstance, indicates the need. It must be reviewed every 12 months from the date of issue, in the light of current practice, to determine whether any revision is necessary.

All copies of S.O.P(s). which are superseded by any revision must be withdrawn from active use, and appropriate change-control records maintained to ensure effective implementation of this requirement.

Figure 11.4B   Complaints and Defect Report Procedure

5. Procedure

5.1 Complaints regarding product quality, reports of defective products, and the like may be received in a number of formats (written or oral), at second or third hand, and from a number of sources, for example:

  a. From within the Company.
  b. From some point within the distribution chain.
  c. From customers, users, or patients.
  d. From a Local, National, or International Regulatory Agency.

5.2 In all cases, the initial receipt of the complaint/defect report, whatever its nature, format, or source, must be recorded **immediately** by the recipient on a "COMPLAINT/DEFECT REPORT RECORD" form (see copy attached). Supplies of these forms will be available to, and held by, all those who are likely to receive such reports.

5.3 Overall responsibility for dealing with Complaints and Defect Reports will rest with ....(A). In the absence of (A) the deputies who will assume this responsibility are ....(B) and/or ....(C)

5.4 Following receipt of a complaint/defect report, and on completion of parts 1 to 9 of the "Complaint/Defect Report Record," the form, together with any relevant correspondence attached, must be passed immediately to (A) or in his/her absence to one of the designated deputies (see 5.3 above).

5.5 (A), or one of the designated deputies, will review the report and decide whether any immediate action is necessary.

5.6 IMMEDIATE ACTION must be taken if the complaint/report is indicative of

  a. An actual or potential hazard to life, health, or well-being of a user or consumer.

  b. A defect that, while not necessarily hazardous, could adversely affect the standing and reputation of our Company.

"Immediate action" will also be considered necessary where the defect renders the product difficult or inconvenient in use.

5.7 Immediate Actions to be considered, and implemented according to the actual or potential seriousness of the known or suspected hazard, include:

Figure 11.4C   Complaints and Defect Report Procedure

S.O.P. No   Page 4 of 5

a. "Freezing" of all relevant Company stocks held on-site.

b. Reviewing distribution records, and requesting all points in the distribution chain to "freeze" all relevant stocks, and not to use them or distribute them further.

c. Initiation of a RECALL.

e. Informing the local or national Regulatory Authority.

5.8 If the decision is to "freeze"(at any level) beyond Company stocks held on the Company site, or to recall, then reference must be made to the RECALL (or" FREEZE") PROCEDURE, S.O.P. No. ...., which must be implemented immediately.

5.9 If the decision is made that it is necessary to freeze Company stocks only, pending further investigation, then those responsible for holding/maintaining in-house stocks should be informed orally immediately, with confirmation in writing as soon as possible. The information conveyed should include

     a. Name of product(s)
     b. Batch number(s)
     c. Nature of defect (if and as appropriate)
     d. Precise action to be taken

N.B. Particular care must be taken over any suspect goods that have been already assembled and are ready for dispatch.

5.10 If the complaint/defect report is not considered to be sufficiently serious as to warrant recall, freeze, or any other emergency action, a letter of acknowledgment should be sent in response to all such reports received.

5.11 Whatever the action decided, arrangements must be made to have the complaint thoroughly investigated. Following these investigations, the remainder of the Complaint/Defect Report Record should be completed, and the person or body responsible for the original complaint informed of the results of the investigations, in writing.

5.12 The completed Complaint/Defect Report Record should be filed, along with all other relevant letters, papers, and reports, and should be held in a filing system, such that they are readily retrievable for review and examination.

5.13 All records of complaints etc. should periodically and formally be reviewed in order to detect any adverse quality trends, "common factors." etc.

Figure 11.4D   Complaints and Defect Report Procedure

S.O.P. No   **Page 5 of 5**

NAME(S), ADDRESS, TELEPHONE NO., CONTACT POINTS of
LOCAL/NATIONAL REGULATORY AUTHORITY

Address:

Telephone Numbers:

> Office Hours ....
> Other Times ....
> Fax. ....

Other Emergency Contact Numbers:

> ....
> ....
> ....

> (All telephone nos. checked as correct, as at (Date) ....)

Figure 11.4E   Complaints and Defect Report Procedure

the first recipients of a complaint or defect report. They should be trained in its use and in the crucial importance of taking all such reports very seriously.

As the Complaints and Defect Report SOP indicates, if the complaints (etc.) procedure leads to a conclusion to recall (or freeze), then the recall (or freeze) procedure (see Figure 11.6) must be implemented. It is vital that this SOP is kept up to date, particularly in regard to internal and external names, addresses, and phone numbers, and that it is regularly shown (by "dummy runs") to be operable *at any time*. (The need to urgently recall does *not* arise only between 08.00 hours to 17.00 hours, Monday through Friday.)

COMPLAINT/DEFECT REPORT RECORD

Page 1 of 2

1. Date Complaint/Report received .... Time ....

2. Received by ....

3. Received from:

    Name ....

    Address:

    Telephone number ....

    Fax number ....

    e-mail ....

4. Names/addresses/phone numbers, etc. of other contacts/persons/organizations:

5. Product(s) involved:

6. Batch/lot numbers:

7. Nature of complaint/report (attach any written correspondence)

8. Have samples been returned for examination? (Give details)

9. Are samples available for collection/examination? (Give details)

Figure 11.5A   Complaint/Defect Report Record

(Complaints/Defect report record (continued))

Page 2 of 2

10. Results of Investigations/Tests (attach other sheets as necessary)

11. Conclusions, and decision on action to be taken

Signed .... Date ....

12. Letter(s) sent to .... Date ....

13. Also considered necessary to inform: Done/Date:

14. Was decision taken to **FREEZE** (beyond own on-site stocks) or **Recall?**

IF "YES," RECALL (or FREEZE) SOP MUST BE IMPLEMENTED IMMEDIATELY.

Signed ....

Date ....

Figure 11.5B   Complaint/Defect Report Record

Phantazpharm Inc. - Standard Operating Procedure

S.O.P. No:	Date Issued:	Supersedes S.O.P. No: New Document	Review Date:	Page 1 of 6

RECALL (or "FREEZE") PROCEDURE

Contents	Page
1. PURPOSE	2
2. SCOPE	2
3. RESPONSIBILITY	2
4. REVISION	2
5. EFFECTIVENESS OF PROCEDURE	2
6. PROCEDURE	3.
7. CONTACT NAMES, TELEPHONE NUMBERS, etc.	6

Written by: Date:	Approved by: Date:	Authorized by: Date:

Figure 11.6A   Recall (or "Freeze") Procedure

S.O.P. No   Page 2 of 6

1. Purpose

To define the procedure to be followed in order to "freeze" or recall products.
**N.B. This S.O.P. should be read and followed in conjunction with
S.O.P. No. .... "COMPLAINTS and DEFECT REPORT PROCEDURE"**

2. Scope

This S.O.P. is to be implemented whenever a decision is taken to recall, or "freeze"
any product(s), at any stage, for whatever reason.

3. Responsibility

Routine responsibility for ensuring that this procedure is implemented as and when
necessary rests with ....

Writing and approval/authorization of this procedure, and ensuring that it is revised
and updated as necessary are the responsibility of ....

4. Revision

This procedure must be re-written, approved, and authorized whenever any change
in method of operation, or any other circumstance, indicates the need. It must be
reviewed every 12 months from the date of issue, in the light of current practice, to
determine whether any revision is necessary.

All copies of S.O.P(s). that are superseded by any revision must be withdrawn
from active use, and appropriate change-control records maintained to ensure
effective implementation of this requirement.

5. Effectiveness of Procedure

This procedure should be practically evaluated from time-to-time (e.g., by performing
internal "dummy runs") to ensure that it is, and remains, capable of **prompt
implementation at any time**, within or outside normal working hours.

Figure 11.6B   Recall (or "Freeze") Procedure

6. Procedure

6.1 An indication that a product, or product-batch, should be RECALLED, or placed on HOLD ("FROZEN") at some stage in the storage and/or distribution chain, may arise from a number of sources, e.g.:

  a. From within the Company.
  b. From some point within the distribution chain.
  c. From customers, users, or patients.
  d. From a Local, National, or International Regulatory Agency.
  e. From suppliers of products, materials, or other goods.
  e. And/or via our own internal COMPLAINTS and DEFECT REPORT RECORD(s) - See S.O.P. No. ....

6.2 Overall responsibility for the coordination of all Recall or "HOLD/FREEZE" actions will rest with ....(A). In the absence of (A) the deputies who will assume this responsibility are ....(B) and/or ....(C) The decision to RECALL/FREEZE will be taken by the "Recall Coordinator" (A or B/C) in discussion with:

  ....
  ....
  ....
  ....

and with the local or ational Regulatory Agency, as necessary or appropriate.

6.3 Immediately when the decision is made to Recall/Freeze a product, or to issue any any other warnings outside the Company about the distribution or use of any product, the following Company Personnel (in addition to (A) &/or the designated deputies) must be informed at once:

  ....
  ....
  ....
  ....

6.4 At the same time, the local or national Regulatory Agency must be informed of the incident and of the action taken or proposed.
(For telephone numbers etc. see Section 7 of this S.O.P., "Contact Names, Telephone Numbers, etc.").

Figure 11.6C   Recall (or "Freeze") Procedure

6.5   The Recall Coordinator must immediately ensure (oral message, confirmed in writing) that all those within the Company concerned with the storage, packing, and distribution of products and materials are aware of the problem and the action to be taken. The information conveyed should include:

> a. Name of product(s)
> b. Batch number(s)
> c. Nature of defect
> d. Action to be taken

6.6   All affected, or potentially affected, products and any other relevant stocks or materials remaining on-site must be securely segregated and effectively quarantined so as to be sure they cannot be dispatched.

6.7   The Recall Coordinator (or his/her deputies) will arrange for the prompt preparation of a list of the recipients of the defective goods, down to the level at which the Recall or Freeze is to take place. He/she will then ensure that all the listed recipients are informed of:

> a. Name of product(s)
> b. Batch number(s)
> c. Nature of defect
> d. Action to be taken

According to the urgency/extent of potential hazard, this information should be conveyed by direct mailing, telephone call confirmed by mail, or fax. IF THE DEFECT IS CONSIDERED TO BE TOO SERIOUS TO PERMIT ANY DELAY, STEPS SHOULD BE TAKEN TO INFORM **ALL** POTENTIAL OR POSSIBLE RECIPIENTS WITHOUT NECESSARILY WAITING FOR THE COLLATION OF FULL DISTRIBUTION DATA.

6.8   If the defect is considered to represent a SERIOUS and IMMEDIATE HAZARD TO USERS or CONSUMERS, then consideration should be given to the use of the media (TV, radio, press) to provide wider awareness of the problem.

6.9   When a Recall has been initiated, arrangements must be made by the Recall Coordinator to receive (or collect) the returned goods, and to ensure that they are securely segregated from other stock, such that there is no possibility that they could be re-distributed.

6.10 The progress of the Recall should be recorded and monitored, so that the

Figure 11.6D   Recall (or "Freeze") Procedure

S.O.P. No   **Page 5 of 6**

quantity of goods returned can be reconciled against the amount produced, distributed, and still held in stock.

6.11 When it is considered that the Recall/Freeze has been effected as completely as possible, a full report will be prepared by the Recall Coordinator (or deputy) detailing

     a. Reason for Recall/Freeze
     b. Results of full investigation into the cause(s) of the defective product
     c. Action taken to Recall etc.
     d. Numerical details of the reconciliation exercise
     e. Action taken to prevent re-occurrence

As required, a copy of this report will be sent to the local or national Regulatory Agency.

6.12 A list of contact names, addresses, and telephone numbers, in and out of normal working hours, follows on the next page. This information should be regularly checked for current accuracy, and amended as necessary,

Figure 11.6E   Recall (or "Freeze") Procedure

S.O.P. No   **Page 6 of 6**

CONTACT NAMES, TELEPHONE NUMBERS, etc.
(For use in connection with Defect Reports, Recalls, etc.)

COMPANY PERSONNEL

    Name: Extension: Home/Out-of-Hours No.:

LOCAL/NATIONAL REGULATORY AGENCY

Address:

Telephone Numbers:

    Office Hours:
    Other Times:
    Fax.:

(All telephone nos. checked as correct, as at (Date ....)

Figure 11.6F   Recall (or "Freeze") Procedure

ANNEX 1 TO CHAPTER 11 (FROM EC GMP GUIDE, ANNEX 11)

COMPUTERISED SYSTEMS

Principle

The introduction of computerised systems into systems of manufacturing, including storage, distribution and quality control does not alter the need to observe the relevant principles given elsewhere in the Guide. Where a computerised system replaces a manual operation, there should be no resultant decrease in product quality or quality assurance. Consideration should be given to the risk of losing aspects of the previous system which could result from reducing the involvement of operators.

Personnel

1. It is essential that there is the closest co-operation between key personnel and those involved with computer systems. Persons in responsible positions should have the appropriate training for the management and use of systems within their field of responsibility which utilises computers. This should include ensuring that appropriate expertise is available and used to provide advice on aspects of design, validation, installation and operation of computerised system.

Validation

2. The extent of validation necessary will depend on a number of factors including the use to which the system is to be put, whether the validation is to be prospective or retrospective and whether or not novel elements are incorporated. Validation should be considered as part of the complete life cycle of a computer system. This cycle includes the stages of planning, specification, programming, testing, commissioning, documentation, operation, monitoring and modifying.

System

3. Attention should be paid to the siting of equipment in suitable conditions where extraneous factors cannot interfere with the system.
4. A written detailed description of the system should be produced (including diagrams as appropriate) and kept up to date. It should describe the principles, objectives, security measures and scope of the system and the main features of the way in which the computer is used and how it interacts with other systems and procedures.
5. The software is a critical component of a computerised system. The user of such software should take all reasonable steps to ensure that it has been produced in accordance with a system of Quality Assurance.
6. The system should include, where appropriate, built-in checks of the correct entry and processing of data.

7. Before a system using a computer is brought into use, it should be thoroughly tested and confirmed as being capable of achieving the desired results. If a manual system is being replaced, the two should be run in parallel for a time, as a part of this testing and validation.

8. Data should only be entered or amended by persons authorised to do so. Suitable methods of deterring unauthorised entry of data include the use of keys, pass cards, personal codes and restricted access to computer terminals. There should be a defined procedure for the issue, cancellation, and alteration of authorisation to enter and amend data, including the changing of personal passwords. Consideration should be given to systems allowing for recording of attempts to access by unauthorised persons.

9. When critical data are being entered manually (for example the weight and batch number of an ingredient during dispensing), there should be an additional check on the accuracy of the record which is made. This check may be done by a second operator or by validated electronic means.

10. The system should record the identity of operators entering or confirming critical data. Authority to amend entered data should be restricted to nominated persons. Any alteration to an entry of critical data should be authorised and recorded with the reason for the change. Consideration should be given to building into the system the creation of a complete record of all entries and amendments (an "audit trail").

11. Alterations to a system or to a computer program should only be made in accordance with a defined procedure which should include provision for validating, checking, approving and implementing the change. Such an alteration should only be implemented with the agreement of the person responsible for the part of the system concerned, and the alteration should be recorded. Every significant modification should be validated.

12. For quality auditing purposes, it should be possible to obtain clear printed copies of electronically stored data.

13. Data should be secured by physical or electronic means against wilful or accidental damage, in accordance with item 4.9 of the Guide. Stored data should be checked for accessibility, durability and accuracy. If changes are proposed to the computer equipment or its programs, the above mentioned checks should be performed at a frequency appropriate to the storage medium being used.

14. Data should be protected by backing-up at regular intervals. Back-up data should be stored as long as necessary at a separate and secure location.

15. There should be available adequate alternative arrangements for systems which need to be operated in the event of a breakdown. The time required to bring the alternative arrangements into use should be related to the possible urgency of the need to use them. For example, information required to effect a recall must be available at short notice.

16. The procedures to be followed if the system fails or breaks down should be defined and validated. Any failures and remedial action taken should be recorded.

17. A procedure should be established to record and analyse errors and to enable corrective action to be taken.

18. When outside agencies are used to provide a computer service, there should be a formal agreement including a clear statement of the responsibilities of that outside agency (see Chapter 7).

19. When the release of batches for sale or supply is carried out using a computerised system, the system should allow for only a Qualified Person to release the batches and it should clearly identify and record the person releasing the batches.

ANNEX 2 TO CHAPTER 11 (US FDA GUIDANCE)

Guidance for Industry
Part 11, Electronic Records; Electronic Signatures
— Scope and Application

Draft Guidance
This guidance is being distributed for comment purposes only.

Comments and suggestions regarding this draft document should be submitted within 60 days of publication in the *Federal Register* of the notice announcing the availability of the draft guidance. Submit comments to Dockets Management Branch (HFA-305), Food and Drug Administration, 5630 Fishers Lane, rm. 1061, Rockville, MD 20852. All comments should be identified with the docket number listed in the notice of availability that publishes in the *Federal Register.*

For questions regarding this draft document please contact (CDER) Joseph C. Famulare, 301-827-8940, part11@cder.fda.gov; (CBER) David Doleski, 301-827-3031, doleski@cber.fda.gov; (CDRH) John Murray, 301-594-4659, jfm@cdrh.fda.gov; (CVM) Vernon D. Toelle, 301-827-0312, vtoelle@cvm.fda.gov; (CFSAN) JoAnn Ziyad, 202-418-3116, jziyad@cfsan.fda.gov; or (ORA) Scott MacIntire, 301-827-0386, smacinti@ora.fda.gov.

U.S. Department of Health and Human Services
Food and Drug Administration
Center for Drug Evaluation and Research (CDER)
Center for Biologics Evaluation and Research (CBER)
Center for Devices and Radiological Health (CDRH)
Center for Food Safety and Applied Nutrition (CFSAN)
Center for Veterinary Medicine (CVM)
Office of Regulatory Affairs (ORA)

February 2003
Compliance

TABLE OF CONTENTS

GUIDANCE FOR INDUSTRY*

PART 11, ELECTRONIC RECORDS; ELECTRONIC SIGNATURES — SCOPE AND APPLICATION

This draft guidance, when finalized, will represent the Food and Drug Administration's (FDA's) current thinking on this topic. It does not create or confer any rights for or on any person and does not operate to bind FDA or the public. An alternative approach may be used if such approach satisfies the requirements of the applicable statutes and regulations.

I. Introduction

This guidance is intended to describe the Food and Drug Administration's (FDA's) current thinking regarding the scope and application of Part 11 of Title 21 of the Code of Federal Regulations; Electronic Records; Electronic Signatures.**

This document provides guidance to persons who, in fulfillment of a requirement in a statute or another part of FDA's regulations to maintain records or submit information to FDA,*** have chosen to maintain the records or submit designated information electronically and, as a result, have become subject to Part 11. Part 11 applies to records in electronic form that are created, modified, maintained, archived, retrieved, or transmitted under any records requirements set forth in Agency regulations. Part 11 also applies to electronic records

---

*   This guidance has been prepared by the Office of Compliance in the Center for Drug Evaluation and Research (CDER) in consultation with the other Agency centers and the Office of Regulatory Affairs at the Food and Drug Administration.

**   62 FR 13430.

*** These requirements include, for example, certain provisions of the Current Good Manufacturing Practice regulations (21 CFR part 211), the Quality System Regulation (21 CFR part 820), and the Good Laboratory Practice for Nonclinical Laboratory Studies regulations (21 CFR part 58).

submitted to the Agency under the Federal Food, Drug, and Cosmetic Act (the Act) and the Public Health Service Act (the PHS Act), even if such records are not specifically identified in Agency regulations (§ 11.1). The underlying requirements set forth in the Act, PHS Act, and FDA regulations (other than Part 11) are referred to in this guidance document as *predicate rules*.

As an outgrowth of its current good manufacturing practice (CGMP) initiative for human and animal drugs and biologics,* FDA is embarking on a re-examination of Part 11 as it applies to all FDA regulated products. We may revise provisions of Part 11 as a result of that re-examination. This guidance explains that, while this re-examination of Part 11 is under way, we will narrowly interpret the scope of Part 11. It also explains that we intend to exercise enforcement discretion with respect to certain Part 11 requirements. We will not normally take regulatory action to enforce compliance with the validation, audit trail, record retention, and record copying requirements of Part 11 as explained in this guidance. However, records must still be maintained or submitted in accordance with the underlying predicate rules.

In addition, we intend to exercise enforcement discretion and will not normally take regulatory action to enforce Part 11 with regard to systems that were operational before August 20, 1997, the effective date of Part 11 (commonly known as existing or legacy systems) while we are re-examining Part 11.

FDA's guidance documents, including this guidance, do not establish legally enforceable responsibilities. Instead, guidances describe the Agency's current thinking on a topic and should be viewed only as recommendations, unless specific regulatory or statutory requirements are cited. The use of the word *should* in Agency guidances means that something is suggested or recommended, but not required.

II. Background

In March of 1997, FDA issued final Part 11 regulations that provided criteria for acceptance by FDA, under certain circumstances, of electronic records, electronic signatures, and handwritten signatures executed to electronic records as equivalent to paper records and handwritten signatures executed on paper. These regulations, which apply to all FDA program areas, were intended to permit the widest possible use of electronic technology, compatible with FDA's responsibility to protect the public health.

After Part 11 became effective in August 1997, significant discussions ensued between industry, contractors, and the Agency concerning the interpretation and implementation of the rule. FDA has (1) spoken about Part 11 at many conferences and met numerous times with an industry coalition and other interested parties in an effort to hear more about potential Part 11 issues; (2) published a compliance policy guide, CPG 7153.17: Enforcement Policy: 21 CFR Part 11; Electronic Records; Electronic Signatures; and (3) published numerous draft guidance documents including the following:

---

* See *Pharmaceutical CGMPs for the 21st Century: A Risk-Based Approach; A Science and Risk-Based Approach to Product Quality Regulation Incorporating an Integrated Quality Systems Approach* at www.fda.gov/oc/guidance/gmp.html.

- Guidance for industry, 21 CFR Part 11; Electronic Records; Electronic Signatures Validation
- Guidance for industry, 21 CFR Part 11; Electronic Records; Electronic Signatures, Glossary of Terms
- Guidance for industry, 21 CFR Part 11; Electronic Records; Electronic Signatures, Time Stamps
- Guidance for industry, 21 CFR Part 11; Electronic Records; Electronic Signatures, Maintenance of Electronic Records
- Guidance for industry, 21 CFR Part 11; Electronic Records; Electronic Signatures, Electronic Copies of Electronic Records

Some statements by Agency staff may have been misunderstood as statements of official Agency policy. Concerns have been raised that some interpretations of the Part 11 requirements would (1) unnecessarily restrict the use of electronic technology in a manner that is inconsistent with FDA's stated intent in issuing the rule, (2) significantly increase the costs of compliance to an extent that was not contemplated at the time the rule was drafted, and (3) discourage innovation and technological advances without providing a significant public health benefit. These concerns have been raised particularly in the areas of Part 11 requirements for validation, audit trails, record retention, record copying, and legacy systems.

In the *Federal Register* of February 4, 2003, we announced the withdrawal of the draft guidance for industry, *21 CFR Part 11; Electronic Records; Electronic Signatures, Electronic Copies of Electronic Records* because we wanted to avoid loss of time spent by industry in an effort to review and comment on the draft guidance when that draft guidance may no longer be representative of FDA's approach under the new CGMP initiative. The other Part 11 draft guidances were left in place because industry had already had the opportunity to review and comment on them. However, in preparing this guidance, FDA has determined that it might cause confusion to leave standing the other Part 11 draft guidance documents on validation, glossary of terms, time stamps, maintenance of electronic records, and CPG 7153.17. Accordingly, FDA is withdrawing those draft guidances and CPG 7153.17 as well as the guidance on electronic copies of electronic records. FDA received valuable public comments on these draft guidances and plans to use that information to inform the Agency's future decision-making with respect to Part 11.

We have now determined that we will re-examine Part 11, and we may revise provisions of that regulation. To avoid unnecessary expenditures of resources to comply with Part 11 requirements that may be revised through a rulemaking, we are issuing this guidance to describe how we intend to exercise enforcement discretion with regard to certain Part 11 requirements during the re-examination of Part 11.

III. Discussion

A. Overall Approach to Part 11 Requirements

As described in more detail below, the approach outlined in this guidance is based on three main elements:

- Part 11 will be interpreted narrowly; we are now clarifying that fewer records will be considered subject to Part 11.
- For those records that we are now clarifying are subject to Part 11, we intend to exercise enforcement discretion with regard to Part 11 requirements for validation, audit trails, record retention, and record copying, in the manner described in this guidance, and in applying Part 11 to systems that were operational before the effective date of Part 11.
- FDA will enforce predicate rule requirements for records that are subject to Part 11.

It is important to note that FDA's exercise of enforcement discretion as described in this guidance, is limited to the specified Part 11 requirements. We intend to enforce all other provisions of Part 11 including, but not limited to, certain controls for closed systems in § 11.10 (e.g., limiting system access to authorized individuals; use of operational system checks; use of authority checks; use of device checks; determination that persons who develop, maintain, or use electronic systems have the education, training, and experience to perform their assigned tasks; establishment of and adherence to written policies that hold individuals accountable for actions initiated under their electronic signatures; and appropriate controls over systems documentation), the corresponding controls for open systems (§ 11.30), and requirements related to electronic signatures (e.g., §§ 11.50, 11.70, 11.100, 11.200, and 11.300). We expect continued compliance with these provisions, and we will continue to enforce them. Furthermore, persons must comply with applicable predicate rules, and records that are required to be maintained or submitted must remain secure and reliable in accordance with the predicate rules.

B. Details of Approach — Scope of Part 11

1. Narrow Interpretation of Scope

We understand that there have been different views expressed about the scope of Part 11. Some have understood the scope of Part 11 to be very broad. We believe that some of those broad interpretations could lead to unnecessary controls and costs and could discourage innovation and technological advances without providing added benefit to the public health. As a result, we want to clarify that the Agency intends to interpret the scope of Part 11 narrowly.

Under the narrow interpretation of the scope of Part 11, with respect to records required to be maintained or submitted, when persons choose to use records in electronic format in place of paper format, Part 11 would apply. On the other hand, when persons use computers to generate paper printouts of electronic records, those paper records meet all the requirements of the applicable predicate rules, and persons rely on the paper records to perform their regulated activities, the *merely incidental* use of computers in those instances would not trigger Part 11. In such instances, FDA would generally not consider persons to be "using electronic records in lieu of paper records" under §§ 11.2(a) and 11.2(b).

2. Definition of Part 11 Records

Under this narrow interpretation, FDA considers Part 11 to be applicable to the following records or signatures in electronic format (Part 11 records or signatures):

- Records that are required to be maintained by predicate rules and that are maintained in electronic format *in place of paper format*. On the other hand, records (and any associated signatures) that are not required to be retained by predicate rules, but that are nonetheless maintained in electronic format, are not Part 11 records.
- Records that are required to be maintained by predicate rules, are maintained in electronic format *in addition to paper format*, and *are relied on to perform regulated activities*.

In some cases, actual business practices may dictate whether you are *using* electronic records instead of paper records under § 11.2(a). For example, if a record is required to be maintained by a predicate rule and you use a computer to generate a paper printout of the electronic records, but you nonetheless rely on the electronic record to perform regulated activities, the Agency may consider you to be *using* the electronic record instead of the paper record. That is, the Agency may take your business practices into account in determining whether Part 11 applies.

Accordingly, we recommend that, for each record required to be maintained by predicate rules, you determine in advance whether you plan to rely on the electronic record or paper record to perform regulated activities. We recommend that your decision be documented (e.g., in a Standard Operating Procedure (SOP)).

- Records submitted to FDA, under the predicate rules (even if such records are not specifically identified in Agency regulations), in electronic format (assuming the records have been identified in the docket as the types of submissions the Agency accepts in electronic format). However, a record that is not itself submitted, but is used in generating a submission, is not a Part 11 record unless it is otherwise required to be maintained by a predicate rule and it is maintained in electronic format.
- Electronic signatures that are intended to be the equivalent of handwritten signatures, initials, and other general signings required by predicate rules.

C. Approach to Specific Part 11 Requirements

1. Validation

The Agency intends to exercise enforcement discretion regarding the specific Part 11 requirements for validation of computerized systems (§ 11.10(a) and corresponding requirements in § 11.30). Persons must still comply with all applicable predicate rule requirements for validation (e.g., 21 CFR 820.70(i)).

Even if there is no predicate rule requirement to validate a system in a particular instance, it may nonetheless be important to validate the system to ensure the accuracy and reliability of the Part 11 records contained in the system. We suggest that your decision to validate such systems, and the extent of validation, be based on predicate rule requirements to ensure the accuracy and reliability of the records contained in the system. We recommend that you base your approach on a justified and documented risk assessment and a determination of the potential of the system to affect product quality and safety and record integrity. For instance, a word processor used only to generate SOPs would most likely not need to be validated.

For further guidance on validation of computerized systems, see FDA's guidance for industry and FDA Staff *General Principles of Software Validation* and also industry guidance such as the *GAMP 4 Guide* (See References).

2. Audit Trail

The Agency intends to exercise enforcement discretion regarding the specific Part 11 requirements related to computer-generated, time-stamped audit trails (§ 11.10 (e), (k)(2) and any corresponding requirement in § 11.30). Persons must still comply with all applicable predicate rule requirements related to documentation of, for example, date (e.g., § 58.130(e)), time, or sequencing of events.

Even if there are no predicate rule requirements to document, for example, date, time, or sequence of events in a particular instance, it may nonetheless be important to have audit trails or other physical, logical, or procedural security measures to ensure the trustworthiness and reliability of the records. We recommend that your decision on whether to apply audit trails, or other appropriate measures, be based on the need to comply with predicate rule requirements, a justified and documented risk assessment, and a determination of the potential impact on product quality and safety and record integrity. We suggest that you apply appropriate controls based on such an assessment. Audit trails are particularly important where the users are expected to create, modify, or delete regulated records during normal operation.*

3. Legacy Systems

The Agency intends to exercise enforcement discretion with regard to legacy systems that otherwise met predicate rule requirements prior to August 20, 1997, the effective date of Part 11. This means that the Agency will not normally take regulatory action to enforce compliance with any part 11 requirements. However, all systems must comply with all applicable predicate rule requirements and should be fit for their intended use.

4. Copies of Records

The Agency intends to exercise enforcement discretion with regard to the specific Part 11 requirements for generating copies of records (§ 11.10 (b) and any corresponding requirement in § 11.30). You should provide an investigator with reasonable and useful access to records during an inspection. All records held by you are subject to inspection in accordance with predicate rules (e.g., §§ 211.180(c),(d) and 108.35(c)(3)(ii)).

---

* Various guidance documents on information security are available (see References).

We recommend that you supply copies of electronic records by

- Producing copies of records held in common portable formats where records are kept in these formats
- Using established automated conversion or export methods, where available, to make copies in a more common format (including PDF)

In each case, we recommend that you ensure that the copying process used produces copies that preserve the content and meaning of the record. If you have the ability to search, sort, or trend Part 11 records, copies provided to the Agency should provide the same capability if it is technically feasible. You should allow inspection, review, and copying of records in a human readable form, on your site, using your hardware and software, following your established procedures and techniques for accessing those records.

5. Record Retention

The Agency intends to exercise enforcement discretion with regard to the Part 11 requirements for the protection of records to enable their accurate and ready retrieval throughout the records retention period (§ 11.10 (c) and any corresponding requirement in § 11.30). Persons must still comply with all applicable predicate rule requirements for record retention and availability (e.g., §§ 211.180(c),(d), 108.25(g), and 108.35(h)).

We suggest that your decision on how to maintain records be based on predicate rule requirements and that you base your decision on a justified and documented risk assessment and a determination of the value of the records over time.

FDA normally does not intend to object if you decide to archive required records in electronic format to nonelectronic media such as microfilm, microfiche, and paper, or to a standard electronic file format, such as PDF. Persons must still comply with all predicate rule requirements, and the records themselves and any copies of the required records should preserve their content and meaning. In addition, paper and electronic record and signature components can co-exist (i.e., a hybrid situation) as long as predicate rule requirements are met and the content and meaning of those records are preserved.*

REFERENCES

Food and Drug Administration References

1. *Glossary of Computerized System and Software Development Terminology* (Division of Field Investigations, Office of Regional Operations, Office of Regulatory Affairs, FDA 1995) (http://www.fda.gov/ora/inspect_ref/igs/gloss.html)
2. *General Principles of Software Validation; Final Guidance for Industry and FDA Staff* (FDA, Center for Devices and Radiological Health, Center for Biologics Evaluation and Research, 2002) (http://www.fda.gov/cdrh/comp/guidance/938.html)

---

* Examples of hybrid situations include combinations of paper records and electronic records, paper records and electronic signatures, or handwritten signatures executed to electronic records.

3. *Guidance for Industry, FDA Reviewers, and Compliance on Off-The-Shelf Software Use in Medical Devices* (FDA, Center for Devices and Radiological Health, 1999) (http://www.fda.gov/cdrh/ode/guidance/585.html)
4. *Pharmaceutical c GMPs for the 21st Century: A Risk-Based Approach; A Science and Risk-Based Approach to Product Quality Regulation Incorporating an Integrated Quality Systems Approach* (FDA 2002) (http://www.fda.gov/oc/guidance/gmp.html)

Other U.S. Federal References

5. NIST Special Publication SP800-30: *Risk Management Guide for Information Technology Systems* (National Institute of Standards and Technology, U.S. Department of Commerce, 2002) (http://csrc.nist.gov/publications/nistpubs/800-30/sp800-30.pdf)

Industry References

6. The Good Automated Manufacturing Practice (GAMP) Guide for Validation of Automated, GAMP 4 (ISPE/GAMP Forum, 2001) (http://www.ispe.org/gamp/)
7. ISO/IEC 17799:2000 (BS 7799:2000) Information technology — Code of practice for information security management (ISO/IEC, 2000)

ANNEX 3 TO CHAPTER 11

FDA
Guidance for Industry

Investigating Out of Specification (OOS) Test Results
for Pharmaceutical Production

Draft Guidance
This guidance document is being distributed for comment purposes only.

Comments and suggestions regarding this draft document should be submitted within 60 days of publication of the *Federal Register* notice announcing the availability of the draft guidance.

Submit comments to Dockets Management Branch (HFA-305), Food and Drug Administration, 5230 Fishers Lane., rm. 1061, Rockville, MD 20852. All comments should be identified with the docket number listed in the notice of availability that publishes in the *Federal Register.*

Additional copies of this draft guidance document are available from the Drug Information Branch, Division of Communications Management, HFD-210, 5600 Fishers Lane, Rockville, MD 20857, (Tel) 301-827-4573, or from the Internet at http://www.fda.gov/cder/guidance/index.htm.

For questions on the content of the draft document, contact C. Russ Rutledge, (301) 594-2455.

U.S. Department of Health and Human Services
Food and Drug Administration
Center for Drug Evaluation and Research (CDER)

September 1998

TABLE OF CONTENTS

DRAFT — NOT FOR IMPLEMENTATION

I. Introduction

This guidance for industry provides the Agency's current thinking on how to evaluate suspect, or out of specification (OOS), test results. For purposes of this document, the term

OOS results includes ***all*** suspect results that fall outside the specifications or acceptance criteria established in new drug applications, official compendia, or by the manufacturer.

This guidance applies to laboratory testing during the manufacture of active pharmaceutical ingredients, excipients, and other components and the testing of finished products to the extent that current good manufacturing practices (CGMP) regulations apply (21 CFR parts 210 and 211). Specifically, the guidance discusses how to investigate suspect, or OOS results, including the responsibilities of laboratory personnel, the laboratory phase of the investigation, additional testing that may be necessary, when to expand the investigation outside the laboratory, and the final evaluation of all test results.

II. Background

FDA considers the integrity of laboratory testing and documentation records to be of fundamental importance during drug manufacturing. Laboratory testing, which is required by the CGMP regulations (§ 211.165), is necessary to confirm that components, containers and closures, in-process materials, and finished products conform to specifications, including stability. Testing also supports analytical and process validation efforts. General CGMP regulations covering laboratory operations can be found in part 211, subparts I (Laboratory Controls) and J (Records and Reports). These regulations provide for the establishment of scientifically sound and appropriate specifications, standards, and test procedures that are designed to ensure that components and containers of drug products conform to the established standards. Section 211.165(f) of the CGMP regulations specifies that products that fail to meet established standards and other relevant quality control criteria will be rejected.

III. Identifying and Assessing OOS Test Results

FDA regulations require that an investigation be conducted whenever an OOS test result is obtained. The purpose of the investigation is to determine the cause of the OOS Even if a batch is rejected based on an OOS result, the investigation is necessary to determine if the result is associated with other batches of the same drug product or other products. Batch rejection does not negate the need to perform the investigation. The regulations require that a written record of the investigation be made including the conclusions of the investigation and follow-up (211.192).

To be meaningful, the investigation should be thorough, timely, unbiased, well-documented, and scientifically defensible. The first phase of such an investigation should include an initial assessment of the accuracy of the laboratory's data, **Before** test solutions are discarded, whenever possible. This way, hypotheses regarding laboratory error or instrument malfunctions may be tested using the same test solutions. If this initial assessment indicates that no errors were made in the analytical process used to arrive at the data, a complete failure investigation should follow.

A. Responsibility of the Analyst

The first responsibility for achieving accurate laboratory testing results lies with the analyst who is performing the test. The analyst should be aware of potential problems that could occur during the testing process and should watch for problems that could create OOS results.

In accordance with the CGMP regulations (§ 211.160 (b)(4)), the analyst should ensure that only those instruments meeting established specifications are used and that all instruments are properly calibrated. Certain analytical methods have system suitability requirements, and systems not meeting such requirements should not be used. For example, in chromatographic systems, reference standard solutions may be injected at intervals throughout chromatographic runs to measure drift, noise, and repeatability. If reference standard responses indicate that the system is not functioning properly, all of the data collected during the suspect time period should be properly identified and should not be used. The cause of the malfunction should be identified and corrected before a decision is made whether to use any data prior to the suspect period.

Before discarding test preparations or standard preparations, analysts should check the data for compliance with specifications. When unexpected results are obtained and no obvious explanation exists, test preparations should be retained and the analyst should inform the supervisor. An assessment of the accuracy of the results should be started immediately. If errors are obvious, such as the spilling of a sample solution or the incomplete transfer of a sample composite, the analyst should immediately document what happened. Analysts should not knowingly continue an analysis they expect to invalidate at a later time for an assignable cause (i.e., analyses should not be completed for the sole purpose of seeing what results can be obtained when obvious errors are known). These same responsibilities extend to analysts at contract testing laboratories.

B. Responsibilities of the Supervisor

Once an OOS result has been identified, the supervisor's assessment should be objective and timely. There should be no preconceived assumptions as to the cause of the OOS result. Data should be assessed promptly to ascertain if the results may be attributed to laboratory error, or whether the results could indicate problems in the manufacturing process. An immediate assessment could include re-examination of the actual solutions, test units, and glassware used in the original measurements and preparations, which would allow more credibility to be given to laboratory error theories.

The following steps should be taken as part of the supervisor's assessment:

1. Discuss the test method with the analyst; confirm analyst knowledge of and performance of the correct procedure.
2. Examine the raw data obtained in the analysis, including chromatograms and spectra, and identify anomalous or suspect information.
3. Confirm the performance of the instruments.
4. Determine that appropriate reference standards, solvents, reagents, and other solutions were used and that they meet quality control specifications.
5. Evaluate the performance of the testing method to ensure that it is performing according to the standard expected based on method validation data.
6. Document and preserve evidence of this assessment.

The assignment of a cause for OOS results will be greatly facilitated if the retained sample preparations are examined promptly. Hypotheses regarding what might have happened (e.g., dilution error, instrument malfunction) can be tested. Examination of the retained solutions can be performed as part of the laboratory investigation.

Examples:

- Solutions can be re-injected as part of an investigation where a transient equipment malfunction is suspected. This could occur if bubbles were introduced during an injection on a chromatographic system, which other tests indicated was performing properly. Such theories are difficult to prove. However, a reinjection can provide strong evidence that the problem should be attributed to the instrument, rather than the sample or its preparation.
- For release rate testing of certain specialized dosage forms, where possible, examination of the dosage unit tested might determine whether it was damaged in a way that affected its performance. Such damage would provide evidence to invalidate the OOS test result, and a retest would be indicated.
- Further extraction of a dosage unit can be performed to determine whether it was fully extracted during the original analysis. Incomplete extraction could invalidate the test results and should lead to questions regarding validation of the test method.

It is important that each step in the investigation be fully documented. The supervisor should ascertain not only the reliability of the individual value obtained, but also the significance these OOS results represent in the overall quality assurance program. Supervisors should be especially alert to developing trends. Laboratory error should be relatively rare. Frequent errors suggest a problem that might be due to inadequate training of analysts, poorly maintained or improperly calibrated equipment, or careless work. Whenever laboratory error is identified, the firm should determine the source of that error and take corrective action to ensure that it does not occur again. To ensure full compliance with the CGMP regulations, the manufacturer also should maintain adequate documentation of the corrective action. In summary, when clear evidence of laboratory error exists, laboratory testing results should be invalidated. When evidence of laboratory error remains unclear, a failure investigation should be conducted to determine what caused the unexpected results. It should not be assumed that failing test results are attributable to analytical error without performing and documenting an investigation. Both the initial laboratory assessment and the following failure investigation should be documented fully.

IV. Investigating OOS Test Results

When the initial assessment does not determine that laboratory error caused the OOS result and testing results appear to be accurate, a full-scale failure investigation using a predefined procedure should be conducted. The objective of such an investigation should be to identify the source of the OOS result. Varying test results could indicate problems in the manufacturing process, or result from sampling problems. Such investigations present a challenge both to employees and to management and should be given the highest priority. The investigation should be conducted by the quality control unit and should involve all other departments that could be implicated, including manufacturing, process development, maintenance, and engineering. Other potential problems should be identified and investigated.

The records and documentation of the manufacturing process should be fully investigated to determine the possible cause of the OOS results.

A. General Investigational Principles

A failure investigation should consist of a timely, thorough, and well-documented review. The written record should reflect that the following general steps have been taken.

1. The reason for the investigation has been clearly identified.
2. The manufacturing process sequences that may have caused the problem should be summarized.
3. Results of the documentation review should be provided with the assignment of actual or probable cause.
4. A review should be made to determine if the problem has occurred previously.

5. Corrective actions taken should be described. The general review should include a list of other batches and products possibly affected and any required corrective actions taken including any comments and signatures of appropriate production and quality control personnel regarding any material that may have been reprocessed after additional testing.

B. Laboratory Phase of an Investigation

A number of practices are used during the laboratory phase of an investigation. These include: (1) retesting a portion of the original sample, (2) testing a specimen from the collection of a new sample from the batch, (3) resampling testing data, and (4) using outlier testing.

1. Retesting

Part of the investigation may involve retesting of a portion of the original sample. The sample used for the retesting should be taken from the same homogeneous material that was originally collected from the lot, tested, and yielded the OOS results. For a liquid, it may be from the original unit liquid product or composite of the liquid product; for a solid it may be an additional weighing from the same sample composite that had been prepared by the analyst.

Situations where retesting is indicated include investigating testing instrument malfunctions or to identify a possible sample handling integrity problem, for example, a suspected dilution error. Generally, retesting is neither specified nor prohibited by approved applications or by the compendia. Decisions to retest should be based on the objectives of the testing and sound scientific judgment. Retesting should be performed by an analyst other than the one who performed the original test.

The CGMP regulations require the establishment of specifications, standards, sampling plans, test procedures, and other laboratory control mechanisms (§ 211.160). The establishment of such control mechanisms for examination of additional specimens for commercial or regulatory compliance testing must be in accordance with "predetermined guidelines or sampling strategies" (USP 23, General Notices and Requirements, p.9). Some firms have used a strategy of repeated testing until a passing result is obtained (testing into compliance), then disregarding the OOS results without scientific justification. Testing into compliance is objectionable under the CGMPs. The number of retests to be performed on a sample should be specified in advance by the firm in the SOP. The number may vary depending upon the variability of the particular test method employed, but should be based on scientifically sound, supportable principles. The number should not be adjusted depending on the results obtained. The firm's predetermined testing procedures should contain a point at which the testing ends and the product is evaluated. If, at this point, the results are unsatisfactory, the batch is suspect and must be rejected or held pending further investigation (§ 211.165(f)).

In the case of a clearly identified laboratory error, the retest results would substitute for the original test results. The original results should be retained, however, and an explanation recorded. This record should be initialed and dated by the involved persons and include a discussion of the error and supervisory comments.

If no laboratory or statistical errors are identified in the first test, there is no scientific basis for invalidating initial OOS results in favor of passing retest results. All test results, both passing and suspect, should be reported and considered in batch release decisions.

2. Resampling

While retesting refers to analysis of the original sample, resampling involves analyzing a specimen from the collection of a new sample from the batch. The establishment of control mechanisms for examination of additional specimens for commercial or regulatory compliance testing should be in accordance with predetermined procedures and sampling strategies (§ 211.165(c)).

In some cases, when all data have been examined, it may be concluded that the original sample was prepared improperly and was therefore not representative of the batch (§ 211.160(b)(3)). A resampling of the batch should be conducted if the investigation shows that the original sample was not representative of the batch. This would be indicated, for example, by widely varied results obtained from several aliquots of the original composite (after determining there was no error in the performance of the analysis). Resampling should be performed by the same qualified, validated methods that were used for the initial sample. However, if the investigation determines that the initial sampling method was in error, a new accurate sampling method must be developed, qualified, and documented (§§ 211.160 and 165(c)).

3. Averaging

Averaging test data can be a valid approach, but its use depends upon the sample and its purpose. For example, in an optical rotation test, several discrete measurements are averaged to determine the optical rotation for a sample, and this average is reported as the test result. If the sample can be assumed to be homogeneous (i.e., an individual sample preparation designed to be homogeneous), using averages can provide a more accurate result. In the case of microbiological assays, the USP prefers the use of averages because of the innate variability of the biological test system.

Reliance on averages has the disadvantage of hiding variability among individual test results. For this reason, unless averaging is specified by the test method or adequate written investigation procedures, all individual test results should be reported. In some cases, a statistical treatment of the variability of results should be reported. For example, in a test for dosage form content uniformity, the standard deviation (or relative standard deviation) is also reported.

Averaging also can conceal variations in the different portions of the sample. For example, the use of averages is inappropriate when performing powder blend/mixture uniformity or dosage form content uniformity determinations. In these cases, the testing is intended to measure variability within the product, and the individual results should be reported. It should be noted that a test might consist of replicates to arrive at a result. For instance, an HPLC assay result may be determined by averaging the peak responses from a number of consecutive, replicate injections from the same preparation (usually 2 or 3). The assay result would be calculated using the peak response average. This determination is considered one test and one result. This is a distinct difference from the analysis of different portions from a lot, intended to determine variability within the lot. The use of replicates should be included in the written, approved, test methodology. Unexpected variation in replicate determinations should trigger investigation and documentation requirements (21 CFR 211.192).

In some cases, a series of assay results may be a part of the test procedure. If some of the results are OOS and some are within specification and all are within the documented variation of the method, the passing results should be given no more credence than the failing results, in the absence of documented evidence that analytical error had occurred.

Relying on test data averaging in such a case can be particularly misleading. For example, in an assay with a given range of 90 to 110 percent, test results of 89 percent, 89 percent, and 92 percent would produce an average of 90 percent even though two of the assay values represent failing results.

To use averaged results for assay reporting, all test results should conform to specifications. Although the above average of 90 percent may be useful in terms of an overall assessment of process capabilities, the individual assay results indicate nonconformance because two of the three results are outside of the range. A low assay value should also trigger concerns that the batch was not formulated properly because the batch must be formulated with the intent to provide not less than 100 percent of the labeled or established amount of active ingredient (21 CFR 211.101(a)). The above example does not necessarily require the manufacturer to fail the batch, but indicates that an immediate investigation should be conducted for batch disposition decisions.

4. Outlier Tests

The CGMP regulations require that statistically valid quality control criteria include appropriate acceptance and/or rejection levels (§ 211.165(d)). On rare occasions, a value may be obtained that is markedly different from the others in a series obtained using a validated method. Such a value may qualify as a statistical outlier. An outlier may result from a deviation from prescribed test methods, or it may be the result of variability in the sample. It should never be assumed that the reason for an outlier is error in the testing procedure, rather than inherent variability in the sample being tested.

Outlier testing is a statistical procedure for identifying from an array those data that are extreme. The possible use of outlier tests should be determined in advance. This should be written into SOPs for data interpretation and be well documented. The SOPs should include the specific outlier test to be applied with relevant parameters specified in advance. The SOPs should specify the minimum number of results required to obtain a statistically significant assessment from the specified outlier test.

For biological assays having a high variability, an outlier test may be an appropriate statistical analysis to identify those results that are statistically extreme observations. The USP describes outlier tests in the section on Design and Analysis of Biological Assays (USP 23, p. 1705). In these cases, the outlier observation is omitted from calculations. The USP also states that "arbitrary rejection *or* retention of an apparently aberrant response can be a serious source of bias…the rejection of observations solely on the basis of their relative magnitudes is a procedure to be used sparingly" (USP 23, p. 1705).

For validated chemical tests with relatively small variance, and if the sample being tested can be considered homogeneous (for example, an assay of composited dosage form to determine strength), an outlier test is only a statistical analysis of the data obtained from testing and retesting. It will not identify the cause of an extreme observation and, therefore, should not be used to invalidate the data. An outlier test may be useful as part of the evaluation of the significance of that result for batch evaluation, along with other data. Outlier tests have no applicability in cases where the variability in the product is what is being assessed, such as for content uniformity, dissolution, or release rate determinations. In these applications, a value perceived to be an outlier may in fact be an accurate result of a nonuniform product.

V. Concluding the Investigation

To conclude the investigation, the results should be evaluated, the batch quality should be determined, and a release decision should be made. The SOPs should be followed in arriving at this point. Once a batch has been rejected, there is no limit to further testing to determine the cause of the failure so that a corrective action can be taken.

A. Interpretation of Investigation Results

An OOS result does not necessarily mean the subject batch fails and must be rejected. The OOS result should be investigated, and the findings of the investigation, including retest results, should be interpreted to evaluate the batch and reach a decision regarding release or rejection (§ 211.165). In those instances where an investigation has revealed a cause, and the suspect result is invalidated, the result should not used to evaluate the quality of the batch or lot. Invalidation of a discrete test result may be done only upon the observation and documentation of a test event that can reasonably be determined to have caused the OOS result. In those cases where the investigation indicates an OOS result is caused by a factor affecting the batch quality (i.e., an OOS result is confirmed), the result should be used in evaluating the quality of the batch or lot. A confirmed OOS result indicates that the batch does not meet established standards or specifications and should result in the batch's rejection, in accordance with § 211.165(f), and proper disposal.

For inconclusive investigations — in cases where an investigation (1) does not reveal a cause for the OOS test result and (2) does not confirm OOS result — the OOS result should be retained in the record and given full consideration in the batch or lot disposition decision.

Statistical treatments of data should not be used to invalidate a discrete chemical test result. In very rare occasions and only after a full investigation has failed to reveal the cause of the OOS result, a statistical analysis may be valuable as one assessment of the probability of the OOS result as discordant, and for providing perspective on the result in the overall evaluation of the quality of the batch. Records must be kept of complete data derived from all tests performed to ensure compliance with established specifications and standards (21 CFR 211.194).

B. Reporting

For those products that are the subject of applications, regulations require submitting within three working days a field alert report (FAR) of information concerning any failure of a distributed batch to meet any of the specifications established in an application (21 CFR 314.81(b)(1)(ii)). OOS test results not invalidated on distributed batches or lots for this class of products are considered to be one kind of "information concerning any failure" described in this regulation. This includes OOS results that are considered to be discordant and of low value in batch quality evaluation. In these cases, an FAR should be submitted.

12

RETURNED AND SALVAGED DRUG PRODUCTS

US CGMPS

Subpart K — Returned and Salvaged Drug Products

Sec. 211.204 Returned drug products

Returned drug products shall be identified as such and held. If the conditions under which returned drug products have been held, stored, or shipped before or during their return, or if the condition of the drug product, its container, carton, or labeling, as a result of storage or shipping, casts doubt on the safety, identity, strength, quality or purity of the drug product, the returned drug product shall be destroyed unless examination, testing, or other investigations prove the drug product meets appropriate standards of safety, identity, strength, quality, or purity. A drug product may be reprocessed provided the subsequent drug product meets appropriate standards, specifications, and characteristics. Records of returned drug products shall be maintained and shall include the name and label potency of the drug product dosage form, lot number (or control number or batch number), reason for the return, quantity returned, date of disposition, and ultimate disposition of the returned drug product. If the reason for a drug product being returned implicates associated batches, an appropriate investigation shall be conducted in accordance with the requirements of Sec. 211.192. Procedures for the holding, testing, and reprocessing of returned drug products shall be in writing and shall be followed.

Sec. 211.208 Drug product salvaging

Drug products that have been subjected to improper storage conditions including extremes in temperature, humidity, smoke, fumes, pressure, age, or radiation due to natural disasters, fires, accidents, or equipment failures shall not be salvaged and returned to the marketplace. Whenever there is a question whether drug products have been subjected to such conditions, salvaging operations may be conducted only if there is (a) evidence from laboratory tests and assays (including animal feeding studies where applicable) that the drug products meet all applicable standards of identity, strength, quality, and purity and (b) evidence from inspection of the premises that the drug products and their associated packaging were not subjected to improper storage conditions as a result of the disaster or accident. Organoleptic examinations shall be acceptable only as supplemental evidence that the drug products meet appropriate standards of identity, strength, quality, and purity. Records including name, lot number, and disposition shall be maintained for drug products subject to this section.

EC GMP GUIDE

Chapter 3 Premises and Equipment

3.23 Segregated areas should be provided for the storage of rejected, recalled, or returned materials.

Chapter 5 Production

Rejected, recovered and returned materials

5.61 Rejected materials and products should be clearly marked as such and stored separately in restricted areas. They should either be returned to the suppliers or, where appropriate, reprocessed or destroyed. Whatever action is taken should be approved and recorded by authorised personnel.

5.62 The reprocessing of rejected products should be exceptional. It is only permitted if the quality of the final product is not affected, if the specifications are met and if it is done in accordance with a defined and authorised procedure after evaluation of the risks involved. Record should be kept of the reprocessing.

5.63 The recovery of all or part of earlier batches which conform to the required quality by incorporation into a batch of the same product at a defined stage of manufacture should be authorised beforehand. This recovery should be carried out in accordance with a defined procedure after evaluation of the risks involved, including any possible effect on shelf life. The recovery should be recorded.

5.64 The need for additional testing of any finished product which has been reprocessed, or into which a recovered product has been incorporated, should be considered by the Quality Control Department.

5.65 Products returned from the market and which have left the control of the manufacturer should be destroyed unless without doubt their quality is satisfactory; they may be considered for re-sale, re-labelling or recovery in a subsequent batch only after they have been critically assessed by the Quality Control Department in accordance with a written procedure. The nature of the product, any special storage conditions it requires, its condition and history, and the time elapsed since it was issued should all be taken into account in this assessment. Where any doubt arises over the quality of the product, it should not be considered suitable for re-issue or re-use, although basic chemical reprocessing to recover active ingredient may be possible. Any action taken should be appropriately recorded.

Sec. 211.208 of the US cGMPs is concerned with the treatment of products returned from the market (for whatever reason) and with the possibility of salvaging products that may have been damaged by improper storage, or a variety of extreme conditions. Whether this applies to products that have suffered such conditions within or outside the company, or both, is not made clear.

Both the two regulatory documents are cautionary, but perhaps the EC GMP Guide is more strongly so. It stresses that "rejected products" (and it is implicit that this means products that have been manufactured in-house and have failed to comply with specification when tested) and "products returned from the market and which have left the control of the manufacturer" should not be considered suitable for reissue, unless there can be no doubt about their quality, "although basic chemical reprocessing to recover active ingredient may be possible." It hardly needs to be added that, in the circumstances described, it is very difficult to be entirely free of doubts over quality.

It needs to be remembered that:

a. In the case of a batch of product, rejected in-house, any attempt to effect a recovery of such a batch (for example, by adding it to a subsequent batch or batches) must be regarded as a significant deviation from the manufacturing process as given in the original registration submission. It could well alter the bioavailabilty characteristics of the product. The same would apply to grinding, regranulating, and recompressing a batch of tablets found to be too hard or too soft. This could result in a tablet with very different disintegration, dissolution, and absorption characteristics.

b. There is very little chance of being certain that a product that has left the control of a manufacturer has not suffered some inadvertent, or even intentional, damage. The small economic advantage that *might* accrue from redistributing returned goods is hardly worth the risk reissuing a dangerously defective product. Laboratory testing is unlikely to detect a small proportion of units that have been dangerously tampered with.

Thus, any form of recovery or reworking of products or materials that have been rejected, or which have been subjected to conditions that may have affected their quality, or which have been returned from the market or from wholesalers, should only be undertaken after the most careful consideration of all relevant events and circumstances. The key question to be constantly asked, and answered, is "can we be absolutely sure that the product resulting from the proposed recovery or rework will comply, in every respect, to the relevant standards of quality, purity, and safety?" Unless the answer is an unqualified "yes," then the rejected, returned, or questionable product should be destroyed, and any plans to recover or rework should be abandoned.

In the event that any recovery or rework *is* undertaken, the reasons and authorization for so doing, the rework or recovery process adopted, and the grounds for considering that the resultant product complies with all relevantly applicable quality standards must be carefully and thoroughly documented.

13

STERILE PRODUCTS MANUFACTURE — BASIC PRINCIPLES

Perhaps a little surprisingly, the US cGMPs do not have a section, or subsection on specific cGMPs for sterile products manufacture. There are no more than a few references, made almost in passing, in other contexts. For example:

US CGMPS

Subpart C — Buildings and Facilities

Sec. 211.42 Design and construction features

...

(c) Operations shall be performed within specifically defined areas of adequate size. There shall be separate or defined areas or such other control systems for the firm's operations as are necessary to prevent contamination or mix-ups during the course of the following procedures:

...

(10) Aseptic processing, which includes as appropriate:
  (i) Floors, walls, and ceilings of smooth, hard surfaces that are easily cleanable;
  (ii) Temperature and humidity controls;
  (iii) An air supply filtered through high-efficiency particulate air filters under positive pressure, regardless of whether flow is laminar or nonlaminar;
  (iv) A system for monitoring environmental conditions;
  (v) A system for cleaning and disinfecting the room and equipment to produce aseptic conditions;
  (vi) A system for maintaining any equipment used to control the aseptic conditions.

Subpart I — Laboratory Controls

Sec. 211.165 Testing and release for distribution

> ...
>
> (a)   For each batch of drug product, there shall be appropriate laboratory determination of satisfactory conformance to final specifications for the drug product, including the identity and strength of each active ingredient, prior to release. Where sterility and/or pyrogen testing are conducted on specific batches of shortlived radiopharmaceuticals, such batches may be released prior to completion of sterility and/or pyrogen testing, provided such testing is completed as soon as possible.
>
> (b)   There shall be appropriate laboratory testing, as necessary, of each batch of drug product required to be free of objectionable microorganisms.

Note: In the above quote from the UScGMPs, the requirements for "design and construction features," in Sec. 211.42 9(c)(10) refer specifically, and apparently only, to facilities used for aseptic processing. It is, however, doubtful if any person or body — including the FDA — would disagree that they apply also to premises used for the manufacture of terminally sterilized sterile products.

In contrast, the EC GMP Guide devotes an entire, quite substantial Annex (1) to the Manufacture of Sterile Medicinal Products. This is shown, in Annex 1 to this chapter, in its most recently revised form (issued June 2003, operational from September 2003). It must be commented that this particular EC Annex has frequently been considered to be unsatisfactory by industrial experts, both in terms of its content, and in the imprecision and ambiguity of its expression. This recently issued revision has done nothing to ease the dissatisfaction and confusion; rather it has increased it.

In 1987, the FDA issued a Guideline on Sterile Drug Products Produced by Aseptic Processing. Recently, the FDA has issued a Draft for Comment entitled Guidance for Industry — Sterile Drug Products Produced by Aseptic Processing — Current Good Manufacturing Practice. It is stated that "this revision updates and clarifies the 1987 guidance" and that "when finalized (it) will represent ... FDA's current thinking on this topic." All involved in the manufacture of aseptically processed sterile products should ensure that they are aware of the content of this guidance document when it is finalized.

What now follows is a general discussion of the concepts, principles and practices of sterile products manufacture.

BASIC PRINCIPLES

Sterile products are very significantly different from nonsterile products. The requirement for a product to be sterile raises a whole range of manufacturing, control, and quality issues, additional to those that are relevant to nonsterile products. The same quality and GMP considerations that apply to *non*sterile products apply equally to sterile products. But the attainment and maintenance of the sterile state imposes additional quality assuring demands. That is, the special requirements (ethical, professional, and regulatory) for sterile products manufacture are additional to, rather than separate from, those that apply to drug products in general. Before looking further at those special GMP requirements it is necessary to establish some basic concepts and principles and to look at the various methods of sterile products manufacture.

DEFINITION OF STERILITY

As a first step, it is necessary to examine what is meant by the terms "sterile" and "sterility."

"Sterility" is so central a concept in the context of products used for parenteral administration (and similarly critical applications) that it is both surprising and alarming that there has been so much variation, laxity, and confusion in the use of the word.

A definition that has been offered so many times it is now impossible to trace its origins is:

(a) A sterile product is one that is free from living microorganisms

And its equivalent:

(b) A sterile product is one that is free from viable microorganisms

It could be pointed out that these two closely related definitions would permit the presence of goldfish, tadpoles, newts, and other *macro*organisms.

Another common definition is:

(c) Sterility is the absence of organisms able to reproduce themselves

This latter definition has recently been restated, with some variation, by Agalloco and Akers[1] as:

(d) Sterility is defined as the absence of micro organisms having the ability to reproduce

Definition d would permit the presence of macroorganisms, and both c and d the presence of living microorganisms, still potentially dangerous, but which had had their reproductive processes inhibited (for example, by sublethal doses of radiation).

A totally uncompromised, and uncompromising, definition[2] is:

(e) Sterility: The complete absence of living organisms (Note: The state of sterility is an absolute — there are no degrees of sterility)

The note added to this definition might well be considered to be a pedantic redundancy, but it did serve to emphasize the *absolute,* nonrelative nature of sterility, at a time (1983) when there was still a not uncommon belief that it was possible that things could be almost sterile, or nearly sterile.

There have, indeed, been those who seem to find such an absolute definition *too* uncompromising — and who prefer a concept of a state of near or almost sterility, where apparently the odd organism, here and there, is acceptable.

The first official statement of this type appeared in an amendment to the Nordic Pharmacopoeia in 1970

> Sterile drugs must be prepared and sterilized under conditions which aim at such a result that in one million units there will be no more than one living micro -organism.

This same broad general line was also followed by the United States Pharmacopoeia (USP). For example, USP XXI declared:

> (It is)… generally accepted that…injectable articles or… devices purporting to be sterile…(when autoclaved) attain a 10^{-6} microbial survivor probability, i.e., assurance of less than one chance in a million that viable organisms are present in the sterilized article or dosage form..

The British Pharmacopoeia (BP) in 1988 adopted a similar (but only an *approximately* similar) stance in considering sterility to be:

> …a theoretical level of not more than one living micro-organism in 10^6 containers in the final product.

This concept of a "theoretical" level of microorganisms is a strange one, for surely here, of all places, we should be thinking of what is, or is not, *really* (not "theoretically") present. It also needs to be noted that there are subtle but very real differences between these two pharmacopoeial definitions. The USP considered sterility to be a less-than-one-in-a-million chance that one (or any number?) of containers, are contaminated with one or any number of organisms, whereas BP (in 1988) permitted, to qualify as sterile no more than one organism in only one out of a million containers. The potential problems (for patients) need to be considered before any unquestioning acceptance of such definitions. Consider, for example, a Large Volume Parenteral infusion (LVP). As real life events have tragically illustrated,[3] in such products, even normally nonpathogenic organisms can kill. In many such products, one organism today can become many millions tomorrow. More than 100 million units of LVP solutions are administered annually throughout the world. Is it acceptable that one in every million of those may contain organisms? If it is, then logically we must also be prepared to accept 100-plus unnecessarily dead patients per year.

Although more recent editions of USP and BP have evidenced a degree of backing away from the "less-than-one-in-a-million" concept, possibly as a result of a dawning realization of the potentially serious practical consequences of adopting such a position, it is unfortunately true that those earlier, questionable concepts have tended to linger in the general consciousness.

The current edition of the USP (2000) states:

> Within the strictest definition of sterility, a specimen would be deemed sterile only when there is complete absence of viable organisms from it. However, this absolute definition cannot currently be applied to finished compendial articles because of limitations in testing. Absolute sterility cannot be practically demonstrated… The sterility of a lot purported to be sterile is therefore defined in probability terms, where the likelihood of a contaminated unit or article is acceptably remote.

What is, in fact, "acceptably remote" is not explicitly stated in this passage. It is perhaps implicit in a later comment in the same volume:

> It is generally accepted that terminally sterilized injectable articles or critical devices purporting to be sterile when processed in the autoclave attain a 10^{-6} microbial survivor level, i.e., assurance of less than one chance in one million that viable micro-organisms are present in the sterilized article or dosage form. With heat-stable articles, the approach often is to considerably exceed the critical time necessary to achieve the 10^{-6} microbial survivor probability (over-kill). However, with an article where an extensive heat exposure may have a damaging effect, it may not be feasible to employ the overkill approach. In this latter instance the development of the sterilization cycle depends heavily on knowledge of the microbial burden of the product, based on examination over a suitable time period of a substantial number of lots of the pre-sterilized product.

This very last clause is distinctly questionable, in suggesting as it does the idea that a sort of "mean pre-sterilization bioburden" figure should be established by the examination of a finite number of batches over a finite period of time. In fact, the sound approach is to determine and record the bioburden in/on each lot, or load, *each and every time*, immediately prior to the sterilization process, in order to ensure that it complies with a pre-established limit. It is also to be noted that the entire latter compendial statement applies only to "articles … *processed in the autoclave*." (Author's emphasis)

The current BP (2002) unequivocally states "sterility is the absence of viable organisms." However, it adds later in the same appendix (XVIII):

> The achievement of sterility within any one item in a population submitted to a sterilization process cannot be guaranteed, nor can it be demonstrated…

It is also worthy of note that BP 2002, Appendix XVIII also, crucially, states:

> Wherever possible a process in which the product is sterilized in its final container (terminal sterilization) is chosen.

One of the more extreme and perhaps most potentially unsafe views on what constitutes sterility appears in the statement:[4]

> Out of a batch of one million units only one container may contain an organism (Statistically maximally 3 at a 95% confidence level)."[4]

Of all the "quantitative/probability" definitions of sterility, the one that perhaps gives the most comfort (if it means what it says) is the one implied in an FDA Compliance Guideline,[5] which requires, as one of a number of preconditions for parametric release, that the sterilization process has been:

> …Validated to achieve…. bioburden reduction to 10^{0}, with a minimum safety factor of an additional six logarithm reduction.

What this *seems* to be saying (and one hopes that this is a correct interpretation) is that, to be considered satisfactory, a sterilization process must be designed and validated to achieve a reduction of the bioburden in the entire load to no more than 1 (= 10^0), with an additional overkill factor of 10^6. This would seem to be an entirely different proposition to permitting an organism (or organisms) in one out of every million containers.

The "European" definition[6] presents a paradox:

> *Sterility* is the absence of living organisms. The conditions of the sterility test are given in the European Pharmacopoeia.

With the first clause, there can be no argument. It is then totally undermined by the immediately following implication that the "sterility test" has relevance to the establishment of a state of sterility throughout a batch of product, which is complete nonsense, and dangerous nonsense at that.

Most, if not all, of the statements on sterility so far appear to be based on an assumption, explicit or implied, that the method used to achieve that state is some form of heat-sterilization process. A further FDA Guide[7] gives the impression that its author believes that moist heat is the *only* method:

> *Sterilization*: The use of steam and pressure to kill any bacteria that may be able to contaminate that environment or vessel.

This definition may also be criticized on a number of other counts, not least the implications that the pressure kills bacteria, that sterilization can only be achieved by *killing* bacteria, and that bacteria are the only organisms that must be destroyed or removed.

There are, of course, methods of sterilization other than by steam. How has sterility been defined in relation to these? What, for example, of filtration with subsequent aseptic filling? An early official statement on this subject appeared in a WHO document[8] published in 1973. It reads:

> The operations where liquid preparations are filled should be checked. This may be done, e.g., at least twice a year, by filling not less than 1000 containers with nutrient medium.... and incubating. If the containers... show a contamination rate above 0.3% some countries do not consider the procedure acceptable. (WHO, Sterility of Biologicals, 1973)

Space does not permit an analysis of the dire statistical implications for human morbidity of this statement, which established, in "official" terms, the concept of the "media fill," or "process simulation test." One can only hope that one never requires an injection in any of those countries (if they still exist) that *do* consider that a contamination rate above 0.3% is acceptable.

Nevertheless, the US Parenteral Drug Association's Monograph No. 2 on Validation of Aseptic Filling for Solution Drug Products[9] commented that the WHO 1973 level of 0.3% is "widely accepted," but "a manufacturer should strive for a contamination level of less than 0.1%."

This last document has been replaced by a PDA Technical Report No. 22 on Process Simulation Testing of Aseptically Filled Products,[10] where under Interpretation of Results and Acceptance Criteria it is stated that "... the ultimate goal for the number of positives in any process simulation test should be zero. A sterile product is, after all, one which contains no viable organisms." The report adds, "there are, however, numerous technical problems in achieving this goal," and "the selection of acceptance criteria for aseptic processing validation is the central issue to be resolved in the conduct of process simulation tests." Disappointingly, or perhaps inevitably, the report then offers little, if anything, in the way of quantitative advice on how to solve the "technical problems," or to resolve "the central issue."

Thus, in the field of aseptic processing, positions seem to shift from considerations of acceptable contamination levels of 3 in 1000, to 1 in 1000, to more recently expressed views that the target should be no more than 1 in 10,000. And these aims have to be compared with the 1 in a million, for heat sterilization processes, which appears to be acceptable, in some form or other, to some authorities. Is sterility therefore to be considered as a moving target?

Sufficient examples have been given, to demonstrate a range of variable (and indeed, often ambiguous) views on what is meant by the word "sterility." In so many statements on the subject, there is an element of compromise. Sterility tends to be regarded by some, very wrongly, as a *conditional*, rather than an *absolute* state.

What is the problem? Why this indecision and ambiguity over so fundamental an issue? The answer is a very simple one. It is that there is a *fundamental flaw* at the heart of much thinking and writing about sterility, and that flaw resides in confusion between the nature of a concept or a state, and the probability of the existence of that state.

The point was well made, in 1977, in a paper by Brown and Gilbert[11]:

> The concept of sterility is absolute. Whether or not a product is sterile is inevitably a matter of probability.

If this distinction were universally noted and adopted, the problem would cease to exist. The *only possible* definition of sterility is the uncompromised, unconditional, and absolute one given in the original UK GMP Guide.[2] The question of the existence of such an absolute, negative state must, inevitably, be a matter of probability, not of absolute certainty. But there is nothing odd or new about that. We have long been aware that we do not live in a grand, simple, predictable, deterministic, Newtonian universe. We inhabit a quantum uncertainty, *probabilistic* universe (or, at least for the present we think we do). In such a universe, the question of the existence of the absolute state of sterility is, of course, a matter of probability, just as the existence (or chances of happening) of any other state, event, thing or occurrence is a matter of probability (not certainty).

This does not preclude our *aiming* to achieve this (or any other) absolute state. In the case of *sterility*, our concern should be over whether we have in fact achieved that state at an acceptable level of probability, and it is not unreasonable to suggest that what may be regarded as an acceptable level of probability could well be considered to vary according to circumstances.

Compare, for example, two different types of terminally heat-sterilized product:

a. A small volume (say, 0.5 or 1.0ml) injection of a *non*growth supporting liquid, intended for intramuscular or subcutaneous injection

And:

b. A Large Volume Parenteral (LVP), say, 1 liter of (growth supporting) Dextrose/Saline solution for intravenous infusion.

That both should be sterile, there can be no question. However, one could well consider that the *level of assurance of the probability of attainment of that state* is more critical in the latter than the former.

WHY STERILITY?

Why are some products required to be sterile? The most important, simple, and obvious, reason is that, with certain routes of administration, there are special dangers of causing serious infections if they are not.

In the case of orally administered products such as tablets, capsules, syrups, etc., the body has natural defenses against at least low levels of contamination, although it is obviously not a good idea that even these products should be contaminated with dangerous organisms. Products required to be given in, or through, more sensitive areas must be sterile.

Injected products bypass the body's natural defenses against microorganisms, and the consequences of injecting even slightly contaminated products can be very serious. Injection of products contaminated with microorganisms can cause, and *has* caused, death. The same applies to things like devices, instruments, and implants that are intended to be inserted into blood vessels, muscles, or other parts of the body.

For the same reason, products intended for use in the eye (drops, ointments, lotions), for application to wounds, sores, or broken skin (liquids, creams, ointments, or dressings), or used in surgical procedures to irrigate body cavities must be sterile. It is also usually considered best that eardrops should be sterile, and some consider that nosedrops should be sterile as well.

In addition, materials, equipment, containers, closures, etc. used in the manufacture of sterile products, and for use in some microbiological and pathological laboratory procedures, will need to be sterile.

There is thus a wide range of uses, and routes of administration, of sterile products and materials. For parenteral products alone, there are a number of different injection routes, e.g., subcutaneous, intradermal, intramuscular, intravenous, intrathecal, intra-articular, intracardial, intraperitoneal, intracisternal, peridural.

It will be apparent that parenteral routes of administration, all other things being equal, present a significantly higher level of potential patient-risk than, for example, oral administration, and the question inevitably arises, "Why inject if this is such a potentially dangerous route of administration?" There are a number of possible reasons for choosing this route. They include:

1. If the active substance in the product is destroyed when it is taken by mouth. (For example, some substances are inactivated by the digestive substances in the gut, yet retain their activity when injected into the bloodstream or into the muscles.)

2. When very rapid action is required, for example in an emergency, after injury, or in the case of a severe infection. (Action following an injection is usually much quicker than when a product is swallowed.)
3. When it is necessary to target the part of the body where the action of the medicine is required to be more accurately sited than is usually possible with products taken by mouth (for example, injection into the heart, brain or spinal canal).
4. When the patient is unable to take the medicine in any other way, for example, he or she cannot swallow or is unconscious.

STERILIZATION — FUNDAMENTAL CONCEPTS

Not infrequently there is talk of sterilization as if it were one single, discrete type of operation (sterilization is sterilization is sterilization, as it were). This is just not so. There are a number of different methods of sterilization. They include:

Heat (steam or dry heat)
Radiation (e.g., gamma ray or electron beam)
Gas (e.g., ethylene oxide)
Filtration (with subsequent aseptic handling)

These are not mere variations on a basic theme. These processes are all very different from one another. Each has its own technology, mode of action, and application. It is no exaggeration to state that the technological difference between the manufacture of sterile products using (a) steam sterilization and (b) filtration with aseptic processing is as great, if not greater, as the difference between the manufacture of tablets and the manufacture of ointments. Thoughtlessly to lump together all possible types of sterile manufacture as if they were all essentially the same is but one of a number of possible errors of judgment.

A feature that *is* common to all the types of processes is, obviously, the objective of making a product that is, in fact, sterile. Concomitantly, the products should also, in many cases, be free of particles and of pyrogens (or bacterial endotoxins). The attributes of freedom from organisms, from nonviable particles, and from pyrogenic substances may be said to be interconnected. Some microorganism species can release toxic metabolites, or endotoxins, which can render parenteral products highly dangerous to patents. From a human point of view, the most significant of these are pyrogens. They largely consist of polyliposaccharide components of gram-negative bacterial cell walls. They are relatively heat-stable and can be present when the bacteria have been destroyed or removed. They can cause acute febrile reactions when introduced directly into the bloodstream.

Another fundamental point, which is common to all types of sterilization, and which it is absolutely vital to grasp is that *sole reliance cannot be placed on the sterilization process alone, in isolation, to achieve sterility.* Much depends on:

■ The microbial condition of the materials, or articles, as they are presented to the sterilization process

■ How they are prepared and handled *before* the actual sterilization, by whom, and under what conditions
■ The preestablished validity of the sterilization process itself, and the careful control of that process during the sterilization
■ What happens after the sterilization process to confirm its efficacy and to prevent product recontamination

The achievement of sterility requires the application of disciplines and techniques additional to those required in the manufacture of other products. As has been noted, the requirement that a product should be sterile imposes additional, not merely alternative, demands.

A further fundamental issue, which bears crucially on all types of sterilization processes, is the fragile fallibility of the only end-product test available to us as weak support in the assurance of that most crucial quality characteristic — the *de facto* sterility of a purportedly sterile product. The following (from Russel et al.[12]) shows the chances of passing the standard pharmacopoeial sterility test, for various levels of contamination (statistical probability considerations only; no account was taken of microbiological limitations).:

Number of Units Contaminated	Chance of Passing Test
0.1%	98%
1.0%	82%
5.0%	36%
10.0%	12%

This is hardly confidence-inspiring, even for those who have faith in end-product testing as a determinant of quality. Compared with the sterility test, an assay (for example) to determine the content of active ingredient of a tablet by HPLC is a powerfully quality-assuring tool. The very fact of the sheer poverty of this test must color all thought and action in this field and necessitates more disciplined approaches and higher orders of care and attention. Crucial to success are:

■ The *people* involved and their *training*
■ The *premises* used and the *environmental standards* therein
■ The *equipment* and its *commissioning, cleaning*, and *sterilization*
■ The quality of the *materials* used (including *water*)
■ *Validation* of the sterilization process
■ *In-process control* of the process
■ *In-process control* of the manufacturing environment

Following the actual sterilization process, it is vital to guard against mix-up of sterilized with nonsterile product or recontamination, for example, by the air entering a sterilizer, or by water used for spray cooling a sterilized load.

METHODS OF STERILIZATION

A distinction may be drawn between two main approaches to the manufacture of sterile products — *terminal sterilization* and processes where the sterilization is not "terminal" and where some form of *aseptic processing* must follow the actual sterilization stage.

The two approaches may be compared as follows:

A terminal sterilization process usually involves:

1. Taking clean, low bioburden (but not necessarily sterile) bulk product, containers, and closures
2. Filling and sealing the product in what we may term in a generalized sense a "high quality environment"
3. Sterilizing the filled and sealed product

In contrast, aseptic production involves:

1. Taking *previously sterilized bulk* product, containers, and closures
2. Filling and sealing the product in an "*extremely* high quality environment," protecting all the while against recontamination

Thus, in comparison with terminal sterilization, aseptic processing:

1. Has more variables
2. Usually involves more than one type of sterilization process (e.g., dry heat for glass containers, steam for rubber closures, filtration for a fluid product)
3. Is very environment-sensitive
4. Is very operator-dependent

Another important distinction between terminal heat sterilization, and filtration followed by aseptic filling, and one which has perhaps not received the emphasis it requires, is that terminal heat sterilization is a singular, "once and for all," optimally controllable, batch-wise process that takes place over a finite, relatively restricted period of time. In contrast, the filtration of a liquid and filling it aseptically is a sequential, or serial, process. Although, through temperature variations within the chamber (which should, in any event, have been investigated and controlled in advance), there may be variations in heat input in different parts of the load, it is still reasonable to regard a chamber-load as one (putatively homogeneous) batch, or subbatch. A filtered and aseptically filled product cannot by any means be regarded in the same light. The filtration takes place over a relatively extended period of time. It cannot be asserted that the first runnings through the filter are exactly the same as the last. Similarly, the filling operation continues over an extended period of time — a period during which conditions can change, either through accident, inadvertence, or carelessness. Total homogeneity, throughout the run, of product and conditions, cannot be assumed. Some items, but not all, in the run may become contaminated through operator intervention, coughing or sneezing, hairy arms over open sterilized vials, and the like. This leads to a strange and inexplicable paradox. Although it may be acceptable (in the context of parametric release) to omit sterility testing when products or materials have been terminally sterilized, experts generally hold that the sterility tests should always be performed on samples of aseptically processed products. Yet, it is fundamental to the drawing of conclusions about a population, on the basis of the examination of samples drawn from that population, that it can be assumed to be homogeneous, with any faults or defects evenly or randomly distributed. It is entirely *impossible* to make such an assumption about aseptically processed products.

To recapitulate, the main methods of sterilization may be classified as follows:

Heat — steam or dry
Radiation (e.g., gamma ray or electron beam)
Gas (e.g., ethylene oxide)
Filtration (plus aseptic handling)

As already noted, these processes are all very different from one another. For example, filtration differs from all the others in that sterilization by heat, radiation, or gas *kills* organisms. Filtration *removes* them. There is a difference.

Heat Sterilization

Sterilization by heat is generally considered to be the most reliable method, and the one to choose if possible. The question thus arises — in what circumstances would it *not* be possible to use a heat-sterilization process?

The main reason for choosing not to use a heat-sterilization process is when the product or material cannot stand the heat required and would therefore break down or deteriorate.

Steam Sterilization

Of the two possible forms of heat sterilization, steam sterilization is more effective than dry heat at the same temperature. The reasons for this are the better contact (and thus heat transfer) that the steam provides, and the fact that steam has a greater heat energy content than, say, water or air at the same temperature. Steam has:

■ Sensible heat, that is the heat required to raise the temperature of a mass of water to its boiling point

And
■ Latent heat of vaporization, that is, the additional heat energy absorbed when a liquid (e.g., water) at its boiling point is converted to steam

Both these forms of heat energy are transferred from the steam to the objects being sterilized, and it is heat energy that kills the organisms. Approximately 80% of the total heat energy in saturated steam is the latent heat, and this is released when the steam makes contact with a cooler surface and condenses.

This condensation leads to an immediate contraction of the steam and a localized lower pressure region into which more steam will flow.

Steam sterilization is used, for example, to sterilize aqueous solutions in sealed containers where the steam acts as a heat transfer medium to raise the temperature of the solution to the desired sterilizing temperature. Here, it is crucial that the integrity of the container seals are assuredly validated, to avoid entry of the condensed steam into the containers. Water-wettable articles, such as empty containers, container seals, instruments, and machine and equipment parts are steam sterilized by direct contact. Here, the quality of the steam is crucial. It must be pure, or clean steam, otherwise the contact surfaces will suffer contamination by impurities in the steam. Where such items are steam sterilized, it is essential that precautions are taken to prevent recontamination after the process has been completed. This may be done, for example, by wrapping the item to be sterilized in a material that allows the removal of air and the penetration of steam, but which provides a barrier against entry of microorganisms after the sterilization. Sheets and bags of suitable wrapping material (treated paper, etc.) are available commercially. Often, two layers of the material are used (the double wrapping technique). This enables successive removal of the two layers, while, for example, the sterilized article is passed through a hatchway into a Clean Room, with the inner wrapping removed only when the sterilized item is under some form of protection against recontamination.

Steam sterilization is not a suitable method for the sterilization of sealed containers of oily solutions or suspensions. This is because the special effectiveness of steam sterilization is due to the *moist*, not dry, heat. Oily material in a sealed container may reach the temperature of the steam sterilizing chamber, but the heat will only be dry heat and that (at the temperatures usually used for steam sterilization) will not be sufficient.

It is important to note that boiling water (or steam) at normal atmospheric pressure (that is, water or steam at 100°C) will not kill all organisms. It will kill many, even most, of them. But some microorganisms are extraordinarily resistant to heat, particularly those that form spores, and can survive boiling water for long periods. Therefore, while in certain cases it may be considered safe to drink water that is only lightly contaminated, after boiling, higher temperatures are needed to ensure sterilization. To achieve these higher steam temperatures, it is of course necessary to operate under pressure, that is in autoclaves.

The British Pharmacopoeia states that "the preferred combination of temperature and time is 12°C *maintained throughout the load* (writer's emphasis) during a holding period of 15 minutes," and adds that other combinations of temperature and time may be used, provided they have been shown to achieve the desired result.

Commonly accepted as effective combinations (and the over-pressure that is required to achieve the corresponding steam temperature) are:

Temperature (°C)	Over Pressure (PS1)	Time (mins)
115–118	10	30
121–124	15	15
126–129	20	10
134–138	32	3

(It is generally considered that temperatures below 115°C do not provide a sufficient level of sterility assurance.)

Important points to note are:

■ As the temperature increases the time required reduces significantly.
■ The temperate must be achieved throughout the load for the time required. That is, for example, a temperature of 121°C must be achieved at the coldest part, of the coldest item, in the coldest section of the load, for at least 15 minutes. It is *not* sufficient that at some point, or at a few points only, this temperature is achieved for the specified time.
■ The pressure is only used to achieve the required temperature and contributes nothing to the sterilization process. It is the temperature that must be used to control and monitor the process.

Steam Sterilization: Autoclaves

Autoclaves are sealable pressure vessels, specially designed for use in the sterilization of materials, products, devices, and equipment by steam under pressure. They come in a great range of shapes and sizes, from small bench-top laboratory models, to large industrial autoclaves, free standing or inserted through walls. They may be single-ended (that is, with just one door at one end) or double-ended (that is with a door at each end). The advantage of a double-ended autoclave is that it makes for more effective segregation of sterilized from nonsterilized items, especially if one end is on one side of a wall, and the other end on the other side. Unless very great care is taken, there is always a risk with single-ended autoclaves, of mix-up between items being removed from the chamber after sterilization, and the next (unsterilized) load waiting to go in. Double-ended autoclaves also allow the direct passage of a sterilized load into a clean, or aseptic, area.

At one time, autoclave cycles were controlled manually. Today, in all but the simplest laboratory units, autoclaves (once loaded and with the doors closed) function automatically in accordance with a preset time/temperature cycle. Heat sensors within the chamber detect when the required temperature has been reached. It is essential that there is complete surety that this temperature has been reached throughout the entire load. The load is then held at the required temperature for the predetermined period of time, for example, 15 minutes at 121°C. There then follows a cooling-down period before the door(s) of the chamber can be reopened and the sterilized items unloaded. One reason for this cooling phase is obvious. The load, immediately following sterilization, will be dangerously hot. But there is another very important safety reason. When containers of fluid are being sterilized, a high pressure builds up inside the containers. They do not burst during the process because the pressure inside the containers is balanced by the pressure outside them, within the autoclave chamber. If, however, the autoclave doors are opened (thus allowing the pressure in the chamber to drop suddenly to atmospheric pressure) before the load has cooled, then there could be some very nasty explosions. Fortunately, most modern autoclaves are designed so that the doors *cannot* be opened until the cooling part of the overall cycle is completed.

Thus, there are 3 main stages to an autoclave sterilization cycle:

1. Heating up
2. Holding period
3. Cooling down

There are also a variety of pre-heat-up phases, depending on the type of autoclave, and various measures have also been introduced to speed up the cooling phase, and thus increase autoclave through-put.

Types of Autoclave

When the chamber of an autoclave contains only saturated steam under pressure, there is a direct linear relationship between the temperature of the steam and the pressure. The higher the pressure, the higher the temperature. Elementary though it may seem, it needs to be stressed that the pressure is used solely to obtain the temperature required (microorganisms are killed by the heat and are not crushed to death, as some people still seem to believe) and that the direct relationship between the temperature and the pressure does not hold if there is both steam and air, or some other gas, in the chamber. Worse still, unless special steps are taken, there is a strong possibility that in some parts of the chamber or load there will be "layers" or "pockets" of air within the chamber. Normally, since steam has a lower density than air, it is the steam that will rise to the top of the chamber. This could, and has, caused serious problems. The steam and air could well be at different temperatures, and items that are surrounded by air (and not steam) would only be subjected to *dry* heat, which is far less effective (at a given temperature, for a given time) than steam.

This *is* a serious problem. Patients have been killed due to administration of IV infusions that failed to be sterilized because of a layer of air in an autoclave.

A major feature of autoclave design concerns the measures taken to overcome the problem of air in the chamber. Various types, which tackle this problem in various different ways, include:

Downward Displacement Autoclaves When using earlier and simpler autoclave models, it is necessary to displace the air in the chamber, through a pressure-regulated drain, or vent, at the bottom of the chamber, before the sterilization cycle properly begins. This displacement is effected by allowing steam to enter at the top of the chamber for what is called a period of "free steaming." Because the steam displaces and forces the air out through the drain at the bottom, this type of autoclave has been called a "downward displacement" autoclave.

Vacuum Purged (or Assisted) Autoclaves Vacuum purged, or assisted, autoclaves are autoclaves fitted with vacuum pumps to remove the air, from the bottom of the chamber, before or as the steam enters at the top. This can have an additional useful function. At the end of a sterilization cycle, the pump can be used again to evacuate the chamber, remove the steam, and thus allow more rapid drying of the contents.

Fan Assisted Autoclaves Another way of overcoming the problem of air in an autoclave chamber is to ensure that there is a complete mixture of the steam and air. That is, it becomes a question not of removing the air, but of mixing it thoroughly. The important point is to ensure that it *is* thoroughly mixed, and that the entire load is held at the required temperature for the required time. It is thus even more critical that the sterilization cycle is controlled via internal temperature, and not pressure. Various designs of internal fans, and similar devices, have been used to ensure this mixing.

Porous Load Autoclaves In addition to the problem of layering of air and steam, there is the possibility of trapped pockets of air. These can occur, for example, within packs of dressings or inside wrapped articles, equipment, and tubing. To overcome this problem, autoclaves used for sterilizing such articles are equipped so they can run a "porous load cycle," the essential feature of which is a repeating, or "pulsing" application of the vacuum — an alternating series of pulling and releasing the vacuum before the heating begins. This vacuum pulsing can also be used to dry the load at the end of the cycle. In a typical porous load cycle, the internal chamber is first reduced to around 2.5 kPa absolute pressure. This is followed by alternating steam injection and evacuation (pulsing) to dilute and remove any residual air. When this has been removed, steam is admitted rapidly so as to reach sterilizing temperature (and pressure), which is maintained for the required time. Finally, there is a rapid evacuation to around 6–7 kPa absolute, to remove the steam and dry the load. Air is then admitted to the chamber through a bacteria-retaining filter.

Air Ballasted Autoclaves When plastics containers (bags, bottles, vials, ampoules) are sterilized in an autoclave, at the temperatures required for sterilization, such containers soften and can distort. When the steam pressure drops at the end of the sterilization phase, the internal pressure can cause the containers to "balloon" or even burst. Some autoclaves are designed to compensate for this, by pumping in air under pressure. This over-pressure within the autoclave chamber prevents the ballooning, etc., and can be held until the temperature has dropped below the softening point of the plastic, when the over-pressure can be reduced. In such autoclaves, initial air removal is usually unnecessary, but a good circulating fan system is essential to ensure thorough mixing of steam and air.

Spray Cooled Autoclaves As a means of reducing the time taken for the load to cool down (and thus increase through-put, and also reduce the possible deleterious effects that prolonged heat may have on products or materials) a number of autoclaves are fitted with a rapid or spray-cooling facility. Following the heating phase, the load is subjected to a fine spray or mist of cold water. This can reduce the time taken to cool a load from a matter of hours to a matter of minutes.

Pumping air into an autoclave chamber for ballasting and spraying water in for cooling can give rise to further potential contamination problems, as can the air that will also enter as the autoclave cools and when the doors are opened. When sterilizing liquid products in well-sealed containers, and if there is no concern about the sterility of the *outsides* of the containers, then there is no problem, *if* there is absolute surety that there is no chance the water or air can enter the cooling container. If there is not this assurance (and serious infections have been caused by contaminated cooling water penetrating faulty vial or bottle seals), or when attempting to sterilize the surfaces of any items that can come into contact with the entering air or water, then this air or water must itself be sterile (and, of course, free from chemical impurities or additives).

Continuous Sterilizers   The need for the large-scale, cost-effective industrial production of sterile fluids has led to the development of continuous sterilizers. These take the form of substantial tower-like structures up to 20 meters or more high. They typically contain three interconnected chambers: a water-filled preheating chamber, a central steam sterilization chamber, and a water-cooling chamber. The hydrostatic pressure in the preheating and cooling chambers seals and counter-balances the steam pressure in the sterilizing chamber. Filled containers move through the chambers on an ascending and descending conveyer belt system. The holding time in each chamber is governed by the speed of the conveyor belt. Sequential spray cooling and drying may also be incorporated in the system.

These different types of autoclave are not all mutually exclusive, and any one autoclave can have more than one of the features discussed. For example, a particular autoclave can be vacuum purged, air ballasted, and spray cooled and fitted to run a porous load cycle if required. It is then possible to select the combination of features required for a specific sterilization process.

Steam Sterilization: Sterilization in Place (SIP)

A traditional method for the connecting of vessels, pipework, tubing, and filter assemblies has been to separately sterilize each component part and then to assemble them together using aseptic technique, that is by making "aseptic connections." This sort of approach is not all that practical in large-scale high-through-put production, and it introduces a further element of risk of contamination when the connections are made. A more recent, practical approach is to "Sterilize in Place" (SIP). To apply this technique, it is first necessary to design and install all the manufacturing equipment, pipework, etc. so that it is possible to use SIP — sealable stainless steel vessels, capable of withstanding steam pressure, stainless steel pipework, all connected -up by sanitary fittings and so on. The complete assembly (e.g., sealed pressure mixing tank with stirrer, connected to pump, connected to filter assembly, connected to filling machine, etc.) is assembled in a clean, but not necessarily sterile, state, and the whole is sterilized internally by feeding in steam under pressure, at the required temperature for the required time. The entire system must be designed so it is possible to do this, with entry point(s) for the steam, drainage points for condensate, and places to insert temperature sensing probes. SIP as a "bolt-on modification" is neither a practical nor sensible proposition.

Dry Heat

This is a process used in, for example, hot air ovens or sterilizing tunnels.

Because dry heat is less effective than steam, higher temperatures and longer exposure times are required. Again, it is vital that *all* parts of *all* items being sterilized reach at least the required temperature for at least the required time.

Generally accepted as effective time/temperature combinations for dry heat sterilization are:

Temperature (°C)	Time (minutes)
180	30
170	60
160	120

Other time/temperature combinations are possible and acceptable if they are demonstrably capable of achieving the desired and intended effect.

Dry heat sterilization is used for articles that are thermostable, but that are either moisture sensitive or impermeable to steam. It is used for equipment, containers (e.g., glass) and other packaging components, dry chemical substances (i.e., BPCs or APIs that can stand the heat), and for nonaqueous solutions and suspensions. Oils, fats, waxes, and silicone lubricants can be sterilized by dry heat. Dry heat thus can be used to sterilize oily injections, implants, ointment bases, certain surgical dressings, and absorbable gelatin sponges. Dry heat temperatures of 250°C and above can be used to both sterilize and depyrogenate glass containers and other glassware.

The most common form of dry heat sterilizer is the *hot air oven*. These are usually constructed of stainless steel, with electrical heating elements placed around the internal wall of the insulated chamber, and arranged so as to minimize localized overheating. Heat is delivered to the load mainly by convection and radiation, and fans are fitted for efficient internal air circulation and optimum heat distribution. To this end, shelves within the chamber are constructed of perforated steel or of steel mesh. Internal temperature is thermostatically controlled, and monitored by thermocouples.

Overall cycle times in hot air ovens can be lengthy, and when, say, glass vials are being sterilized for aseptic filling, the vial-sterilization process is decoupled from the filling process. The filling phase must wait for the availability of a batch of sterilized vials. There is also the contamination-risking handling that is necessary in transporting the vials from the oven to the filling line. These problems, of course, can be, and are, regularly overcome. A more efficient and effective, solution (where production volume justifies the cost) is the on-line sterilization tunnel, which has the additional advantage that it can also depyrogenate.

Clean vials are loaded onto the conveyor, which passes through the tunnel from outside the aseptic filling room. The conveyor belt transports the vials through the tunnel, where they are exposed to high temperature, sterile-filtered air. Before they leave the tunnel, and onto the aseptic filling line, the vials are cooled rapidly by a sterile filtered air flow. Some designs utilize internal infrared heaters in addition to the filtered air flows. Differential pressures, and the internal air flows prevent entrance of contamination into the filling room, from the entrance of the tunnel.

Radiation Sterilization

Radiation sterilization can be used to sterilize materials that are heat sensitive but able to withstand the relatively high radiation levels required. The most common forms of radiation employed are gamma rays and accelerated electron beams. Other forms of radiation have been suggested, for example ultraviolet light, which cannot be recommended for product sterilization due to its poor penetrating power and its extensive absorption by glass, plastics, particles, and turbid liquids. Recently the use of pulsed "pure bright light" as a sterilant has been proposed. Initial results look very promising, but more independent work needs to be done (Dunn[13]). The use of high frequency microwave continuous sterilizers has also been proposed (Ebara et al.[14]).

The usual source of gamma radiation is colbalt-60, although some irradiation plants use caesium-137. Electron beams are generated by high-energy electron accelerators. Radiation sterilization is not the sort of process that is usually carried out in a standard pharmaceutical, or device, manufacturing facility. Specialized plants, with elaborate precautionary systems, are necessary to protect both operators and the environment. Thus radiation sterilization tends very largely to be an operation contracted out by manufacturers. Radiation sterilization has been performed as both a batch and a continuous process. The most usually preferred method is a continuous process, in which a conveyor system takes the individual items (containers, sealed cartons, bags, etc.) through the irradiation chamber in such a manner as to ensure that the orientation of the load in relation to the source varies so all parts of the load receive the same dose. The dose delivered depends upon the strength of the source, the distance from the source, the resistance of the material between the articles to be sterilized and the source, and the time of exposure. Exposure time must be adjusted to allow for any decay of the source. In the UK and in most of Europe, a minimum absorbed dose of 25 kGy has normally been accepted as the standard for radiation sterilization. However, as an interesting example of discordance within a "harmonized" community, in Scandinavia doses of up to 35 kGy have been specified. In the US, the Association for the Advancement of Medical Instrumentation has developed dose-setting guidelines based upon the extent and resistance of the presterilization bioburden present in or on the load.

It is necessary to monitor the dose received during radiation sterilization. To this end, dosimeters (usually of the "red perspex" type) are used in sufficient number, and in packages sufficiently close to one another, so as to ensure that in a continuous process there will always be at least two dosimeters in the chamber, exposed to the source. Biological indicators may also be used as an additional control measure.

Sterilization by irradiation can be very effective, especially since it can be used to sterilize products and materials that are already packaged, provided the radiation can penetrate the package. (Radiation is not used to sterilize just the surface of a product, material, device, or article, given that the radiation is able to penetrate that product etc.) It thus provides a means of terminal sterilization without the use of heat, but it needs to be recognized that it can have serious deleterious effects on a number of products, materials, compounds, and containers. Further, expensive and complex plant and equipment is required, such as is usually not found in a normal pharmaceutical or device manufacturing factory.

Some materials, such as natural rubber, styrene (and styrene derivatives, such as ABS), polyethylene, polycarbonate, polysulphones, silicones, cellulose compounds, and nylon show little if any change after gamma irradiation, but many materials can be adversely affected. Some glasses, and plastics (such as polyvinyl chloride, polytetrafluoroethylene, and polypropylene) may be discolored, and this discolorization my continue after the sterilization process has finished. Gas may be liberated; for example, hydrogen chloride from polyvinyl chloride. Mechanical properties (such as increased brittleness or hardness) may be altered. Many chemical compounds can be degraded by having their molecules rearranged, and this effect is greatest in the presence of water. This severely limits the use of radiation for the sterilization of aqueous solutions. However, radiation has been used successfully for the sterilization of a wide range of pharmaceutical products and materials, including enzymes, vitamins, minerals, antibiotics, monoclonal antibodies, and peptides. It can also be used to sterilize some containers and closures. Plasma, tissue, and bone grafts can be sterilized by gamma irradiation. Perhaps, in sheer volume, the major use of gamma radiation is in the sterilization of medical equipment devices, including surgical gowns, hoods and masks, dressings, catheters, syringes, needles, surgical blades, prosthetic implants, and the like.

Gas Sterilization

Many chemical substances are toxic to microorganisms, but only a very limited number of them have any use as sterilizing agents (as distinct from the considerable range of chemical compounds that can be effectively used as disinfectants).

While other substances, in a gaseous state, have been proposed and tried (for example, formaldehyde, propylene oxide, vapor phase hydrogen peroxide), in practice the substance that has been by far the most widely used as a gas sterilant is ethylene oxide (commonly, and unscientifically, abbreviated to "EtO").

Ethylene oxide can only be used to sterilize surfaces. That is, unlike heat or radiation it cannot penetrate through the walls of many containers to the product inside. (It is therefore not possible, for example, to sterilize sealed vials or ampoules of liquid products by EtO, although it will penetrate certain plastic films and bags).

Among the *dis*advantages of ethylene oxide is that it is highly explosive (it is usually diluted with an inert gas, such as a fluorinated hydrocarbon, or carbon dioxide) and toxic, both acutely and chronically. Acute toxic reactions resulting from inhalation include headache, nausea, vomiting, and respiratory and conjunctival irritation. Contact with the skin causes burns and sensitization. Chronic exposure can result in neurological, ocular, and hematological reactions. Ethylene oxide is mutagenic and a suspected human carcinogen. For these and other reasons (for example, the dependence for the success of this method on a number of critical operational parameters, all of which need to be most carefully controlled and monitored), its use is declining. Certainly, where possible, other methods are preferred. Factors that affect the efficacy of ethylene oxide as a sterilant include:

- Concentration of the gas
- Temperature
- Gas pressure
- Humidity (some moisture must be present, but with an excess, activity declines)
- Time of exposure
- Gas distribution and penetration

All of the above must be carefully controlled and monitored.

Failure to attain the required conditions in relation to any single process parameter can result in failure to sterilize the load, in whole or in part. For this reason, it is generally considered necessary to include biological indicators in each load. Furthermore, because of the complex interrelationships between these various factors, there is no universally accepted standard cycle for ethylene oxide sterilization. Cycles used have included ethylene oxide gas concentrations of 250mg/liter to 1500mg/liter, relative humidity from 30 to 90%, temperatures from 30 to 65°C, and exposure times from 1 to 30 hours. Thus, a sterilization cycle must be designed, developed, and validated for each specific product and product load configuration.

Ethylene oxide sterilization is carried out in a purpose-built ethylene oxide sterilizer, with a gas-tight jacketed stainless steel chamber, built to withstand high pressure and vacuum. The product or material is loaded into the chamber in a predefined configuration. This configuration must remain consistently the same as the configuration adopted when the process was validated, otherwise the validation can no longer be considered valid. A vacuum is drawn on the chamber to about 2 kPa, to remove air and thus to facilitate later ethylene oxide penetration. Pulses of steam then enter the chamber to moisturize the load, and to raise it to the required sterilization temperature. Ethylene gas is then admitted through a heated vaporizer. The gas is circulated within the chamber, using internal fans or external recirculation loops, to ensure even internal gas distribution. At the end of the exposure time, the gas mixture is exhausted from the chamber and rendered harmless by acid hydrolysis or catalytic oxidation.

A further problem is the reaction products left as residues after the sterilization. These include such toxic substances as ethylene chlorohydrin and ethylene glycol. Thus, the load must be held, under controlled airflow and temperature conditions, for a further degassing period of up to 10 days. Precautions are, obviously, necessary to prevent microbial recontamination of the load.

Ethylene oxide is mainly used to sterilize dressings, catheters, infusion and giving sets, syringes, prostheses, and similar devices. It can also be used to sterilize some plastic containers and closures, and also some thermolabile powders (provided the required humidity does not present a problem).

Another form of gaseous sterilization that has been developed is sterilization by low temperature steam and formaldehyde (LTSF). Formaldehyde is acutely toxic to human beings. It is also mutagenic, and possibly carcinogenic. It has been widely used to disinfect clean rooms. In admixture with steam, at temperatures between 60 and 80°C, it has been used for a range of medical devices.

Filtration Sterilization

Although all the different sterilization processes so far discussed are significantly different from one another in terms of technology, mode of action, and applications, filtration is "more different" than all the rest. It:

- Removes, rather than kills, organisms
- It is only applicable to fluids (liquids and gases)

Furthermore, it cannot be used as a terminal sterilization process. Any sterile-filtered bulk product that is not going to be subjected to a further terminal sterilization must be filled and sealed into final containers observing special aseptic precautions to prevent recontamination. Thus, a further element of risk of contamination is inevitably involved. Hence, it is generally considered that where it is possible to terminally heat sterilize products (particularly injections), this should be done, with filtration sterilization restricted to use where the product cannot withstand heating. However, more modern and automated techniques such as barrier (or isolator) or blow-fill-seal technology are being increasingly refined and used to significantly reduce the risks of recontamination following sterilization by filtration.

Isolator and BFS Technology

Isolator technology is concerned with the containment and close control of the environment immediately surrounding the work in process. It has two possible main objectives: (a) protecting the product from contamination from the operator(s) and the local environment and (b) protecting operator(s) from any hazardous products or materials. A well-designed, constructed, and operated isolator, avoiding as it can any direct contact between human operators and the immediate product environment, can significantly reduce the risk of microbiological contamination of products and materials processed within them. Problems to be guarded against are the possibility of puncturing of the fabric of an isolator, and of leakage, particularly around any transfer devices or ports. The revised EC "Sterile Annex" considers that, when used for aseptic processing, isolators should be sited in at least a Grade D background operating environment.

Isolator processes should be validated, with particular regard for the quality of the air both inside and outside of the isolator, cleaning and disinfection of the isolator, and the integrity of the isolator and any transfer mechanisms. They should be physically and microbiologically monitored, as for any other aseptic process, with attention being given to leak testing of the isolator and any glove or sleeve system (see Coles[20]).

Blow-fill-seal (BFS) machines are specialist purpose-built pieces of equipment in which containers are formed from a thermoplastic granulate, filled, and then sealed, all in a continuous automated operation. Originally developed for other purposes, they have for some years been available adapted for use in the manufacture of pharmaceutical, device, and other healthcare products, specifically, sterile pharmaceutical products.

In this process, bulk solution prepared under "microbiologically clean" or sterile conditions (as appropriate) is delivered to the machine via a bacteria retaining filter; pipework, filter housings, and machine parts in contact with the product having been previously steam sterilized in place. Filtered compressed air and granules of a plastics material conforming to a predetermined specification and known to be compatible with the product to be filled (usually polyethylene, polypropylene, or polyethylene/polypropylene copolymers) are also supplied to the machine.

Within the machine, the plastics granules are extruded downward under pressure (up to 350 Bar) as a hot, hollow, moldable, plastics tube (termed in the trade a "parison") or tubes. As a result of the high-pressure extrusion process, the parison reaches a temperature of 170 to 230°C. The configuration and internal integrity of the parison is maintained by an internal downward flow of filtered air under pressure. The two halves of a mold close around the parison and seal the base. Simultaneously, the top of the parison is cut free by a hot knife edge. The plastics material is now formed into a container (or containers, as determined by the design of the mold) by vacuum or sterile air pressure.

The container(s), having been transferred to the filling position (still within the mold) are immediately filled with a carefully metered volume of the solution, displacing the sterile air. Both the air and the solution are filtered through microorganism retaining filters immediately before entry into the forming, or formed, container(s).

When the required volume is filled into the container(s), the filling unit is raised, and the containers are sealed automatically. The mold then opens, releasing a package formed, filled, and sealed all in the one continuous, automatic cycle, which takes a matter of seconds (usually around 10 to 25 seconds, depending on the volume filled). Meanwhile, parison extrusion continues, and the cycle repeats. The filled and sealed units will usually require "cropping" or "deflashing" of excess plastic.

In versions of these machines adapted for aseptic manufacture, the filling cycle occurs in an internal, sterile, filtered air-flushed environment (the "air shower"). The machines can also be used to fill suspensions, ointments, creams, and liquids other than aqueous solutions, although with such products it is not necessarily always possible to employ the final aseptic product filtration facility.

"Multiblock" versions of these machines permit the formation of a number (or set) of containers at each pass, from one parison by one mold.

Some impressive claims have been made for the advantages of blow-fill-seal (Jones et al., 1995,[15] Leo 1990,[16] Sharp 1988,[17] 1988,[18] and 1990[19]) and isolator technology (Coles[20]) in providing greater confidence of sterility in aseptic processing. Not, it would seem without justice, although some regulatory people appear to be skeptical about these technologies (Hargreaves 1990[21]).

In essence, the process of filtration sterilization is a simple one. A liquid is forced, under moderate pressure, through a filter in the form of a membrane or a cartridge, the filter itself and the assembly in which it is mounted having first been sterilized. Various types, sizes, grades, and porosities of filter are available from a number of specialist manufacturers, and for effective sterilization a pore size of not more than 0.22 micrometers (micron) is required. Given the right filter for the job, properly assembled, fitted, and sterilized, and with confirmation that the filter is not damaged before use (or has not become damaged in use), a very high level of assurance of the sterility of the liquid *as it emerges through the filter* is possible, especially if the fluid is passed through two filters, connected in series. The main problem is ensuring that the liquid does not become recontaminated after the filtration (e.g., in the filling process). And that is not easy, particularly when and where people are involved. Aseptic filling involving human operators is a highly skilled operation, requiring excellent aseptic technique.

Sterilizing grade filters have been made from sintered glass or sintered metal, but the most commonly used filter media today are polymeric membranes, either as sheets or discs, or as one or more layers of the filter material supported on an inert and relatively rigid matrix to form a "cartridge filter." A wide range of types, sizes, and porosity of such filters are available from specialist suppliers. Filter manufacture and supply is a relatively large, competitive business, and filter manufacturers are generally more than ready to provide details on their range, and to give advice on applications and to offer technical backup.

It needs to be noted that although filters are designated as having "nominal" or "absolute" pore-size ratings (e.g., of 0.22 micrometers), it is not correct to view a filter as a sort of sieve plate, pierced by holes all of identical size (of, say, 0.22 micrometers). The pores in a membrane, in fact, consist of a range of sizes, characterized by a pore-size distribution. It is thus preferable to refer to a given type and make of filter in terms its organism-retentive properties (numbers and dimensions of organisms). A number of filter manufacturers provide this information, and in the US, HIMA guidelines[22] have been issued, which specify, for a sterilizing grade filter, a minimum challenge of 10^7 *Pseudomonas diminuta* (0.5 x 1.0–4.0 micrometers) per square centimeter of filter surface, with no passage of these organisms through the filter, into the filtrate.

The following briefly summarizes the different major methods of sterilization, their applications, advantages, and disadvantages:

STEAM

Heating in an Autoclave by steam under pressure or Sterilization in Place (SIP) Steam is effective at lower temperatures for shorter times than dry heat, because of its latent heat in addition to its sensible heat — kills, but does not remove organisms
Typical temperature/times 121°C for 15 minutes or 134°C for 3 minutes (other time/temperature combinations are possible)

Applications

- Aqueous preparations in sealed containers
- Equipment, instruments
- Dressings
- Manufacturing vessels, pipework, tubing

- Filter assemblies
- Surgical gowns, drapes, and dressings made of cellulosic materials
- Some medical devices

Advantages

- A process for *terminal* sterilization
- Can provide a very high margin of safety
- Can destroy viruses as well as bacteria, etc.
- Relatively short process time

Disadvantages

- Unsuitable for materials or articles damaged by heat
- Only suitable for water-wettable materials or sealed aqueous solutions (i.e., *not* oils or powders)
- Expensive (capital and energy costs)

DRY HEAT

Hot air ovens or sterilizing tunnels
Less effective than steam (at a given temperature) — kills but does not remove organisms
Typical temperature/times: 180°C for 30 minutes 160°C for 120 minutes

Applications

- Nonaqueous products (e.g., oils, powders) in sealed containers
- Glass and metal articles, equipment, parts etc.
- Some plastics
- Oils, fats, and waxes
- Silicone lubricants

Advantages

- Can be used for terminal sterilization
- Kills viruses
- Can be used for depyrogenation
- Can be used for nonaqueous and nonwettable materials
- Can be less damaging to metals than steam

Disadvantages

- Can only be used for items that can withstand the necessary heat input
- Unsuitable for most dressings, plastics, and rubber materials

RADIATION

Exposure to gamma radiation or high speed electrons
Typical required dose 25 kGy

Applications

- Some heat-sensitive materials, but not those that are radiation sensitive
- Dressings
- Surgical instruments
- Some chemical compounds

Advantages

- No heat effects
- Can be used for terminal sterilization
- Can be accurately controlled

Disadvantages

- Expensive specialist plant and equipment
- Most manufacturers have to contract out
- Elaborate safety precautions necessary
- Has deleterious effects on a number of chemicals, products, and containers (e.g., plastics)

ETHYLENE OXIDE

Usually used as a mixture of ethylene oxide with an inert gas (carbon dioxide, CFCs). Concentration , pressure, humidity, temperature, and nature of material and its packaging all influence effectiveness as a sterilant.

Applications

- Heat-sensitive materials
- Rubber and plastics
- Dressings and fabrics

Advantages

- No heat damage
- Effective at low temperatures
- Effective against many organisms

Disadvantages

- Toxic
- Explosive
- Expensive
- Slow
- Needs time for dispersal of residual gas and reaction products

FILTRATION

Fluid is passed through a sterile filter (membrane or cartridge) with a pore size of 0.22 micrometers, or one known to have equivalent bacteria-retaining properties

Applications

- Liquids and solutions that cannot withstand heat treatment
- Gasses

Advantages

- No heat effect
- Removes dead, as well as living, organisms
- Clarifies as well as sterilizes

Disadvantages

- Not a terminal process
- Requires additional aseptic handling and filling following sterilization
- Will not remove or inactivate viruses
- Requires skilled techniques
- Some substances may be absorbed by filter
- Cannot be used for suspensions
- Filter-integrity tests must be performed
- Very operator and environment sensitive

REFERENCES

1. Agalloco, J. and Akers, J., A critical look at sterility assurance, *Eur J Parent Sci,*; 7(4) , 97–103, 2002.
2. Sharp, J., ed., Guide to Good Pharmaceutical Manufacturing Practice, Her Majesty's Stationery Office, London, 1983.
3. Clothier, C., Report of the committee appointed to inquire into the circumstances, including the production, which led to the use of contaminated infusion fluids in the Devonport section of Plymouth General Hospital, Her Majesty's Stationery Office, London, 1972.

4. Polderman, J., *Introduction to Pharmaceutical Production*, Novib, The Hague, 1990.
5. US FDA, Compliance Guideline on Parametric Release, US Food and Drug Administration, Rockville, MD, 1993.
6. European Commission, Guide to Good Manufacturing Practice for Medicinal Products, Luxembourg, Office for Official Publications of the EC, 1998, Revised 2002. Reprinted in the MCA Rules and Guidance for Pharmaceutical Manufacturers and Distributors, TSO, London, 2002.
7. US FDA, Guide to the Inspection of Lyophilization of Parenterals, US Food and Drug Administration, Rockville, MD, 1993.
8. WHO, Sterility of Biologicals, World Health Organization report, Geneva, 1973.
9. PDA, Validation of aseptic filling for solution drug products, US Parenteral Drug Association, *Technical Monograph no. 2*, 1980.
10. PDA, Process simulation testing of aseptic filling for solution drug products, Technical Report no. 22, *PDA J Paren Sci Technol,* 50: S1 (Suppl), 1996.
11. Brown, M. and Gilbert, P,. Increasing the probability of sterility of medicinal products, *J Pharm Pharmacog,* 29, 517–523, 1977.
12. Russel, A.D., in Hugo and Russel *Pharmaceutical Microbiology,* Blackwell Scientific, Oxford, 1983.
13. Dunn, J. et al., PureBright pulsed light processing and sterilization, *European Journal of Parenteral Sciences,* 3(4), 105–114, 1998.
14. Ebara, T. et al., Development and practical application of high-frequency wave (microwave) continuous sterilizer, *European Journal of Parenteral Sciences,* 3(2), 39–47, 1998.
15. Jones, D., Topping, P., and Sharp, J., Environmental microbial challenges to an aseptic blow-fill-seal process, *PDA J. Pharm. Science and Tech.,* 49(3), 226–234, 1995.
16. Leo, F., Chapter in *Aseptic Pharmaceutical Manufacturing — Technology for the 1990s,* Olson and Groves eds., Interpharm Press, 1990.
17. Sharp, J., Manufacture of sterile pharmaceutical products using blow-fill-seal technology, *PJ ,* 239, 1987.
18. Sharp, J., Validation of a new form-fill-seal installation, *Manf. Chemist,* February 22, 1988.
19. Sharp, J., Aseptic validation of a form-fill-seal installation: Principles and practices, *J. Parent. Science and Tech.,* 44/5, 1990.
20. Coles, T., *Isolation Technology — a Practical Guide*, Interpharm Press, Illinois, 1998.
21. Hargreaves, P., Good manufacturing practice in the control of contamination, in Denyer, S. and Baird, R. *Guide to Microbiological Control in Pharmaceuticals*, Ellis Horwood Ltd., New York, 1990.
22. Health Industries Manufacturers Association, Microbiological evaluation of filters, HIMA Document No. 3, Vol 14., Washington, D.C., 1982.

ANNEX 1 TO CHAPTER 13

ANNEX 1 to EC GMP Guide

Manufacture of Sterile Medicinal Products

Principle

The manufacture of sterile products is subject to special requirements in order to minimise risks of microbiological contamination, and of particulate and pyrogen contamination. Much depends on the skill, training and attitudes of the personnel involved. Quality Assurance is particularly important, and this type of manufacture must strictly follow carefully established and validated methods of preparation and procedure. Sole reliance for sterility or other quality aspects must not be placed on any terminal process or finished product test.

Note: This guidance does not lay down detailed methods for determining the microbiological and particulate cleanliness of air, surfaces etc. Reference should be made to other documents such as the EN/ISO Standards.

General

1. The manufacture of sterile products should be carried out in clean areas entry to which should be through airlocks for personnel and/or for equipment and materials. Clean areas should be maintained to an appropriate cleanliness standard and supplied with air which has passed through filters of an appropriate efficiency.

2. The various operations of component preparation, product preparation and filling should be carried out in separate areas within the clean area. Manufacturing operations are divided into two categories; firstly those where the product is terminally sterilized, and secondly those which are conducted aseptically at some or all stages.

3. Clean areas for the manufacture of sterile products are classified according to the required characteristics of the environment. Each manufacturing operation requires an appropriate environmental cleanliness level in the operational state in order to minimise the risks of particulate or microbial contamination of the product or materials being handled.

In order to meet "in operation" conditions these areas should be designed to reach certain specified air-cleanliness levels in the "at rest" occupancy state. The "at-rest" state is the condition where the installation is installed and operating, complete with production equipment but with no operating personnel present. The "in operation" state is the condition where the installation is functioning in the defined operating mode with the specified number of personnel working. The "in operation" and "at rest" states should be defined for each clean room or suite of clean rooms.

For the manufacture of sterile medicinal products 4 grades can be distinguished.

Grade A : The local zone for high risk operations, e.g., filling zone, stopper bowls, open ampoules and vials, making aseptic connections. Normally such conditions are provided by a laminar air flow work station. Laminar air flow systems should provide a homogeneous air speed in a range of 0.36 – 0.54 m/s (guidance value) at the working position in open clean room applications.

The maintenance of laminarity should be demonstrated and validated.

A uni-directional air flow and lower velocities may be used in closed isolators and glove boxes.

Grade B : For aseptic preparation and filling, this is the background environment for the grade A zone.

Grade C and D: Clean areas for carrying out less critical stages in the manufacture of sterile products.

The airborne particulate classification for these grades is given in the following table:

Grade	maximum permitted number of particles/m³ equal to or above (a)			
	at rest (b)		in operation (b)	
	0.5 μm (d)	5 μm	0.5 μm (d)	5 μm
A	3 500	1 (e)	3 500	1 (e)
B (c)	3 500	1 (e)	350 000	2 000
C (c)	350 000	2 000	3 500 000	20 000
D (c)	3 500 000	20 000	not defined (f)	not defined (f)

Notes:

(a) Particle measurement based on the use of a discrete airborne particle counter to measure the concentration of particles at designated sizes equal to or greater than the threshold stated. A continuous measurement system should be used for monitoring the concentration of particles in the grade A zone, and is recommended for the surrounding grade B areas. For routine testing the total sample volume should not be less than 1 m³ for grade A and B areas and preferably also in grade C areas.

(b) The particulate conditions given in the table for the "at rest" state should be achieved after a short "clean up" period of 15-20 minutes (guidance value) in an unmanned state after completion of operations. The particulate conditions for grade A "in operation" given in the table should be maintained in the zone immediately surrounding the product whenever the product or open container is exposed to the environment. It is accepted that it may not always be possible to demonstrate conformity with particulate standards at the point of fill when filling is in progress, due to the generation of particles or droplets from the product itself.

(c) In order to reach the B, C and D air grades, the number of air changes should be related to the size of the room and the equipment and personnel present in the room. The air system should be provided with appropriate terminal filters such as HEPA for grades A, B and C.

(d) The guidance given for the maximum permitted number of particles in the "at rest" and "in operation" conditions correspond approximately to the cleanliness classes in the EN/ISO 14644-1 at a particle size of 0.5 μm.

(e) These areas are expected to be completely free from particles of size greater than or equal to 5 μm. As it is impossible to demonstrate the absence of particles with any statistical significance the limits are set to 1 particle/m³. During the clean room qualification it should be shown that the areas can be maintained within the defined limits.

(f) The requirements and limits will depend on the nature of the operations carried out.

Other characteristics such as temperature and relative humidity depend on the product and nature of the operations carried out. These parameters should not interfere with the defined cleanliness standard.

Examples of operations to be carried out in the various grades are given in the table below. (see also par. 11 and 12)

Grade	Examples of operations for terminally sterilized products. (see par. 11)
A	Filling of products, when unusually at risk
C	Preparation of solutions, when unusually at risk. Filling of products
D	Preparation of solutions and components for subsequent filling

Grade	Examples of operations for aseptic preparations. (see par. 12)
A	Aseptic preparation and filling.
C	Preparation of solutions to be filtered.
D	Handling of components after washing.

The particulate conditions given in the table for the "at rest" state should be achieved in the unmanned state after a short "clean up" period of 15–20 minutes (guidance value), after completion of operations. The particulate conditions for grade A in operation given in the table should be maintained in the zone immediately surrounding the product whenever the product or open container is exposed to the environment. It is accepted that it may not always be possible to demonstrate conformity with particulate standards at the point of fill when filling is in progress, due to the generation of particles or droplets from the product itself.

4. The areas should be monitored during operation, in order to control the particulate cleanliness of the various grades.
5. Where aseptic operations are performed monitoring should be frequent using methods such as settle plates, volumetric air and surface sampling (e.g., swabs and contact plates). Sampling methods used in operation should not interfere with zone protection. Results from monitoring should be considered when reviewing batch documentation for finished product release. Surfaces and personnel should be monitored after critical operations.

Additional microbiological monitoring is also required outside production operations, e.g., after validation of systems, cleaning and sanitisation.

Recommended limits for microbiological monitoring of clean areas during operation:

Grade	air sample cfu/m³	settle plates (diam.90mm), cfu/4 hours (b)	contact plates (diam.55mm), cfu/plate	glove print 5 fingers cfu/plate
		Recommended limits for microbial contamination(a)		
A	< 1	< 1	< 1	< 1
B	10	5	5	5
C	100	50	25	–
D	200	100	500	–

Notes:

(a) These are average values.

(b) Individual settle plates may be exposed for less than 4 hours.

6. Appropriate alert and action limits should be set for the results of particulate and microbiological monitoring. If these limits are exceeded operating procedures should prescribe corrective action.

Isolator technology

7. The utilisation of isolator technology to minimise human interventions in processing areas may result in a significant decrease in the risk of microbiological contamination of aseptically manufactured products from the environment. There are many possible designs of isolators and transfer devices. The isolator and the background environment should be designed so that the required air quality for the respective zones can be realised. Isolators are constructed of various materials more or less prone to puncture and leakage. Transfer devices may vary from a single door to double door designs to fully sealed systems incorporating sterilisation mechanisms.

The transfer of materials into and out of the unit is one of the greatest potential sources of contamination. In general the area inside the isolator is the local zone for high risk manipulations, although it is recognised that laminar air flow may not exist in the working zone of all such devices.

The air classification required for the background environment depends on the design of the isolator and its application. It should be controlled and for aseptic processing it should be at least grade D.

8. Isolators should be introduced only after appropriate validation. Validation should take into account all critical factors of isolator technology, for example the quality of the air inside and outside (background) the isolator, sanitisation of the isolator, the transfer process and isolator integrity.

9. Monitoring should be carried out routinely and should include frequent leak testing of the isolator and glove/sleeve system

Blow/fill/seal technology

10. Blow/fill/seal units are purpose built machines in which, in one continuous operation, containers are formed from a thermoplastic granulate, filled and then sealed, all by the one automatic machine. Blow/fill/seal equipment used for aseptic production which is fitted with an effective grade A air shower may be installed in at least a grade C environment, provided that grade A/B clothing is used. The environment should comply with the viable and non viable limits at rest and the viable limit only when in operation. Blow/fill/seal equipment used for the production of products which are terminally sterilized should be installed in at least a grade D environment.

Because of this special technology particular attention should be paid to, at least the following: equipment design and qualification, validation and reproducibility of cleaning-in-place and sterilisation-in-place, background cleanroom environment in which the equipment is located, operator training and clothing, and interventions in the critical zone of the equipment including any aseptic assembly prior to the commencement of filling.

Terminally sterilized products

11.   Preparation of components and most products should be done in at least a grade D environment in order to give low risk of microbial and particulate contamination, suitable for filtration and sterilisation. Where the product is at a high or unusual risk of microbial contamination, (for example, because the product actively supports microbial growth or must be held for a long period before sterilisation or is necessarily processed not mainly in closed vessels), then preparation should be carried out in a grade C environment.

Filling of products for terminal sterilisation should be carried out in at least a grade C environment.

Where the product is at unusual risk of contamination from the environment, for example because the filling operation is slow or the containers are wide-necked or are necessarily exposed for more than a few seconds before sealing, the filling should be done in a grade A zone with at least a grade C background. Preparation and filling of ointments, creams, suspensions and emulsions should generally be carried out in a grade C environment before terminal sterilisation.

Aseptic preparation

12.   Components after washing should be handled in at least a grade D environment. Handling of sterile starting materials and components, unless subjected to sterilisation or filtration through a micro-organism-retaining filter later in the process, should be done in a grade A environment with grade B background.

Preparation of solutions which are to be sterile filtered during the process should be done in a grade C environment; if not filtered, the preparation of materials and products should be done in a grade A environment with a grade B background.

Handling and filling of aseptically prepared products should be done in a grade A environment with a grade B background.

Prior to the completion of stoppering, transfer of partially closed containers, as used in freeze drying should be done either in a grade A environment with grade B background or in sealed transfer trays in a grade B environment.

Preparation and filling of sterile ointments, creams, suspensions and emulsions should be done in a grade A environment, with a grade B background, when the product is exposed and is not subsequently filtered.

Personnel

13. Only the minimum number of personnel required should be present in clean areas; this is particularly important during aseptic processing. Inspections and controls should be conducted outside the clean areas as far as possible.

14. All personnel (including those concerned with cleaning and maintenance) employed in such areas should receive regular training in disciplines relevant to the correct manufacture of sterile products. This training should include reference to hygiene and to the basic elements of microbiology. When outside staff who have not received such training (e.g., building or maintenance contractors) need to be brought in, particular care should be taken over their instruction and supervision.

15. Staff who have been engaged in the processing of animal tissue materials or of cultures of micro-organisms other than those used in the current manufacturing process should not enter sterile-product areas unless rigorous and clearly defined entry procedures have been followed.

16. High standards of personal hygiene and cleanliness are essential. Personnel involved in the manufacture of sterile preparations should be instructed to report any condition which may cause the shedding of abnormal numbers or types of contaminants; periodic health checks for such conditions are desirable. Actions to be taken about personnel who could be introducing undue microbiological hazard should be decided by a designated competent person.

17. Changing and washing should follow a written procedure designed to minimise contamination of clean area clothing or carry-through of contaminants to the clean areas.

18. Wristwatches, make-up and jewellery should not be worn in clean areas.

19. The clothing and its quality should be appropriate for the process and the grade of the working area. It should be worn in such a way as to protect the product from contamination.

The description of clothing required for each grade is given below:

Grade D: Hair and, where relevant, beard should be covered. A general protective suit and appropriate shoes or overshoes should be worn. Appropriate measures should be taken to avoid any contamination coming from outside the clean area.

Grade C: Hair and where relevant beard and moustache should be covered. A single or two-piece trouser suit, gathered at the wrists and with high neck and appropriate shoes or overshoes should be worn. They should shed virtually no fibres or particulate matter.

Grade A/B: Headgear should totally enclose hair and, where relevant, beard and moustache; it should be tucked into the neck of the suit; a face mask should be worn to prevent the shedding of droplets. Appropriate sterilized, non-powdered rubber or plastic gloves and sterilized or disinfected footwear should be worn. Trouser-legs should be tucked inside the footwear and garment sleeves into the gloves. The protective clothing should shed virtually no fibres or particulate matter and retain particles shed by the body.

20. Outdoor clothing should not be brought into changing rooms leading to grade B and C rooms. For every worker in a grade A/B area, clean sterile (sterilized or adequately sanitised) protective garments should be provided at each work session. Gloves should be regularly disinfected during operations. Masks and gloves should be changed at least for every working session.

21. Clean area clothing should be cleaned and handled in such a way that it does not gather additional contaminants which can later be shed. These operations should follow written procedures. Separate laundry facilities for such clothing are desirable. Inappropriate treatment of clothing will damage fibres and may increase the risk of shedding of particles.

Premises

22. In clean areas, all exposed surfaces should be smooth, impervious and unbroken in order to minimise the shedding or accumulation of particles or micro-organisms and to permit the repeated application of cleaning agents, and disinfectants where used.

23. To reduce accumulation of dust and to facilitate cleaning there should be no uncleanable recesses and a minimum of projecting ledges, shelves, cupboards and equipment. Doors should be designed to avoid those uncleanable recesses; sliding doors may be undesirable for this reason.

24. False ceilings should be sealed to prevent contamination from the space above them.

25. Pipes and ducts and other utilities should be installed so that they do not create recesses, unsealed openings and surfaces which are difficult to clean.

26. Sinks and drains should be prohibited in grade A/B areas used for aseptic manufacture. In other areas air breaks should be fitted between the machine or sink and the drains. Floor drains in lower grade clean rooms should be fitted with traps or water seals to prevent back-flow.

27. Changing rooms should be designed as airlocks and used to provide physical separation of the different stages of changing and so minimise microbial and particulate contamination of protective clothing. They should be flushed effectively with filtered air. The final stage of the changing room should, in the at-rest state, be the same grade as the area into which it leads. The use of separate changing rooms for entering and leaving clean areas is sometimes desirable. In general hand washing facilities should be provided only in the first stage of the changing rooms.

28. Both airlock doors should not be opened simultaneously. An interlocking system or a visual and/or audible warning system should be operated to prevent the opening of more than one door at a time.

29. A filtered air supply should maintain a positive pressure and an air flow relative to surrounding areas of a lower grade under all operational conditions and should flush the area effectively. Adjacent rooms of different grades should have a pressure differential of 10 - 15 pascals (guidance values). Particular attention should be paid to the protection of the zone of greatest risk, that is, the immediate environment to which a product and cleaned components which contact the product are exposed. The various recommendations regarding air supplies and pressure differentials may need to be modified where it becomes necessary to contain some materials, e.g., pathogenic, highly toxic, radioactive or

live viral or bacterial materials or products. Decontamination of facilities and treatment of air leaving a clean area may be necessary for some operations.

30. It should be demonstrated that air-flow patterns do not present a contamination risk, e.g., care should be taken to ensure that air flows do not distribute particles from a particle-generating person, operation or machine to a zone of higher product risk.

31. A warning system should be provided to indicate failure in the air supply. Indicators of pressure differences should be fitted between areas where these differences are important. These pressure differences should be recorded regularly or otherwise documented.

Equipment

32. A conveyor belt should not pass through a partition between a grade A or B area and a processing area of lower air cleanliness, unless the belt itself is continually sterilized (e.g., in a sterilising tunnel).

33. As far as practicable equipment, fittings and services should be designed and installed so that operations, maintenance and repairs can be carried out outside the clean area. If sterilisation is required, it should be carried out, wherever possible, after complete reassembly.

34. When equipment maintenance has been carried out within the clean area, the area should be cleaned, disinfected and/or sterilized where appropriate, before processing recommences if the required standards of cleanliness and/or asepsis have not been maintained during the work.

35. Water treatment plants and distribution systems should be designed, constructed and maintained so as to ensure a reliable source of water of an appropriate quality. They should not be operated beyond their designed capacity. Water for injections should be produced, stored and distributed in a manner which prevents microbial growth, for example by constant circulation at a temperature above 70°C.

36. All equipment such as sterilisers, air handling and filtration systems, air vent and gas filters, water treatment, generation, storage and distribution systems should be subject to validation and planned maintenance; their return to use should be approved.

Sanitation

37. The sanitation of clean areas is particularly important. They should be cleaned thoroughly in accordance with a written programme. Where disinfectants are used, more than one type should be employed. Monitoring should be undertaken regularly in order to detect the development of resistant strains.

38. Disinfectants and detergents should be monitored for microbial contamination; dilutions should be kept in previously cleaned containers and should only be stored for defined periods unless sterilized. Disinfectants and detergents used in Grades A and B areas should be sterile prior to use.

39. Fumigation of clean areas may be useful for reducing microbiological contamination in inaccessible places.

Processing

40. Precautions to minimise contamination should be taken during all processing stages including the stages before sterilisation.

41. Preparations of microbiological origin should not be made or filled in areas used for the processing of other medicinal products; however, vaccines of dead organisms or of bacterial extracts may be filled, after inactivation, in the same premises as other sterile medicinal products.

42. Validation of aseptic processing should include a process simulation test using a nutrient medium (media fill). Selection of the nutrient medium should be made based on dosage form of the product and selectivity, clarity, concentration and suitability for sterilisation of the nutrient medium. The process simulation test should imitate as closely as possible the routine aseptic manufacturing process and include all the critical subsequent manufacturing steps. It should also take into account various interventions known to occur during normal production as well as worst case situations. Process simulation tests should be performed as initial validation with three consecutive satisfactory simulation tests per shift and repeated at defined intervals and after any significant modification to the HVAC-system, equipment, process and number of shifts. Normally process simulation tests should be repeated twice a year per shift and process. The number of containers used for media fills should be sufficient to enable a valid evaluation. For small batches, the number of containers for media fills should at least equal the size of the product batch. The target should be zero growth but a contamination rate of less than 0.1% with 95% confidence limit is acceptable. The manufacturer should establish alert and action limits. Any contamination should be investigated.

43. Care should be taken that any validation does not compromise the processes.

44. Water sources, water treatment equipment and treated water should be monitored regularly for chemical and biological contamination and, as appropriate, for endotoxins. Records should be maintained of the results of the monitoring and of any action taken.

45. Activities in clean areas and especially when aseptic operations are in progress should be kept to a minimum and movement of personnel should be controlled and methodical, to avoid excessive shedding of particles and organisms due to over-vigorous activity. The ambient temperature and humidity should not be uncomfortably high because of the nature of the garments worn.

46. Microbiological contamination of starting materials should be minimal. Specifications should include requirements for microbiological quality when the need for this has been indicated by monitoring.

47. Containers and materials liable to generate fibres should be minimised in clean areas.

48. Where appropriate, measures should be taken to minimise the particulate contamination of the end product.

49. Components, containers and equipment should be handled after the final cleaning process in such a way that they are not recontaminated.

50. The interval between the washing and drying and the sterilisation of components, containers and equipment as well as between their sterilisation and use should be minimised and subject to a time-limit appropriate to the storage conditions.

51. The time between the start of the preparation of a solution and its sterilisation or filtration through a micro-organism-retaining filter should be minimised. There should be a set maximum permissible time for each product that takes into account its composition and the prescribed method of storage.

52. The bioburden should be monitored before sterilisation. There should be working limits on contamination immediately before sterilisation which are related to the efficiency of the method to be used. Where appropriate the absence of pyrogens should be monitored. All solutions, in particular large volume infusion fluids, should be passed through a micro-organism-retaining filter, if possible sited immediately before filling.

53. Components, containers, equipment and any other article required in a clean area where aseptic work takes place should be sterilized and passed into the area through double-ended sterilizers sealed into the wall, or by a procedure which achieves the same objective of not introducing contamination. Non-combustible gases should be passed through micro-organism retentive filters.

54. The efficacy of any new procedure should be validated, and the validation verified at scheduled intervals based on performance history or when any significant change is made in the process or equipment.

Sterilisation

55. All sterilisation processes should be validated. Particular attention should be given when the adopted sterilisation method is not described in the current edition of the European Pharmacopoeia, or when it is used for a product which is not a simple aqueous or oily solution. Where possible, heat sterilisation is the method of choice. In any case, the sterilisation process must be in accordance with the marketing and manufacturing authorisations.

56. Before any sterilisation process is adopted its suitability for the product and its efficacy in achieving the desired sterilising conditions in all parts of each type of load to be processed should be demonstrated by physical measurements and by biological indicators where appropriate. The validity of the process should be verified at scheduled intervals, at least annually, and whenever significant modifications have been made to the equipment. Records should be kept of the results.

57. For effective sterilisation the whole of the material must be subjected to the required treatment and the process should be designed to ensure that this is achieved.

58. Validated loading patterns should be established for all sterilisation processes.

59. Biological indicators should be considered as an additional method for monitoring the sterilisation. They should be stored and used according to the manufacturers instructions, and their quality checked by positive controls. If biological indicators are used, strict precautions should be taken to avoid transferring microbial contamination from them.

60. There should be a clear means of differentiating products which have not been sterilized from those which have. Each basket, tray or other carrier of products or components should be clearly labelled with the material name, its batch number and an indication of whether or not it has been sterilized. Indicators such as autoclave tape may be used, where

appropriate, to indicate whether or not a batch (or sub-batch) has passed through a sterilisation process, but they do not give a reliable indication that the lot is, in fact, sterile.

61. Sterilisation records should be available for each sterilisation run. They should be approved as part of the batch release procedure.

Sterilisation by heat

62. Each heat sterilisation cycle should be recorded on a time/temperature chart with a sufficiently large scale or by other appropriate equipment with suitable accuracy and precision. The position of the temperature probes used for controlling and/or recording should have been determined during the validation, and where applicable also checked against a second independent temperature probe located at the same position.

63. Chemical or biological indicators may also be used, but should not take the place of physical measurements.

64. Sufficient time must be allowed for the whole of the load to reach the required temperature before measurement of the sterilising time-period is commenced. This time must be determined for each type of load to be processed.

65. After the high temperature phase of a heat sterilisation cycle, precautions should be taken against contamination of a sterilized load during cooling. Any cooling fluid or gas in contact with the product should be sterilized unless it can be shown that any leaking container would not be approved for use.

Moist heat

66. Both temperature and pressure should be used to monitor the process. Control instrumentation should normally be independent of monitoring instrumentation and recording charts. Where automated control and monitoring systems are used for these applications they should be validated to ensure that critical process requirements are met. System and cycle faults should be registered by the system and observed by the operator. The reading of the independent temperature indicator should be routinely checked against the chart recorder during the sterilisation period. For sterilisers fitted with a drain at the bottom of the chamber, it may also be necessary to record the temperature at this position, throughout the sterilisation period. There should be frequent leak tests on the chamber when a vacuum phase is part of the cycle.

67. The items to be sterilized, other than products in sealed containers, should be wrapped in a material which allows removal of air and penetration of steam but which prevents recontamination after sterilisation. All parts of the load should be in contact with the sterilising agent at the required temperature for the required time.

68. Care should be taken to ensure that steam used for sterilisation is of suitable quality and does not contain additives at a level which could cause contamination of product or equipment.

Dry heat

69. The process used should include air circulation within the chamber and the maintenance of a positive pressure to prevent the entry of non-sterile air. Any air admitted should be passed through a HEPA filter. Where this process is also intended to remove pyrogens, challenge tests using endotoxins should be used as part of the validation.

Sterilisation by radiation

70. Radiation sterilisation is used mainly for the sterilisation of heat sensitive materials and products. Many medicinal products and some packaging materials are radiation-sensitive, so this method is permissible only when the absence of deleterious effects on the product has been confirmed experimentally. Ultraviolet irradiation is not normally an acceptable method of sterilisation.

71. During the sterilisation procedure the radiation dose should be measured. For this purpose, dosimetry indicators which are independent of dose rate should be used, giving a quantitative measurement of the dose received by the product itself. Dosimeters should be inserted in the load in sufficient number and close enough together to ensure that there is always a dosimeter in the irradiator. Where plastic dosimeters are used they should be used within the time-limit of their calibration. Dosimeter absorbances should be read within a short period after exposure to radiation.

72. Biological indicators may be used as an additional control.

73. Validation procedures should ensure that the effects of variations in density of the packages are considered.

74. Materials handling procedures should prevent mix-up between irradiated and non-irradiated materials. Radiation sensitive colour disks should also be used on each package to differentiate between packages which have been subjected to irradiation and those which have not.

75. The total radiation dose should be administered within a predetermined time span.

Sterilisation with ethylene oxide

76. This method should only be used when no other method is practicable. During process validation it should be shown that there is no damaging effect on the product and that the conditions and time allowed for degassing are such as to reduce any residual gas and reaction products to defined acceptable limits for the type of product or material.

77. Direct contact between gas and microbial cells is essential; precautions should be taken to avoid the presence of organisms likely to be enclosed in material such as crystals or dried protein. The nature and quantity of packaging materials can significantly affect the process.

78. Before exposure to the gas, materials should be brought into equilibrium with the humidity and temperature required by the process. The time required for this should be balanced against the opposing need to minimise the time before sterilisation.

79. Each sterilisation cycle should be monitored with suitable biological indicators, using the appropriate number of test pieces distributed

throughout the load. The information so obtained should form part of the batch record.

80. For each sterilisation cycle, records should be made of the time taken to complete the cycle, of the pressure, temperature and humidity within the chamber during the process and of the gas concentration and of the total amount of gas used. The pressure and temperature should be recorded throughout the cycle on a chart. The record(s) should form part of the batch record.

81. After sterilisation, the load should be stored in a controlled manner under ventilated conditions to allow residual gas and reaction products to reduce to the defined level. This process should be validated.

Filtration of medicinal products which cannot be sterilized in their final container

82. Filtration alone is not considered sufficient when sterilisation in the final container is possible. With regard to methods currently available, steam sterilisation is to be preferred. If the product cannot be sterilized in the final container, solutions or liquids can be filtered through a sterilefilter of nominal pore size of 0.22 micron (or less), or with at least equivalent micro-organism retaining properties, into a previously sterilized container. Such filters can remove most bacteria and moulds, but not all viruses or mycoplasmas. Consideration should be given to complementing the filtration process with some degree of heat treatment.

83. Due to the potential additional risks of the filtration method as compared with other sterilization processes, a second filtration via a further sterilized micro-organism retaining filter, immediately prior to filling, may be advisable. The final sterile filtration should be carried out as close as possible to the filling point.

84. Fibre shedding characteristics of filters should be minimal.

85. The integrity of the sterilized filter should be verified before use and should be confirmed immediately after use by an appropriate method such as a bubble point, diffusive flow or pressure hold test. The time taken to filter a known volume of bulk solution and the pressure difference to be used across the filter should be determined during validation and any significant differences from this during routine manufacturing, should be noted and investigated. Results of these checks should be included in the batch record. The integrity of critical gas and air vent filters should be confirmed after use. The integrity of other filters should be confirmed at appropriate intervals.

86. The same filter should not be used for more than one working day unless such use has been validated.

87. The filter should not affect the product by removal of ingredients from it or by release of substances into it.

Finishing of sterile products

88. Containers should be closed by appropriately validated methods. Containers closed by fusion, e.g., glass or plastic ampoules should be subject to 100% integrity testing. Samples of other containers should be checked for integrity according to appropriate procedures.

89. Containers sealed under vacuum should be tested for maintenance of that vacuum after an appropriate, pre-determined period.

90. Filled containers of parenteral products should be inspected individually for extraneous contamination or other defects. When inspection is done visually, it should be done under suitable and controlled conditions of illumination and background. Operators doing the inspection should pass regular eye-sight checks, with spectacles if worn, and be allowed frequent breaks from inspection. Where other methods of inspection are used, the process should be validated and the performance of the equipment checked at intervals. Results should be recorded.

Quality control

91. The sterility test applied to the finished product should only be regarded as the last in a series of control measures by which sterility is assured. The test should be validated for the product(s) concerned.

92. In those cases where parametric release has been authorised, special attention should be paid to the validation and the monitoring of the entire manufacturing process.

93. Samples taken for sterility testing should be representative of the whole of the batch, but should in particular include samples taken from parts of the batch considered to be most at risk of contamination, e.g.:

a. for products which have been filled aseptically, samples should include containers filled at the beginning and end of the batch and after any significant intervention,

b. or products which have been heat sterilized in their final containers, consideration should be given to taking samples from the potentially coolest part of the load.

14

GMP AND QUALITY ASSURANCE
IN STERILE PRODUCTS
MANUFACTURE

A fundamental prerequisite for the manufacture of *quality* sterile products, as with other pharmaceutical products, is the research and development of products that are demonstrably efficacious and acceptably safe. The additional requirement for a sterile product is that a suitable, *fit for purpose*, sterilization process has been established and validated as part of the product development phase. There is no point whatsoever in developing a product that can be shown to be splendidly efficacious and safe from an "ordinary" pharmacological or toxicological aspect, but which, if intended for injection, cannot be reliably sterilized when manufactured on a commercial scale.

It must always be remembered that mere passing of a sterility test is far from being an adequate determinant of sterility. Further, the need is not only for a valid sterilization process, vital though that is. What happens both before and after the sterilization process must also be regarded as crucially important. Every effort must be made to ensure that the items to be sterilized present the lowest possible microbial challenge to the sterilization process. The reason for this is to (a) reduce the risk of failure in part of the load or batch and (b) to guard against the presence of dangerous levels of pyrogenic substances that will not be destroyed or removed by most sterilization processes. The standard, or pharmacopoeial, sterilization cycles are not designed to ensure the sterilization of very highly contaminated material, and careful control and monitoring of presterilization bioburden is essential. Furthermore, it is both senseless and dangerous to competently sterilize a product by applying a validated process to a low bioburden product and then allow it to become recontaminated. This can perhaps most readily happen when a product is filled or packaged after bulk sterilization. It can also happen when an autoclave load is spray cooled, or air is admitted to a sterilization chamber, and there is a failure of the container seals, or if there are cracks, pinholes, or tears in the containers. Much, therefore, needs to be known about container quality and integrity, and about the microbial quality of the cooling water or the admitted air. There is also the possibility of what might be termed "macrocontamination" — a mix-up between sterilized and nonsterilized items as a consequence of inadequate segregation between lots leaving a sterilizer and those waiting to be loaded into it. It has happened.

Earlier in this book were set out the various factors that need to be considered before a product — any product — is released for sale or supply. These same factors apply just as much to sterile as to nonsterile products, but in addition, there are a number of further matters to be considered. These may be set down as follows:

a. Have the products or materials been handled and processed in premises, and under environmental conditions, appropriate to the manufacture of the type of sterile product concerned?

b. Have the personnel involved (at all levels, including those concerned with cleaning, maintenance, monitoring, and testing) been properly and thoroughly trained in the disciplines relevant to sterile products manufacture? (This training should include hygienic practices, at least the elements of microbiology, and an inculcation of a thorough awareness of the consequences to patients of failure to do the job properly.)

c. Have all persons entering sterile products manufacturing areas been, at all times, appropriately clad in the protective clothing as prescribed for the area?

d. Has the product, or material, been subjected to the prescribed validated sterilization procedure, in precise conformity with all laid-down process parameters, including loading patterns? Is there evidence (including automatic, and any manual, sterilizer records) available that this was, in fact, so? For example:

    i. For autoclaves and hot air ovens, time/temperature (and pressure, if relevant) recorder charts

    ii. For a radiation sterilization, evidence that the required radiation dose has been delivered to the load, either by means of examination of the exposed dosimeters themselves, or on the basis of unquestionably authoritative certification by person or persons who have, in turn, examined the exposed dosimeters

    iii. For an ethylene oxide sterilization, evidence of the quality of the incoming supply of ethylene oxide, load preconditioning records (time, temperature, relative humidity), sterilization cycle records (time, temperature, pressure, humidity, gas concentration), biological indicator incubation results, and degassing records (time, temperature)

    iv. For a filtration sterilization and aseptic fill:
      – Identification of the filter used
      – Evidence of sterilization of filter system before use
      – Results of filter integrity test(s)
      – Prefiltration bioburden data

- Details, and evidence, of sterilization of product, containers and other package components, before filling
- Aseptic filling area environmental monitoring data
- Sterility test results
- Evidence of aseptic filling validation

PREMISES AND SERVICES FOR STERILE PRODUCTS MANUFACTURE

Clean Rooms

Much of the manufacture and filling of sterile products is carried out in "clean rooms," or at least within confined areas that have a defined and controlled level of microbial (viable) and particulate contamination.

A clean room, in this special sense, is not merely a room that is clean. It is a room that is classifiable into one or other of several classes, or grades, of clean-ness.

The concept and design of clean rooms was first developed (in the early 1960s) for the microelectronic and aerospace industries. In these industries, it is important to protect microelectronic components against even the finest particles. Viable contaminants, as such, are of no special significance to micro-circuits, only in so far as bacteria, etc. are themselves particles.

Because of this origin in industries where product *sterility* is not the aim, classifications of the various classes of clean room are usually based upon the number and size of the particles (purely as *particles*) permitted per unit volume of the air in the room. Many published clean room standards also have specifications for humidity, temperature, lighting, and air pressure. It was only later that the pharmaceutical and related industries adopted the concept, and then (in some cases) added to it certain permissible levels of microbial (or viable) contamination. Even so, a lot of what is said and written on this subject, in relation to pharmaceutical manufacturing, still sounds or reads as if it were based on a premise that it is the *inanimate* particles that are crucially important. In fact (all other things being equal), the presence or absence of *viable* contamination in a parenteral product is a quality and patient safety issue that is even more critical than the presence or absence of nonviable contamination. It may be said, however, that although there has been a lack of convincing demonstration of a direct linear relationship between the numbers of nonviable and the numbers of viable particles in a given volume of air, it is entirely reasonable to suppose that where there are low levels of nonviable particles, there will concomitantly be low levels of microorganisms. Microorganisms are not usually found floating freely in air, but most characteristically are to be found associated with, or "rafting upon" particles or droplets.

The unit of measure used to define Clean Rooms is the micrometer (μm = 0.001 mm), very commonly termed (albeit inaccurately) a "micron." The first official published Standard for Clean Rooms was the United States Federal Standard 209: Clean Room and Work Station Requirements, Controlled Environment. This standard has gone through a number of revisions over the years since the 1960s — 209B, 209C, 209D, 209E, etc. — but the basic idea behind the US classification has remained the same. It is also the easiest official clean room standard to grasp and remember. It is based on permitted numbers, per cubic foot of air, of particles of a size 0.5 mμ and larger.

There are three classes of US Federal Standard 209 Clean Room, which are particularly relevant to sterile products manufacture: Class 100, Class 10,000, and Class 100,000, thus:

US Federal Standard 209 E Clean Room Standards

	Max. Permitted number of Particles per Cubic Foot of Air in Room	
	0.5 mμ and larger	*5.0 mμ and larger*
Class 100	100	0
Class 10,000	10,000	65
Class 100,000	100,000	700

Note: From issue "E" onward, the US Federal Standard 209 introduced (in addition to, and *not* in replacement of, the original classification based on particles per cubic foot) a parallel series based on particles *per cubic meter*. These all have the prefix "M." It is necessary to point-out that this "M" should be taken to stand for "metric," and to warn against the confusion that has arisen in some quarters by an erroneous assumption that this "M" stands for "microbial." It does not.

This US Standard defines a number of other clean room parameters and conditions, such as temperature, humidity, air pressure, operator clothing and behavior, and the instruments and devices used to measure and count particles in the air, but to date it has made no reference to permitted levels of microorganisms.

Following the lead given by the US Federal Standard 209, a number of national and other bodies have produced Standards for Clean Rooms, in essence, much like the US Standard but with changes in the nomenclature used for the various classes, or *grades*. This tends to make things seem more confused than they really are. An essential difference is that most clean room standards of European origin are based on numbers of particles per cubic *meter* of air, rather than per cubic *foot*.

For example, in 1976, British Standard (BS) No. 5295, Environmental Cleanliness in Enclosed Spaces, was published. This defined four main classes of clean room. In sterile products manufacturing areas, normally only the first three of them are of interest or concern, Class 1, Class 2, and Class 3. While at first sight these may look different, they are in fact closely similar to the Federal Standard's Class 100, Class 10,000, and Class 100,000, respectively.

What British Standard 5295 did was express the permitted number of particles in each class in terms of a cubic meter of air, not per cubic foot.

The BS 5295 (1976) figures for the 3 Classes can be summarized as follows:

	Max. Permitted No. of Particles per Cubic Meter of Air in Room			
	0.5 mμ (or larger)	*1.0 mμ (or larger)*	*5.0 mμ (or larger)*	*10 mμ (or larger)*
CLASS 1	3000	N/A	0	0
CLASS 2	300,000	N/A	2000	30
CLASS 3	N/A	1,00,000	20,000	4,000

Now, for example, 100 per cubic foot is *roughly* equivalent to 3,500 per cubic meter, and 10,000 per cubic foot is more-or-less the same as 350,000 per cubic meter. So, in the end, the specifications for the different classes of clean room in BS 5295 (1976) are quite close to the US Standard, i.e., the US Standard was recalculated in terms of a cubic *meter* of air, and then rounded down to make it a little tighter.

CLEAN ROOM STANDARDS AND GMP GUIDELINES

The US cGMPs (for finished pharmaceuticals) make no mention of clean rooms, in the specific sense, and offer no quantitative standards, neither for viable nor nonviable particles. Subpart C, Buildings and Facilities, includes, under Sec. 211.46 (Ventilation, air filtration, air heating, and cooling), the following:

(a)   Adequate ventilation shall be provided.
(b)   Equipment for adequate control over air pressure, micro-organisms, dust, humidity, and temperature shall be provided when appropriate..........
(c)   Air filtration systems, including prefilters and particulate matter air filters, shall be used when appropriate on air supplies to production areas......

The US Drug Product cGMPs do not contain any statement on environmental standards any more specific than that.

The 1987 FDA Guideline on Sterile Drug Products Produced by Aseptic Processing[1] is somewhat more specific. It does not, however, refer to "clean rooms" as such, but it does draw a distinction between a "Critical Area … in which the sterilized dosage form, containers and closures are exposed to the environment" and a "Controlled Area where unsterilised product, in-process materials, and container/closures are prepared."

According to this FDA guideline (currently under revision), air in a critical area, "in the immediate proximity of exposed sterilized containers/closures and filling/closing operations is of acceptable quality when it has a per-cubic-foot particle count of no more than 100 in a size range of 0.5 micron and larger (Class 100) when measured not more than one foot away from the work site, and upstream of the air flow…." In a critical area the air "should also be of a high microbial quality. A incidence of no more than one colony forming unit per 10 cubic feet is considered as attainable and desirable."

This FDA guideline considers that, in a controlled area, the air "is generally of acceptable particulate quality if it has a per-cubic-foot particle count of not more than 100,000 in a size range of 0.5 micron and larger (Class 100,000) when measured in the vicinity of the exposed articles during periods of activity. With regard to microbial quality, an incidence of no more than 25 colony forming units per 10 cubic feet is acceptable."

The 3rd edition of the UK GMP Guide (1983) set out, probably for the first time in any official GMP guideline, its own Basic Environmental Standards for the Manufacture of Sterile Products. These were based on the British Standard 5295 (1976), but with the addition of a series of maximum permitted levels of viable organisms per cubic meter of air. These standards were summarized in a table, which, in addition to giving levels for particles that are similar to those of BS 5295 (1976), also gives figures for:

Air filter efficiency
Air changes (per hour) in the room
Viable organisms per cubic meter

There is also a cross-reference to other Classifications, including BS 5295 and US Fed. 209 and a note that air pressures should "always be highest in the area of greatest risk," and that "air pressure differentials between rooms of successively higher to lower risk should be at least 1.5 mm (0.06 inch) water gauge" (i.e., approximately equivalent to a 15 Pa air pressure differential). This exemplifies the concept of the air pressure "cascade." Air pressure in sterile manufacturing areas should generally be higher than in the "world," or factory, outside. Furthermore, the pressure should be highest where a sterile product is, or is liable to be, exposed (i.e., is at greatest risk from contamination — cf . the FDA "Critical Area"), with the pressure becoming successively lower in the various interconnecting rooms, as the risk decreases. The intention is that sterile manufacturing areas should always be flushed by filtered air in a way that will most efficiently sweep contamination *away* from the more to the less critical zones.

The UK GMP classification dropped the term "class" and refers instead to "grades." There are also grades "1/A" and "1/B." Grade 1/A refers to required conditions where product is manipulated at, or under, a special unidirectional air-flow (Laminar Air Flow, or LAF) work station, and grade 1/B to high standard general room conditions, in which, for example, such an LAF work station would be sited. This standard also made it clear that, except for 1/A, it is intended that the various specified *particle* levels should be achieved when the room is unmanned and no work is in progress, but should be recoverable within a short period after personnel have left. The assumption was that it is not possible to retain these conditions generally when people are working in the room and also that the filtered air supply is running continuously, and thus rapidly flushes out the contamination caused by people. However, as is made very clear, it is vital that the specified conditions should be maintained at all times "in the zone immediately surrounding the product whenever the product is exposed." That is another important concept in Sterile Pr oducts Manufacture — the provision of a general high-standard Clean Room environment, with additional localized pr otection where product is or could be exposed.

There are a number of other published clean room classifications. They all tend to be similar, but use different labels for the different classes or grades.

The EC GMP Guide's "Airborne particulate classification" system, as it appears in the most recent revision of the EC GMP Guide's Annex 1, Manufacture of Sterile Medicinal Products (September 2003) is shown in its entirety in Annex 1 to the previous chapter. This revised EC Annex 1 also includes a table of "Recommended limits for microbial contamination."

THE STERILE PRODUCTS MANUFACTURING AREA, OR "SUITE"

Before we turn to the layout etc. of a sterile suite ther e is an *absolutely fundamental* point, that cannot be stressed too often:

All Sterile Products must be manufactured under carefully controlled and monitored conditions, and sole reliance should NOT be placed on any terminal process or test for assurance of the microbial and particulate quality of the end-product. (UK GMP Guide 1983)

The British GMP Guide might well now be considered to be superseded, but this particular statement has timeless relevance. The essential point is that great care must be taken to protect the product from contamination *throughout the entire manufacturing process*, and it is not sufficient merely to rely on a final filtration or sterilization to "clean things up." It certainly *cannot* be hoped that if it fails to do so, then end-product testing, particularly by the notably fallible sterility test, will detect any problems. That is why sterile products are manufactured in specially designed *sterile manufacturing areas*, or "*suites*" of one or more *clean rooms*.

A clean room, as originally conceived, may be defined thus:

A clean room is an enclosed space with quantitatively specified control of:

> Particles
> Temperature
> Pressure
> Humidity

– constructed with non-porous surfaces which are easy to clean and maintain, with controlled access via air-locks, and operated in accordance with procedures designed to keep contamination below a defined low level."

Notes:

a.   Although it is customary to classify clean rooms on the basis of a single parameter — the room air particulate level — the term "clean room" implies a number of other important factors.

b.   The definition, as given above, does indeed refer to a clean room "as originally conceived." It was only when this engineering concept was applied to the manufacture of pharmaceuticals, medical devices, and other healthcare products that standards for permitted microbial levels were grafted on to it.

There are a few other expressions that also need to be defined:

Air Lock: This is an enclosed space with a door at each end, which is placed between rooms (for example, between two different grades, or classes, of clean rooms) in order to control the air-flow between the rooms when they need to be entered. It is usual to arrange, by mechanical, electromagnetic, or electronic devices (or, less desirably, by visible or audible warning systems) that both doors are not open at the same time. (That would defeat the whole object of the air lock). Air locks may be intended either for the passage of people or of materials.

Pass-Through Hatch: This is a sort of mini air lock, inserted in a wall between two different rooms, usually at bench height. It serves the same purpose as an air lock, and is usually used for the passage of materials or smaller items of equipment. Again, the two doors should not be opened (or openable) at the same time.

Double-Ended Sterilizer: This is a sterilizer that has a door at each end. When inserted through a wall to a clean room, it means that items can be loaded into the sterilizer on one side of the wall and (following sterilization) can be taken out on the other. There are two main advantages. Use of a double-ended sterilizer guards against the contamination of a clean room, which could happen if goods were transported in or out by other means, and it also helps to prevent mix-ups between sterilized and unsterilized materials.

Sterilizing Tunnel: This is an enclosed conveyor system on which, for example, ampoules and vials can be conveyed into a filling room while (during their journey through the tunnel) they are subjected to sufficient heat to sterilize them and, usually also, to destroy pyrogens.

LAYOUT OF A STERILE PRODUCTS SUITE

A basic problem when writing on this topic is what system (or nomenclature) to use to denote the different air classes, grades, or levels. What system will be most readily comprehensible to the widest range of readers? The US cGMPs do not set down specific clean room standards, and the US Federal Standard 209E does not state limits for microbial contamination. The classification and terminology of the latest "sterile products" annex to the EC GMP Guide has, as we have noted, a number ambiguities and imprecisions. In the discussion that follows, therefore, such terms as "grade" and "class" have been abandoned, and the different levels of air quality have simply been designated as (A), (B), (C), and (D).

Thus, for these present purposes:

(A) Refers to the conditions at or under an operational laminar air flow work station, — approximately equivalent to US Federal Standard Class 100, with the additional requirement of less than 1 cfu/m^3.

(B) Approximately equivalent to Class 100, with the additional requirement of not more than 10 cfu/m^3.

(C) Approximately equivalent to Class 10,000, with the additional requirement of not more than 100 cfu/m^3.

(D) Approximately equivalent to Class 100,000, with the additional requirement of not more than 200cfu/m^3.

Figure 14.1 illustrates a basic general idea of a *sterile products suite*. The drawing is not to any scale, it does not indicate true relative sizes, and a number of things (e.g., doors) are omitted. It is merely intended to indicate the relationships between different areas and activities and how they all fit together.

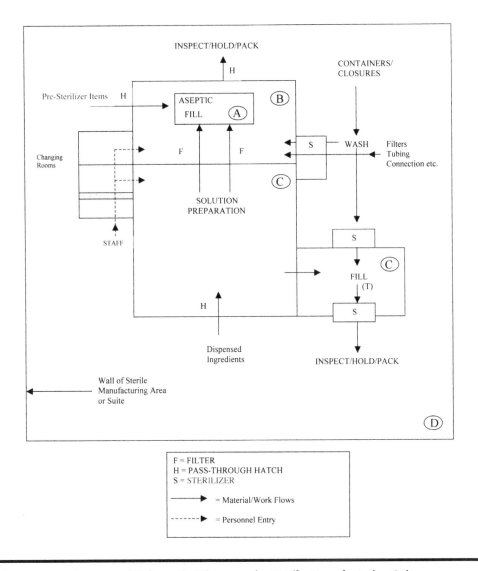

Figure 14.1 Conceptual/Schematic Diagram of a Sterile Manufacturing Suite

Points to note are:

1. The sterile products manufacturing suite should be separate and apart from other manufacturing areas, — either a separate building, or a discrete, walled-off part of a more general manufacturing site. Entry should only be permitted to authorized personnel via controlled, preferably air-locked, entry doors.
2. The environmental air quality within the general area (that is, the area outside the clean rooms, but within the overall suite) need not necessarily conform to the higher standards, although this will depend on the type of work that is carried out. Nevertheless, the air and all surfaces should at least be very clean by any normal (or "domestic") standards — (D) would normally be appropriate for the air quality.

3. The essential heart of the suite is the set of Clean Rooms. Three are shown in the example -
   - A solution preparation room — (C)
   - A filling room for terminally sterilized products ("fill [T]") — (C)
   - A room for aseptic filling and any other aseptic manipulations — (B) with local LAF protection (A)

The material and work flows are as follows:

- Dispensed ingredient for the manufacture of bulk solutions and suspensions, etc. are passed via a pass-through hatch, or an air lock, into the solution preparation room.
- When manufactured, bulk solutions *intended for aseptic filling* are then pumped, via sterilizing grade filters, to the aseptic fill room (filters on the aseptic side; other methods are possible).
- In the aseptic fill room, the sterile filtered solution is filled into containers that have previously been washed, rinsed, and passed into the aseptic filling room through a double-ended sterilizer or through a sterilizing tunnel. Following filling, the filled sterile product then passes out from the filling room via a hatch, for inspection, labeling, and further packaging.

It is essential in the type of aseptic filling process outlined above that the containers are clean and sterile at the time they are filled. There is an alternative form of ampoule, which is completely sealed as purchased. In use, it is usually assumed that the inside of the ampoule is clean and sterile and, therefore, does not need washing, etc. (Indeed, the inside cannot be washed!) Nevertheless, the *outsides* of these ampoules must also be clean and sterile before they pass into the aseptic filling room.

The route for bulk solutions that are intended to be filled and then terminally sterilized is as follows:

- The bulk manufactured solution is transferred to *"filling room [T]"* — either by pump or by use of a mobile sealed mixing vessel, wheeled through the door (not shown), to be filled into clean, washed, rinsed, and dried, and preferably, but not necessarily, sterilized containers. The filled containers then pass out of the filling room [T] through a double-ended sterilizer, in which the product, sealed in its containers, is terminally sterilized.

So, as the diagram indicates, once a bulk solution has been prepared, there are two alternative ways (depending on the nature of the product) by which it can be further processed, and these illustrate the two main, and different, approaches to making sterile products:

1. Filling and sealing the product into its final container and then sterilizing it (terminal sterilization)
2. Sterilizing a product at some earlier, bulk stage and then carrying out further processing, filling, and sealing into sterile containers, using aseptic techniques and taking aseptic precautions

Room Standards for Different Operations

Note: In what follows (A), (B), etc. should be interpreted as indicated above.

The aseptic fill room should be (B), and both the solution preparation room and the filling room [T] should be (C). In addition, in the aseptic fill room, there should certainly be additional localized high-quality filtered air, provided by LAF units (A) protecting the point of fill, and wherever else the product is exposed. It is also at least desirable that there should be additional localized filtered air protection at the place(s) where product is filled in the filling room [T].

It is not usually considered necessary to have (A), (B), or (C) conditions in rooms or areas used for the *initial* preparation of containers, caps and closures, items of equipment, components, etc. prior to washing and sterilization, always provided that precautions are taken to prevent recontamination once such items have been cleaned or sterilized.

However, even the "general areas" within a sterile products suite should at least be *very clean* by everyday standards, and some sterile product manufacturers do in fact specify, at least (D) for their more general areas.

Where any sterilized product, or container, is exposed (or could become exposed) in the room, then the room environment (background) should be (B) when the room is unmanned , and with (A) conditions maintained at critical points (critical for the product, that is) when the room is manned and operational. This applies not only to aseptic filling of previously sterilized liquids, but also to activities such as:

■ Mixing different previously sterilized ingredients to form a bulk product that will not (or cannot) then be sterilized. (An example would be the incorporation of a sterile powder into a sterile ointment base to form an eye ointment.)
■ Capping of vials of freeze-dried material.

For the preparation of bulk solutions and other products that are to be subject to later sterilization, and for the final cleaning (before sterilization) of some components and items of equipment, (C) conditions are required.

Changing Rooms

The possible ways of passing things in and out of these clean rooms include passage via:

Air Locks
Pass-through hatches
Double-ended sterilizers
Sterilizing tunnels

The method adopted will depend on the nature of the product or material, and the nature of the process.

The other important things that have to be got in and out of clean rooms are the people who work in them. And when it is recalled that it is people who are the most likely source of contamination, it becomes clear why special arrangements have to be made for their entry to and exit from clean rooms.

This is done through special changing rooms. The schematic diagram (Figure 14.1) shows one possible arrangement in the form of a single, *3-stage* changing room. The idea is that workers enter through a door from the general area, already clad in some form of clean protective clothing. This first stage is separated from the second by what is usually called a *"step-over barrier."* In fact it would be more correct to call it a *"sit-over barrier,"* for it is usually more like a solid bench, about seat height, extending to the floor (no gaps underneath) and completely crossing the room. Staff entering the room sit on this bench while changing their shoes for clean room footwear, one foot at a time. As the first foot is changed it is swung over the bench, to be followed by the other when that second shoe has been changed. The object is to make sure the dirty shoes do not touch the floor in the second stage, and that the new, clean footwear does not touch the dirtier floor in the first stage.

In the second stage, a first scrub-up and a further change takes place, into special *clean* (but not necessarily sterile) protective garments, and then staff can enter the solution preparation and other (C) rooms.

However, before entry into an *aseptic fill room*, a third stage is necessary, with a change into more thoroughly protective (protective of the product against contamination from the wearer, that is) *sterile* garments.

Notice that, since there are doors:

On first entry to the changing room
Between the second stage and solution preparation
Between the second and third stages
Between the third stage and aseptic fill

changing rooms also act as air locks for the passage of people. (And the same considerations regarding interlocking of doors apply.)

There are many possible variations on this simple arrangement. For example:

- Some arrangements have separate rooms, or routes, for entrance and for exit, and this (laudably) is becoming an increasing trend
- Some have an entirely separate changing room for just an aseptic room
- Some have separate male and female changing rooms, and others, by careful timing, manage to operate a "unisex" system

Air Supply

As already discussed, the air supplied to various rooms must be of quantitatively specified quality. Air from the outside world, or from the rest of the factory is neither suitable nor acceptable.

Hence, no windows (or certainly no *openable* windows) to the outside are permissible. It follows that the air to the various rooms must be a forced supply, delivered via ventilation trunking, through filters designed and tested to ensure that air that has passed through them is of the required quality (that is, contains no more than the specified number of particles, or organisms, per cubic foot or cubic meter).

While it is common for coarser prefilters to be used "upstream," in order to reduce the clogging of the final filters, it is important that the final air filters should be fitted at, or as closely as possible to, the point of entry of the air into a room. This is usually at various places in the ceiling. It is also usual to design clean rooms so that it is possible to change these filters, as and when necessary, without having to do so from within the room itself (that is, for example, by making the change from above a "false" or suspended ceiling). This is to avoid contaminating the room with particles from the "dirty" side of the filter, and to avoid more personnel than absolutely necessary having to enter the room.

Another important feature is that this filtered air should be supplied under pressure, so as to make the whole area at a positive pressure in relation to surrounding areas and, as necessary, give a pressure differential *between* different rooms within the suite. As we have seen, the whole idea is that there should be a "flushing out" of the suite, from the more to the less critical areas, and then to the outside.

It is generally considered that there should be a pressure differential between rooms of successively higher to lower risk of at least 15 Pa (approximately 1.5 mm water gauge = 0.06 inches water gauge), although some authorities consider that a pressure differential of 10 Pa between a one classified area and another adjacent one of lower classification is acceptable, provided that the differential between the classified areas and adjacent unclassified areas is at least 15 Pa. Thus, the air pressure in the aseptic fill room should be higher than in solution preparation, and the pressure in both should be higher than in the changing rooms, which should in turn be at least 15 Pa higher than in the general unclassified area. To give the idea, our simple diagram of a sterile suite can be redrawn to show just the air-flow patterns (See Figure 14.2).

Air can flow from one room to another through air locks, hatches, ports for conveyors or tubing, or through low-level grilles specially installed to aid the overall air pressure balance.

Pressure sensing devices (water gauges, manometers) must be installed to show the pressure differentials between rooms, and there should be audible or visible warning systems that sound, or display, alarm signals if the air supply fails and the required pressure drops.

Designing and installing air supply systems is a highly specialized business, as is the balancing of air pressures so as to achieve the correct differentials and flows. Most usually clean rooms are constructed and installed, under contract, by specialists in this field. Careful selection of an appropriate specialist is crucial to success (some *are* better than others). Just as important is a clear, precise definition of just exactly what is wanted, and close and careful monitoring of the project, as the contract proceeds. When the installation has been completed and commissioned, it should be "handed over," complete with a certificate of conformity to specification. This should contain certification of at least:

1. Air filter integrity — all air inlet filters tested to confirm filter and seal integrity, and conformity to the specified standard. Filter efficiency can

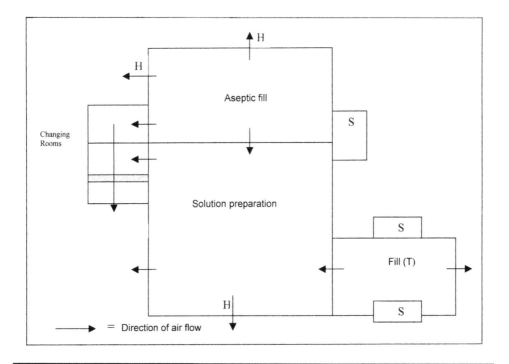

Figure 14.2   Air Flow in Sterile Manufacturing Suite (cf. Figure 14.1)

be tested by introducing particulate contamination upstream (PCU) of the filter, and measuring the particulate contamination level downstream (PCD). Percent efficiency is then determined as: (PCU – PCD)/PCU x 100.

2. Air velocity — measured by anemometer to determine air velocity (m/s) at the internal filter face of each air inlet.

3. Air change rate — calculated for each clean room from the air velocity, and the internal volume of the room.

4. Air particle counts — as measured in each clean room, in terms of the number of particles (of specified size) per cubic meter, or cubic foot, at the positions and heights specified in the standard against which the clean rooms were constructed and commissioned.

In addition there should be reports of checks on air-flow patterns, room pressure differentials, lighting levels, heating and humidity.

Once the system has been installed and the suite is operational, it is necessary to continue to check and monitor air filter (and seal) integrity and efficiency and that the air pressures and flows remain as required and as specified. It is also necessary to check air-flow rates at filter faces, and room air change rates.

The quality of the air, and surfaces, within the rooms must also be regularly monitored by:

■ Total particle counters
■ Air samples (for viable organisms)

- "Settle plates" (petri dishes of nutrient gel) to check microbial deposition on surfaces
- Surface swabs and contact plates
- "Finger dabs"

There will, of course, be people working in these clean rooms. Therefore, there needs to be heating or cooling of the air to ensure the right level of comfort, particularly for operators clothed in the special clean room garments, which can make them uncomfortably hot. It is important that operators do not get too warm, since the more they sweat, the more particles and organisms they will shed.

Total particles per unit volume of air may be determined by drawing a sample of air (of known volume) through a gridded filter membrane (capable of retaining particles of at least the size under investigation, e.g., of at least 0.5 micron) and then examining the membrane under a microscope for the size and numbers of particles. This is the reference method for demonstrating compliance with a number of official standards. However, more often used are the various commercial brands of optical particle counters. In these, the air sample is passed through a light beam, with a light-scattering detection device, the resulting signal being electronically processed to display or print out particle counts at different size ranges. To detect the finest particles, some models use laser beams.

Microbial levels in a clean room may be determined by use of:

- Settle plates
- Air samplers
- Surface sampling
- "Finger dabs"

Settle plates are petri dishes containing sterile nutrient agar. The plates are most usually 90 mm diameter (surface area approx. 0.006 m^2), although plates of 140 mm diameter (approx. 0.015 m^2 surface area) have been used. It is thus necessary when reporting settle-plate results (and when establishing standards for settle-plate counts) to state both the size of the plate(s) and the time of exposure. For valid comparisons to be made between results obtained using different plate sizes, the results should be expressed as cfus/100 cm^2/hour.

The most commonly used growth medium is Tryptone Soya, but others such as Rose Bengal and Sabouraud's medium have been used for specialized purposes. The pates are exposed in the room for a predefined (and subsequently recorded) time, and then incubated for a specified time at a specified temperature, and then examined. The number of colony-forming units (cfu) and the types (or species) of microorganisms found are recorded.

It has been argued that settle plates do not give a measure of the concentration of microorganisms in the *air* in a room. This is true, but it may equally be argued that, in providing a direct measure of the organisms that are *depositing from* the air and onto surfaces (or into containers), they do provide an indication of what the sterile product manufacturer really requires to know — the likely microbial contamination entering into, or onto, products (Whyte[2]).

At one time, it was considered that settle plates were only suitable for exposure for relatively short periods (1/2 to 1 hour), on the grounds that longer exposure times cause the agar to dry-out, and thus fail to culture the organisms that deposit on it. This view is no longer generally held, given that the agar has not been poured too thinly. If the agar layer fills at least half the petri dish, plates may be exposed, to good effect for several hours (Whyte and Niven,[3] Russell et al.[4]).

Air samplers are commercially available in a number of different types:

a. *Cascade samplers* — Where the air being sampled is drawn through perforations onto a stack of agar plates, separated from one another by perforated plates, the perforations in a plate decreasing in diameter from top to bottom of the stack. Organisms attached to the larger particles in the air sample are deposited on the top agar plate. Those attached to the smaller particles cascade down through the perforations until they impact on an agar plate lower down. After the specified time the agar plates are incubated and examined, as for settle plates.

b. In a *slit to agar sampler*, the air sample is drawn through slits and impinges on a rotating agar plate. The dimensions of the slits, their distance from the agar surface, and the air flow rate are adjusted to give optimum capture of the contaminants in the air. The agar plate is incubated and examined as before.

c. *Single sieve to agar sampler* — Here, the air is drawn through a perforated disc so as to impinge on an agar plate, which is then incubated and examined as in the other methods.

d. In a *centrifugal sampler*, air is drawn into the sampling head by the blades of a rotating fanlike impeller. This directs the air onto an agar strip fitted around the inner circumference of the sampling head. After a preset sampling time, the agar strip is incubated and examined.

e. *Filtration* — Here, air is drawn, at a controlled rate for a specified time, through a microorganism-retaining membrane filter, which is then placed on an agar plate, incubated and examined for organism growth.

f. *Liquid impingement* — Here, the air is drawn through an aqueous medium contained in an aspirator bottle, for a specified time. The liquid is then filtered through a membrane, which is then placed on an agar plate and incubated.

In all these air sampling methods, knowledge of the sampling rate, the time period over which the sample was taken, and of the number of cfus counted after incubation will enable the determination of the number of viable organisms present in unit volume (1 cubic meter or 1 cubic foot) of air in the room.

Surface sampling. Surfaces of walls, floors, and work and equipment surfaces can be sampled using moistened sterile cotton swabs, which are then "streaked out" on an agar plate, which is then incubated. Alternatively, and more conveniently, except for less accessible surfaces, contact (or "Rodac") plates can be pressed lightly onto flat surfaces and incubated. The colonies can then be counted to derive a quantitative estimate of surface contamination levels. Following the application of a swab, or a contact plate, the relevant surface area should be wiped with a disinfectant wipe.

Finger dabs. Although "finger dabs" do not directly measure the microbial contamination of the air in a clean room, they do give an indication of the contamination picked up by operators from surfaces in the room.

After, or as, they leave the room, operators touch the tips of all digits of both gloved hands onto an agar plate, which is then incubated. (It should go without saying that the gloves should be discarded and fresh ones put on before an operator continues to work.)

A similar technique can be employed as a training exercise, outside the sterile area, by applying operators' ungloved fingers (and indeed noses, ears, or whatever) to an agar plate to provide them with a graphic illustration of the organisms present on the human body surface.

Frequency of Monitoring/Checking of Clean Room Parameters

1. Physical

    a. Room pressure differentials — there should be a continuous automatic manometric measurement, linked to unmistakable visual and audible warning signals, which are triggered whenever pressure drops below the specified level. The manometer gauges should also be visually checked hourly and the reading recorded at least once per day (or per shift). It is essential that the manometers are regularly calibrated.

    b. Air velocity and room air change rates — perform and record every six months.

    c. Air particle counts — perform daily (or batch-wise) in the more critical areas and weekly in the less.

    d. Air filter integrity and efficiency test — Carry out once or twice a year, unless results of in-process physical and microbial monitoring indicate a more urgent need.

Air flow directions and patterns should also be occasionally checked, as convenient. (Clearly, it is not possible to check this by "smoke" tests when the rooms are in operation. It is usually possible to check quite simply the direction in which air is passing through a grille. But beware — air moving rapidly *away* from the surface of a hand can produce a sense of cooling which can be (and indeed has been) misinterpreted as if were air blowing *onto* the skin surface.

It is quite common for sterile products manufacturers to have the more crucial physical parameters examined and independently certified, say, once or twice a year, by external contracting specialists. While there is no absolute necessity to do this, it can be said, with good reason, that in such a critical area of manufacturing, where lives could be at stake, it does add a significant extra layer of assurance.

2. Microbial

While there is something like general agreement on the frequency of monitoring of physical clean room parameters, no such official or general agreement exists in regard to frequency of microbial monitoring. In the circumstances, the following seems to be a reasonable schedule for the different levels of clean room:

- In clean rooms (A) — daily or batch-wise
- In clean rooms (B) — daily
- In clean rooms (C) — weekly
- In clean rooms (D) — weekly, without all microbial monitoring necessarily being carried-out on each occasion

THE STERILE MANUFACTURING AREA — CONSTRUCTION AND FINISHES

The surfaces of all floors, walls and ceilings should be hard, smooth, impervious and unbroken (i.e. no cracks, holes or other damage). There are three good reasons for this:

1. To prevent the shedding of particles from damaged, or poorly finished brick, building block, plaster, etc.
2. To prevent the accumulation of dust, dirt, and microorganisms on, or in, rough or broken surfaces
3. To permit easy and repeated cleaning and disinfection

Various materials have been used for floors, including welded sheet vinyl, terrazzo, and various "poured" resin floors. A variety of basic structural materials are used for walls — bricks, blocks, plastic-coated metal panels, glass reinforced plastics. All are acceptable, provided that the final finish provides a *smooth, impervious, unbroken* surface. Thus if a wall is constructed of brick or structural block, it must be smooth-plastered and then coated with a hard-setting finish (polyurethane, epoxy, etc.), sprayed or painted on.

Welded sheet vinyl is also used as a wall finish, often as a continuation of the same material when it has also been used as a floor surface.

Where windows are installed, they should not be openable. They should be flush-fitted on the controlled (or classified) area side. Where windows are fitted in a dividing wall between two classified areas or rooms, they should be double-glazed so as to present a flush, ledgeless fit on both sides. If communication is necessary between adjacent clean rooms, it should be via "speech panels" (polymeric membranes that transmit sound while maintaining an airtight seal). They can be used back-to-back in double-glazed windows. When installed in the more critical clean rooms, the usual protective grilles should be removed, as they are difficult to clean. Telephone and intercom installations should generally be avoided, certainly in aseptic processing rooms. If they are deemed essential in, for example, a solution preparation room, they should be purpose-designed, flush-mounted, and with easily wipeable touch-sensitive controls.

Ceilings in sterile areas are often "false" or suspended, to allow for the installation of air-supply ducting, and other services, above. It is important that any suspended ceiling is effectively sealed from the room below, to prevent any possible contamination from the space between the false and the real ceiling.

Where floors meet walls and walls meet ceilings, the joins should be coved, so as to avoid sharp corners that are difficult to clean and can harbor dust, dirt, and microorganisms. It is also important that any such coving should be flush to both floor and wall (or wall and ceiling) and *not* create additional ledges. All too often, coving is so badly installed as to create two dirt- and bacteria-collecting corners where previously there was only one.

It is also important that all potentially dust-collecting surfaces should be kept to a minimum. All lights, doors, windows, air-supply and exit grilles, and the like should be flush-fitting with no, or a minimum of, ledges or ridges. Light fittings should be flush-fitted, permanently sealed-in units, with access for lamp replacement from a service void above the false, or suspended, ceiling. Because of the difficulties of cleaning the sliding gear, sliding doors should be avoided. Doors should normally open inward to the higher standard, or more critical, room so that operators can back in, without having to touch the dirtier side with their gloved hands. Doors into, within, and from changing rooms should be fitted with warning systems or interlocks to prevent more than one door being opened simultaneously, thus causing a drop in overall pressure levels. Electromechanical interlocks are perhaps more effective from this point of view, but are a potential serious hazard to operators in the event of fire.

Unless the safety of workers can be assured by other means (e.g., by the installation of sealed "burst panels" in clean room walls) interlocking bolt systems are probably best avoided. Given well-trained and disciplined operators (which all workers in sterile products manufacturing areas should unquestionably be), clear warning light systems actuated by door-mounted microswitches should be more than adequate.

There will, inevitably, be bench and working surfaces, but these must be constructed (with hard, smooth, impervious, and unbroken surfaces) so it is always possible to clean everything thoroughly on, in, under, and around them. There should be easy access around, and under, all furniture and equipment within a clean room. On the whole, wood, and certainly *bare* wood, should not be used as a material of construction within a clean room.

Any pipework, ducting, or conduit supplying air, water, electricity, or gas to a clean room should be supplied via the void above a false ceiling, or sealed within the walls. If pipework or ducting passes through a wall, it should be completely sealed within the wall through which it passes. There should be no pipes set so close to walls as to create difficult-to-clean dirt traps.

Sinks and drains should be excluded from manufacturing areas wherever possible and avoided entirely in clean rooms in which aseptic operations are carried out.

Where drains must be installed, they should be fitted with effective, easily cleanable traps and air breaks to prevent backflow. They must be easy to clean and disinfect — and, of course, they must *be* clean and disinfected.

Where sinks are installed (for example, in changing rooms and in bulk solution preparation areas), they should preferably be made of stainless steel, and be designed, installed, and maintained so as to minimize risks of microbial growth and contamination. In changing rooms, foot controls for the supply of water and antibacterial liquid soap for washing hands, and automatic, warm-air hand dryers are much preferred. Sinks and hand-wash basins, and the drains from them, should be regularly disinfected.

Wherever possible, major items of equipment should either be movable, for the purposes of cleaning and disinfecting both the equipment and the floor or wall area under or adjacent to it, or built into the fabric of the room so as to present only smooth, cleanable surfaces, and no inaccessible gaps and recesses. Where built-in work benches join walls, flexing of the working surface can cause stress and consequent cracking at the junction. The joint can be sealed with a well- and smoothly applied flexible sealant, silicone, or urethane, but probably the better alternative is the use of freestanding, movable, stainless steel tables rather than fixed benchwork.

In general, it can be said (again — but it *is* important) that all surfaces should be smooth, impervious, and unbroken, that there should be *no* uncleanable gaps, cracks, spaces, holes, or recesses and that there should be a minimum of ledges, shelves, cupboards, and equipment.

Unidirectional Air Flow Work Stations

Laminar Air Flow (LAF) is the term applied to a supply of sterile air forced through high efficiency (HEPA) air filters so that it flows, nonturbulently, in just one direction, over the material or product that is in need of special protection. The object is to sweep away (and keep away) any potential microbial or other particulate contamination. The actual filter faces from which the air emerges are usually fitted with side screens, curtains, or shields, which surround, or partially surround, the product, material, or equipment that needs to be protected. Alternatively, they may be obtained already purpose-built into Laminar Flow cabinets. Unidirectional, or laminar, air flow may be either vertical or horizontal.

Laminar air flow cabinets (contained work stations or unidirectional air flow cabinets) are installed where extra protection against contamination is required, within a standard (or "conventional") clean room. A conventional clean room is a clean room that is not a *totally* laminar flow room, but a room where, normally, the air enters through filters installed in the ceiling and exits through grilles placed low in the walls or under doors, etc.

Services for Sterile Products Areas

Services for manufacturing departments generally have already been covered in Chapter 3 of this book, where the various grades of water and the systems for supplying them were discussed. The special requirements for air supply to sterile manufacturing areas have already been covered in this chapter. It is especially important that all services to sterile product areas should be supplied or installed so as not to represent risks of contaminating the product or the manufacturing environment. The FDA's Guideline on Sterile Drugs Produced by Aseptic Processing (1987) also notes that "gases other than ambient air may also be used in controlled areas. Such gases should, if vented to the area, be of the same quality as the ambient air. Compressed air should be free from demonstrable oil vapor."

Validation of Sterilization Processes

All sterilization processes must be validated. See Chapter 15 and Chapter 16 of this book.

Personnel

Even more than in relation to other types of manufacture, the *people* involved are the single most important factor in the manufacture of *sterile* products. The person or persons who manage sterile products departments should have a full understanding of, and experience in, the special techniques, technologies, and disciplines required — and of the underlying physical, chemical, microbiological, and clinical principles. They should be able to impart their knowledge and understanding to their staff, who should also be selected with care. Workers in sterile products areas should be mature (and that does not necessarily mean old), intelligent people who can fully understand not just what they have to do, but also the reasons for doing it. They must have innately high standards of personal hygiene and be readily able to abide by the special disciplines involved. They should also be free from any disease or condition that could represent an abnormal microbiological hazard to the clean room environment and, hence, to the product. These conditions include, in addition to chronic gastrointestinal and respiratory tract diseases, short-term conditions such as colds, acute diarrhea, skin rashes, boils, open superficial injuries, and peeling sunburn. Operators should be required to report any such conditions, and supervisory staff should be on the lookout for them. There should be periodic health checks.

In addition to those who have chronic skin, respiratory, or gut diseases, persons who have allergies to the synthetic fabrics used in clean room clothing, who are abnormally high shedders of skin flakes or dandruff, have nervous conditions resulting in excessive itching, scratching, etc., or who suffer from any degree of claustrophobia are really not fitted to work in clean rooms.

No person who reports that they have a condition that would preclude their working in a clean area should suffer any penalty for doing so. The thought of loss of earnings might well persuade even the most saintly worker to keep dumb about an adverse health condition.

A certain calm resoluteness of character is also most desirable. To be alone, or perhaps be just one of two or three in a clean room, in a full sterile suit with gloves, hood, mask, and possibly goggles, can prove a lonely, depressing, demotivating experience for some temperaments. Conversely, while a cheery, hail-fellow, whistle-while-you-work attitude may be salutary in some areas of human activity, it is entirely inappropriate in a clean room.

To minimize the contamination inevitably caused by the presence of people, the numbers entering and working in clean rooms should be kept to the minimum necessary for effective working. All activities, such as in-process testing and control, visual inspection, and the like, that do not need to be conducted in a clean room should be performed outside it.

All personnel — including cleaning staff and maintenance engineers — required to work in, or otherwise enter, a clean area should be trained in the techniques and disciplines relevant to the safe and effective manufacture of sterile products. This training, which should not be a "one-off" exercise but should be regularly reinforced with refresher training, should include coverage of personal hygiene, the essential elements of microbiology, and the purpose and correct wearing of protective clothing. Operators should be taught to "know the enemy," and practical demonstrations of growing cultures, finger dabs, and the like will help to get the message home. Training should also include a strong motivational element, stressing responsibilities to patients' health and life, which are, quite literally, "in your hands."

Any outside persons, such as building or maintenance contractors, who have not received the training and who need to enter clean areas should only do so under close supervision, and when wearing protective clothing appropriate to the area.

Personnel — Changing and Clothing

Personnel should only enter a clean area via changing rooms, where washing and changing should proceed in strict accordance with a written procedure. The operators should have been trained to follow this procedure, and a copy of it should be clearly displayed on the changing room wall. The procedure should be designed to minimize contamination of the protective clothing, through, for example, contact with the floor on the "dirtier side," or with operators' shoes. Outdoor clothing should not be taken into clean room changing rooms. The assumption should be that outdoor garments have already been removed elsewhere, and that personnel are already clad in the standard "general factory" protective clothing. Wristwatches and jewelry should be removed as part of the changing process. Plain, simple, wedding rings are generally considered to be an exception, which is both reasonable, sympathetic, and expedient — many people finding it impossible (physically or emotionally) to remove their wedding rings. However, the FDA is said not to agree on this point. Cosmetics, other than perhaps simple particle-free, nonshedding creams, should not be worn.

The protective garments, which should include head- and footwear, should be made from textiles specially manufactured so as to shed virtually no fibers or particles and to retain any particles shed by a human body within. They should be comfortable to wear and loose-fitting to reduce abrasion. Fabric edges should be sealed and seams all-enveloping. Unnecessary tucks and belts should be avoided, and there should be no external pockets. The garments should be worn only in the clean areas. A fresh set of clean (and, if necessary, sterilized) protective garments should be provided each time a person enters, or reenters, a clean room. This should be rigorously enforced where aseptic processing is in operation. In other, less critical, clean rooms, it may be possible to relax this requirement and provide fresh garments once per day, if this can be justified on the basis of monitoring results and other control measures. Even so, fresh head- and footwear, and gloves should be provided for each working session.

Protective clothing, following use, should be washed or cleaned (and, as necessary, sterilized), and thereafter handled in such a way as to prevent it gathering contaminants, and to minimize attrition of the fabric. It needs to be recognized that repeated wearing and laundering or cleaning (and sterilization) can cumulatively damage the fabric so that it becomes no longer suitable for use. This is clearly something that needs to be monitored and controlled, and some methods and standards have been published (see, e.g., ASTM ,[5] AS ,[6] AS 2014[7]).

In (C) and (D) standard rooms, clean one-or two-piece trouser suits should be worn, close fitting at the neck, wrists, and ankles, with high necks. Hair, including any facial hair (beard or mustache) should be covered. Trouser-bottoms should be tucked into overshoes or boots, and sleeves into gloves.

In (B) rooms, and when working at (A) work stations, sterilized, nonshedding coverall trouser suits (preferably one piece) should be worn. Headwear should be of the helmet or cowl type and totally enclose the hair and any beard or moustache. It should be completely tucked into the neck of the suit. Footwear should be of the boot, or "bootie" type, totally enclosing the feet. Trouser bottoms should be completely tucked into the footwear. Powder-free rubber or plastic gloves should be worn with the garment sleeves neatly and completely tucked inside the gloves. Gloves should be regularly disinfected (e.g., with a sterile alcoholic spray or foam) during extended operations. Disposable face masks, covering both the nose and mouth, should be worn. They should be discarded at least each time the wearers leave the clean room, and whenever they become soggy. In the latter circumstances, it is of course necessary to leave the clean room to change the mask. Operators should be trained not to touch masks, or any other part of their face, with their hands when in a clean room.

There is a school of thought that holds that, when working in an aseptic processing area, operators should wear close-fitting goggles. Indeed, the US FDA has been known to insist upon it. There are, however, those who would argue that any benefit, in terms of reduction in contamination hazard, is outweighed by the risks introduced by the additional operator discomfort and the misting of the goggle lenses.

When working at contained LAF work stations, operators should always work downstream of the filter face and of any product, material, or equipment that is being processed or manipulated at the work station. In other words, work should be conducted so that any operator-derived contamination is swept in a direction *away* from the work at hand. Hands or arms should not be interposed between the filter face and the product, as this would cause the air stream to sweep contamination from the operator onto the work — the very reverse of what is required.

Instructions to Operators on Entering and Working in Clean Rooms

The following is intended to serve as a checklist, and perhaps as the basis for an SOP, or a training handout:

1. Keep body, hair, face, hands, and fingernails clean.
2. Report any illnesses, cuts, grazes, or respiratory, gut, or skin problems.
3. Follow the written changing and wash-up procedure *exactly*.
4. Check that your protective clothing is worn properly.
5. Do not wear cosmetics, jewelry, or wristwatches.
6. Leave all personal items (wallets, coins, keys, watches, tissues, combs, etc.) in the changing room.

7. Do not take papers, documents or paper materials into clean rooms, unless they have been specifically approved. (Paper, cardboard, and similar materials are great shedders of particles and fibers.)
8. *No* eating, chewing, drinking.
9. Always move gently and steadily.
10. Do *not* move vigorously. *No* fooling about, singing, or whistling.
11. Avoid talking unless absolutely necessary.
12. Avoid coughing or sneezing. If these are unavoidable, leave the clean room. (We spray a lot of fine drops and microbes about when we talk, sing, whistle, cough, sneeze, or splutter.)
13. Do not touch other operators.
14. Avoid scratching, touching nose and mouth, and rubbing hands.
15. Where gloves are worn, regularly disinfect them as instructed.
16. Always check for worn or damaged garments and torn gloves and change them as necessary. (Even a pinhole in a glove could have disastrous consequences for a patient.)
17. Keep garments fully fastened. Do not unfasten or loosen them.
18. Unless there is a special hazard involved, do not pick up dropped items from the floor.
19. When working at a laminar flow work station it is important to ensure that:
    ■ Nothing is placed between the air filter face and the object, material, or product that is being handled and that needs to be protected. (This would disturb the smoothly sweeping flow of unidirectional air that is keeping the vital areas clean.)
    ■ Always work downstream from the air filter face, and do not let your hands or arms come between the item that is being protected and the air filter face. Blowing contamination from you onto a product defeats the whole object, and could be a danger to patients.

In-Process Control of Sterilization Processes

General

It is a statement of the obvious, but it cannot be overstressed that for a sterilization process to be effective, *all* of the material, product, batch, lot, or load must be subjected to the required treatment. For example, 121°C for 15 minutes is generally considered to be an effective time/temperature cycle for a steam sterilization. This does not mean that it is OK if some part or parts of the autoclave load, reach, and maintain a temperature of 121°C for 15 minutes. It means that the coldest part, of the coldest item or unit, in the coldest area within the chamber, must reach a temperature of at least 121°C, and be held at that temperature for at least 15 minutes. If it is necessary, in order to achieve such conditions, for some other parts of the load to reach much higher temperatures for significantly longer times, thus risking product degradation, then something is wrong with either the design of the autoclave, or the loading pattern, or both.

When a heat sterilization cycle is validated, it must be regarded only as the validation of a specified cycle, in a particular chamber, containing a specific product or material, loaded into the chamber in a specified pattern. If the loading pattern is changed, the process can no longer be considered validated.

Materials and products to be sterilized should not carry a high level of microbial contamination (bioburden). Limits should be set on the presterilization bioburden, and lots intended for sterilization should be tested for microbial levels before being subjected to the sterilization process. This does not necessarily mean that sterilization may not proceed until these test results are available. They should, however, be recorded and form part of the data to be considered when the final release/reject decision is made. The whole object is to present the lowest possible microbial challenge to the sterilization process. To this end, it is sound quality practice to pass solutions, particularly large volume parenterals (LVPs) through a bacteria-retaining filter before filling and terminal sterilization.

Various forms of biological and chemical indicators are available for use in connection with sterilization processes. They can show when a sterilization has *failed*, but cannot necessarily demonstrate that the process has been an overall success. Thus, if used alone, they are not acceptable as proof that a sterilization process has been effective. They may be considered as providing no more than backup to the other evidence, which must be available.

Biological indicators (preparations of bacterial cultures, usually spores, of strains known to be resistant to the type of sterilization process under consideration) are less reliable than physical monitoring and control methods, except in the case of ethylene oxide sterilization. If used, strict precautions must be taken to avoid releasing the resistant contamination into clean rooms, and into or onto products.

Chemical indicators are commercially available for heat, ethylene oxide, and radiation sterilization. They can take the form of adhesive tapes or patches, color-spot cards, small sealed tubes, or sachets. They change color or appearance as a result of chemical reactions activated by exposure to the sterilization process, but it is possible for the change to take place before the sterilization time has been completed and, hence (with the exception of the plastic dosimeters used in radiation sterilization), their inadequacy as *proof* of sterilization. The same applies to the preparations of substances that melt at a sterilizing temperature. They can show that a temperature has been reached, but not that it has been held or for how long. Radiation-sensitive colored discs (not to be confused with plastic dosimeters), which change color when exposed to radiation, serve to distinguish between items that have been exposed to radiation and those that have not, but they do not give a reliable indication of successful sterilization.

It is vital that there should be a clear and foolproof method of distinguishing between goods that have been sterilized and those that have not. Double-ended, through-wall sterilizers are a notable aid in preventing this potentially lethal form of mix-up. In any event, each basket, tray, or other carrier of material or product should be clearly labeled with the material name, its lot number, and an unequivocal indication of whether or not it has been sterilized. Autoclave tape is a useful indicator, as long as it is understood that it can only indicate that a lot has been through an autoclave cycle, not that it has been sterilized.

Each *heat sterilization* cycle should be automatically recorded on a temperature/time chart, or by other suitable automatic means, such as a digital printout. The chart scale should be large enough to permit accurate reading of both the time and temperature. The chart or printout should form part of the permanent batch record and thus form part of data evaluated when making the release/reject decision. The temperature thus recorded should be sensed from a probe placed in the coolest part of the loaded chamber, this point having been determined for each type of load processed. Where control of the cycle is automatic, the heat-sensing *control* probe should be independent of the *recorder* probe. (If the same probe was used for both purposes, and it was defective, it could actuate an inadequate cycle, yet still signal an apparently satisfactory one.)

In a *steam sterilization* process, it is important to ensure that the steam used is of a suitable quality (clean steam) and does not contain additives or other substances that could cause chemical contamination of the product, material, or equipment being sterilized.

After the high temperature (sterilizing) phase of a heat sterilization cycle has finished, there is a risk that as the load cools, air entering the chamber, and particularly any water used in spray cooling, could be drawn into, for example, inadequately sealed vials. Air admitted before the chamber doors are opened should be filtered, and water used for spray cooling should be water for injection quality.

In *ethylene oxide* sterilization, direct contact between the gas and the microbial cells is essential for effective sterilization. Organisms occluded in crystals, or coated with other material, such as dried proteinaceous matter, may not be killed. The nature and quantity of any packaging material can also markedly affect the efficacy of the process. Before exposure to the gas, the materials should be brought into equilibrium with the required temperature and humidity. Throughout the cycle, records should be made of the cycle time, temperature, pressure, humidity, gas concentration, and total amount of gas used. These records should form part of the batch record and be used in the final evaluation of the batch for release/reject. Ethylene oxide sterilization is an instance where use of biological indicators should be considered mandatory, rather than merely a possible useful adjunct. The generally recommended organism is *Bacillus subtilis* var. *niger,* deposited on a suitable carrier. The positioning of these indicators should be selected following validation studies to determine those parts of the load most difficult to sterilize. The information derived from the use of these biological indicators should form part of the batch manufacturing record, as evaluated when making the final release/reject decision.

During *gamma irradiation sterilization*, the dose received should be monitored throughout the process by the use of plastic dosimeters inserted in the load in sufficient numbers and inserted in packs sufficiently close together so as to ensure that in a continuous process there are always at least two dosimeters in the load, exposed to the source. The standard red perspex dosimeters give a reproducible, quantitative, dose-related change in absorbance, which should be read as soon as possible after exposure to the radiation. *Electron beam sterilization* is rather more difficult to control. The dosimeters used are usually in the form of PVC films. In both cases, the dosimetry results should form part of the batch record. Biological indictors can be used, but *not* as a proof of sterilization. Radiation-sensitive adhesive colored discs are used, but only (repeat, *only*) as a means of indicating that a package has been exposed to radiation and not as proof of sterilization.

In *filtration sterilization*, which should only be used when it is not possible or practicable to sterilize by other more secure means, nonfiber shedding filters, which are demonstrably capable of removing microorganisms without removing ingredients from the solution or releasing substances into it, should be used. It is often advisable to use a (possibly coarser grade) prefilter to first remove larger particles and thus reduce the load on the sterilizing filter. It should not be necessary to mention it, but the once widely-used asbestos filters must not be used unless there is some absolute necessity, and only when there is complete assurance that any released asbestos fibers will be removed downstream. Because of the potential additional risk of filtration as compared with other sterilization methods, it is sound practice to follow the first sterilization grade filter with a second in series, downstream. This has, on occasions, been decried as pointless "belt and braces," but it is impossible to be too careful when lives may be at stake and no good arguments have been mounted against such a double filtration.

The integrity of the sterilizing (and sterilized) filter assembly, *in situ* (not just the filter in isolation) should be confirmed before use and rechecked after use by such methods as the bubble-point, pressure-hold, or forward-flow tests. Most major filter manufacturers supply automatic integrity-testing equipment, applicable to their own filters and filter housings, and will assist in the selection and operation of integrity test procedures appropriate to specific filters, products, and applications. The time during which a sterile-filtered bulk solution is held, pending filling and sealing in its final container, should be kept to a defined minimum, appropriate to the conditions under which the bulk filtered solution is stored. Any one filter should not normally be used for more than one working day, unless a longer period of use can be justified by sound validation studies.

After Sterilization

Of major importance is the need to avoid recontamination of a sterilized product or material, and the mix-up of sterilized with nonsterilized items. Ethylene oxide sterilization is a special case where it is necessary to hold sterilized material under controlled, ventilated conditions to allow any residual ethylene oxide and its reaction products to diffuse away. This presents additional problems in the prevention of recontamination and mix-up.

As well as the chemical analytical testing to confirm compliance with specification, sterile products also require further testing that is specific to this type of product. This testing includes:

- Examination for particles
- Sterility testing
- Leak detection testing
- (And possibly) pyrogen (or endotoxin) testing

Examination for Particulate Contamination

The EC GMP Guide's revised (2003) Annex 1, Manufacture of Sterile Medicinal Products (see annex to previous chapter) requires that "filled containers of parenteral products should be inspected individually for extraneous contamination or other defects" and adds that when this is a visual inspection, "it should be done under suitable and controlled conditions of illumination and background." It adds that operators engaged in this work should pass regular eyesight tests, with glasses, if normally worn, and be given frequent breaks from inspection (an acknowledgment of the very real problem of the decline in the efficiency of the human inspecting machine as a result of eye, and general, fatigue.) Pharmacopoeias (e.g., British, European, and United States) have also variously set down requirements for the examination of filled parenterals for visible and subvisible particles, albeit not necessarily for the 100% inspection that is, *perhaps*, implied by the EC GMP Guide's use of the term "individually." The question thus arises as to why it is considered necessary, or at least desirable, for parenteral products to be free of visible, and in the case of large volume parenterals (LVPs) subvisible particulate contamination. Although the evidence for adverse clinical effects following injection of particles is equivocal, even conflicting (Barnett,[8] Sharp[9]), there is nevertheless a general (understandable, and probably justified) perception that particles in injections do represent a significant health hazard. At very least, it can be said that a lack of particles conveys a highly desirable image of a clean, "quality" product, indicative of high manufacturing standards. It is also worth recalling that environmental organisms are most commonly *not* to be found floating freely in the air, but are usually to be found associated with (or "rafting on") inanimate particles. It is thus not unreasonable to postulate that presence of particles implies potential presence of microorganisms.

Visual inspection is a fallible process, relying as it must on subjective, hardly quantifiable judgments under conditions that are difficult to standardize. Not only is it of doubtful value, it is also a dreary, time consuming job that most workers would wish to avoid. It is not surprising, therefore that various automated electronic methods have been developed. For a comparative review of the techniques and equipment available see Akers.[10]

Sterility Testing

The severe statistical limitations of the compendial sterility test have been discussed in the preceding chapter. There are also microbiological limitations, in particular the fact that there is no "universal" growth medium upon or in which all forms of microorganisms may be expected to grow. As generally practiced, sterility tests will not detect viruses, protozoa, exacting parasitic bacteria or many thermophillic and psychrophilic bacteria. Furthermore, organisms that have been damaged but not killed by exposure to sublethal levels of "sterilization," may not show up in the standard sterility test, as they may require conditions for growth, in terms of nutrients, temperature, and time, which the test does not provide.

Despite these acknowledged limitations, the test continues to be performed, even by those who would accept that it has little real significance in terms of the quality of the product. This would appear to be due largely to regulatory requirements and to a nervous perception of potential legal implications. GMP guidelines appear to accept the limitations by declaring, for example, that "the sterility test applied to the finished product should only be regarded as the last in a series of control measures by which sterility is assured" and "compliance with the test does not guarantee sterility of the whole batch, since sampling may fail to select non-sterile containers, and the culture methods used have limits to their sensitivity...." The EC Sterile Annex (rev. 2003) considers that samples taken for sterility testing should be "representative of the whole batch, but should in particular include samples taken from parts of the batch considered to be most at risk of contamination." Examples given are of (a) samples taken from the beginning and the end of an aseptic run and "after any significant intervention" and (b) samples from the "potentially coolest part of the load" in a heat sterilization.

There will be those who would consider that it would be difficult to encompass these requirements, within the limitations of the twenty-unit sample that is usually taken, and they would be right. Akers[10] has considered alternative statistical sampling methods.

Leaks and Leak Testing

In the context of this chapter, a leak may be defined as:

> Any break or interruption in the physical structure of a container and/or its seal that would permit the egress of its contents or the ingress of any substance, article or material or contaminant, living or non-living

The Parenteral Society's Technical Monograph No. 3, The prevention and detection of leaks in ampoules, vials and other parenteral containers,[11] rightly lays stress on the primary importance of *preventing* the formation of leaks.

The two main causes of leaks in ampoules are cracks in the glass and faulty sealing. Mechanical cracks can be caused by collision or abrasion of ampoules, with one another, or with or against other objects, during or after filling. In addition, thermal cracks can be caused in the glass through rapid cooling from higher temperatures, for example, by contact of hot glass with cold machine parts. Such thermal cracks may develop immediately, or regions of stress may be induced, which develop into cracks later. Crack-inducing stresses can be caused during the original ampoule-forming operation, or in sterile product manufacture, during heat sealing, ceramic printing, or heat sterilization. Faulty ampoule seals can arise from maladjustment or faulty setting of ampoule filling and sealing machines.

Methods aiding prevention are obvious: at all stages from the original forming of the ampoules to the dispatch of the finished product, careful steps should be taken to prevent impact and attrition of glass against glass, or with or against any other objects. Empty and filled ampoules awaiting further processing should be assembled neatly on their bases and not just loaded haphazardly in basket loads. To prevent thermal stress cracks, contact must be avoided between hot glass and cold metal. Careful attention is necessary to machine adjustment, including flame settings, to avoid faulty sealing. With proper setting, draw sealing is less likely to give rise to faulty seals than tip sealing.

There are a number of traditional methods for leak-testing ampoules. They include various pressure and vacuum tests such as the common dye intrusion (or "dye bath") test, liquid loss, and "blotting paper" tests. These, and other techniques have been (and are) used, and they all have their limitations, even hazards — for example, that of dye solution entering an ampoule through a leak and then escaping subsequent detection.

Although not entirely free of problems and limitations, automated high voltage detection methods are more sensitive, and are not subject to the limitation of traditional methods, such as fallibility of human inspectors and hazards of undetected dye intrusion. They also have the further advantage that they can detect points of weakness, such as areas of thin glass, which at the time of testing are potential, if not actual, leaks.

With glass vials, again the major stress should be upon *prevention*, not merely detection of leaks. Measures to prevent mechanical and thermal stresses and cracks are the same as for ampoules. To minimize leaks arising from dimensional, physical, and chemical inadequacies or incompatibilities, it is crucial that detailed and comprehensive specifications are agreed with suppliers of both vials and closures and that compliance with specification is checked on all incoming deliveries.

In contrast with a fairly general acceptance of the need for 100% leak testing of glass ampoule products, a brief survey carried out in 1991–1992 [11] indicated that 100% leak testing of glass vial products was the exception, rather than the rule. This is clearly unsatisfactory from a patient safety point of view, unless it can be shown that there is little, if any, possibility of leaks in filled and sealed glass vials, which does not appear to be the case. Pressure and vacuum tests can be applied to glass vial products, with the same limitations and problems for ampoules. However, based on the experience of the relatively few manufacturers of glass vial products who have tried the technique, it seems that automated high voltage detection is applicable to glass vial products. Such trials that have been conducted have shown that leaks do occur in production batches of filled glass vials, both in the vial body and in the closure system.

Leaks in large volume parenteral (LVP) plastics containers (bags) can be caused by:

■ Faults in the welding or sealing of the container when it is fabricated from the plastics sheet
■ Inadequate "fit," or sealing, of components (tubes, closures, ports) attached to the bag
■ Mechanical damage caused by contact with sharp or abrasive surfaces during filling, sterilization, and subsequent handling
■ Pinholes or splits occurring during bag printing

The standard method of checking for leaks in LVP bags is 100% visual inspection, perhaps with a gentle manual squeeze, after sterilization, to check for abnormal quantities of liquid between the bag and its outer overwrap. Some manufacturers apply a light, controlled mechanical pressure via a bar or plate, immediately before the units are examined. As ever, there are limitations in the efficacy of the human visual checking machine.

Leaks in plastics blow-fill-seal (BFS) containers are most commonly caused by:

■ Imperfect heat seals
■ Damage inflicted by the scrap- (or flash-) removing cropper
■ Careless handling

Leak test methods that can be used include dye bath pressure and vacuum tests, vacuum and blotting paper test, and mechanical pressure plus visual examination. Such work that has been carried out to date suggests that automated high voltage detection methods are applicable to filled BFS containers and appear to be more sensitive than the traditional methods.

Prefilled syringes and cartridges would clearly seem to represent a serious patient hazard if they have leaks. Somewhat disturbingly, however, it does appear there is virtually no information available on the incidence and causes of leaks, nor on suitable methods of leak detection. This is clearly an area in need of serious attention.

Pyrogen, or Endotoxin, Testing

A literal translation of the term "pyrogen" (derived from the Greek) is "fire-generating." However, the reaction in humans to injection of pyrogens can include chill, shivering, vasoconstriction, dilation of pupils, respiratory depression, hypertension, nausea, and pains in joints and head, in addition to (or as a result of) the "fire," or rapid increase in body temperature, which the term suggests. It is reasonable to assume that a patient receiving an injection is, in most cases, already ill. This additional stress to the system cannot be considered as anything less than highly undesirable.

Some substances, including some active drug-substances (or APIs, e.g., some steroids) and some viruses are pyrogenic *per se*, but in terms of sterile products manufacturing on an industrial scale, the most significant pyrogen is the *bacterial endotoxin* that is derived from the outer cell wall of certain gram-negative bacteria. This substance is a complex, high-molecular-weight lipopolysaccharide, soluble in water, and relatively heat-stable. It can withstand autoclaving, and can pass through the 0.2 micrometer pores in the filters commonly used for sterilization by filtration. Destruction, or removal, of microorganisms will not necessarily destroy pyrogenic endotoxins. There is thus another very good reason for keeping bacterial contamination at the lowest possible level *at all stages* in the manufacturing process, in addition to that of ensuring the lowest possible challenge to the sterilization procedure. It is to reduce the chance of the presence of endotoxins. Prevention is, as ever, far better than later detection.

Pyrogenic contamination can arise at any stage in the manufacturing process. It may be present in starting materials, most notably in the water used to make solutions, hence, the importance of good quality water, produced by well-designed and monitored systems. Pyrogenic contamination can be present on the surfaces of containers. It is unlikely to be present on glass containers, as manufactured, in view of the temperatures at which glass in blown or molded, but it can be introduced by washing and rinsing glass containers with water, which is not pyrogen-free. It can be removed from glass containers by exposure to temperatures of 250°C or above in, for example, a sterilizing and depyrogenating tunnel. Once present in a solution, it is difficult, if not impossible, to remove. The answer is to not let it develop in the first place.

The traditional test for the detection of pyrogenic substances relies on the fact that the febrile response of rabbits resembles that of humans. The solution under test is injected into rabbits, and the rise, if any, in their rectal temperatures is measured over the period of the test. The rabbit test has a number of disadvantages: it is a limit, rather than a quantitative test, it is time-consuming and subject to the variability and vagaries inherent in all biological test methods, and it cannot be used for solutions of substances that, themselves, prompt or inhibit a pyrogenic response.

A method that overcomes these problems, which can be used for quantitative determinations, and is more sensitive at low endotoxin levels is based on a discovery that a lysate of the amoebocytes from the blood of the "horseshoe crab" (*Limulus polyphemus*, found mainly along the northeastern seaboard of the American continent), in contact with bacterial endotoxin, shows turbidity or undergoes clotting (gelation). This is the Limulus Amoebocyte Lysate (LAL) test. LAL test kits are widely available from commercial suppliers. Although, at its most simple, the turbidity and gelling end-point is determined visually, the method has been refined to permit more precise turbidimetric, colorimetric, nephelometric determinations (see Akers[10]).

Parametric Release

The concept (or, it would be more correct to say, the terminology) of parametric release is another fairly recent development. "Release," in this context, refers to approving a batch of product for distribution and sale, or for further processing — and surely this is always done (or should be done) after the consideration of a number of relevant parameters. "So," it could be asked, "what's new?"

The terminology emerged in the early-to-mid 1980s and originally was related solely to the sterility (or otherwise) of terminally heat-sterilized products. That is, it did not originally bear upon other release criteria, or on the release of any other products, sterile or otherwise.

One of the first (if not *the* first) "official" publications on this subject is an FDA Compliance Policy Guide on Parametric Release — Terminally Sterilized Drug Products. This Guide provides the following definition:

> Parametric Release is defined as a sterility release procedure based upon effective control, monitoring and documentation of a validated sterilization process cycle, *in lieu* of release based upon end-product sterility testing.

It is possible to wonder — if "sterility release" may be based on "effective control," etc., in place (i.e., *in lieu*) of a sterility test result, then the inverse corollary is surely implied that if a sterility test *has* been passed, then "effective control, monitoring and documentation of validated sterilization process" is not necessary, which is contrary to all the principles of quality assurance in the manufacture of sterile products, and is thus presumably not what the FDA really meant.

This FDA guideline then goes on to list the actions that must be taken (and documented) as preconditions for parametric release.

In brief, they are given as:

1. Validation of the cycle to achieve a reduction of the known microbial bioburden to $10°$ (sic), with a minimum safety factor of an additional

six logarithm reduction. (Validation studies to include heat distribution, heat penetration, bioburden, and cycle lethality studies.)
2. Validation of integrity of container/closure
3. Pre-sterilization bioburden testing on each lot, pre-sterilization, and checking comparative resistance of any spore-formers found.
4. Inclusion of chemical or biological indicators in each truck-load

At the risk of appearing not merely to be laboring a point, but to be beating it to death, it is difficult to refrain from asking the question — is not this (with the possible exception of the inclusion of biological indicators in *every* load) a list of things that should be done anyway, whether the lot is to be sterility tested or not?

(Note: The "10°" in 1. above is presumably intended to mean 1.)

This form of parametric release provides an indication of the type and range of process parameters that need to be considered before a product may reasonably be released without testing the end product for a specific quality characteristic. In this particular instance, a notably unreliable test procedure (the sterility test) may be abandoned, with at least a theoretical possibility of regulatory approval, in favor of a rigorous concentration of effort on actions that will provide a significantly higher level of assurance of sterility — an excellent notion, in this context, and one that has been adopted (with official approval) by a few sterile products manufacturers. But they are surprisingly few. The reason for this probably lies in a not unfounded fear that, should action be taken for damages in the case of an alleged sterility failure, learned judges will probably consider "passed pharmacopoeial sterility test" a better defense than technical and statistical arguments that they will not understand.

REFERENCES

1. FDA, Guideline on Sterile Drug Products Produced by Aseptic Processing, Center for Drugs and Biologics and Office of Regulatory Affairs, Food and Drug Administration,Rockville, MD, June 1987.
2. Whyte, W., In support of settle plates, *J. Pharm. Sci. Technol.,* 50/4, 201, 1996.
3. Whyte, W. and Niven, L., Airborne bacterial sampling: the effect of dehydration and sampling time, *J. Parenter. Sci. Technol.,* 40, 182, 1986.
4. Russell, M., Goldsmith, J., and Phillips, I., Some factors affecting the efficiency of settle plates, *J. Hosp. Infect.,* 5, 1989, 1984.
5. ASTM, Sizing and Counting Particulate Contamination in and on Clean Room Garments, American Society for Testing Materials, No. ASTM F51-68, 1968.
6. AS 2013 1977, Clean Room Garments, Standards Association of Australia, North Sydney, NSW, 1977.
7. AS 2014 1977, Code of Practice for Clean Room Garments, Standards Association of Australia, North Sydney, NSW, 1977.
8. Barnett, M. Particulate contamination, in *The Pharmaceutical Codex*, 12th edition, Pharmaceutical Press, London, 1994.
9. Sharp, J., Particulate contamination of injections, in *Good Manufacturing Practice — Philosophy and Applications*, Interpharm Press, Illinois, 1991.
10. Akers, M., Particulate matter testing, Pyrogen testing, and Sterility testing, in *Parenteral Quality Control*, Marcel Dekker, New York, 1985.
11. Parenteral Society, Prevention and Detection of Leaks in Ampoules, Vials and Other Parenteral Containers, Technical Monograph No. 3, Swindon, Wilts, UK, 1992.

15

VALIDATION — GENERAL PRINCIPLES

The US cGMPs make four specific references to validation in the contexts, respectively, of computer systems, supplier test results, sampling and testing in-process materials, and sterilization processes. These are:

US CGMPS

Sec. 211.68 Automatic, mechanical, and electronic equipment

(a)   Automatic, mechanical, or electronic equipment or other types of equipment, including computers, or related systems that will perform a function satisfactorily, may be used in the manufacture, processing, packing, and holding of a drug product. If such equipment is so used, it shall be routinely calibrated, inspected, or checked according to a written program designed to assure proper performance. Written records of those calibration checks and inspections shall be maintained.

(b)   Appropriate controls shall be exercised over computer or related systems to assure that changes in master production and control records or other records are instituted only by authorized personnel. Input to and output from the computer or related system of formulas or other records or data shall be checked for accuracy. The degree and frequency of input/output verification shall be based on the complexity and reliability of the computer or related system. A backup file of data entered into the computer or related system shall be maintained except where certain data, such as calculations performed in connection with laboratory analysis, are eliminated by computerization or other automated processes. In such instances a written record of the program shall be maintained along with appropriate *validation* data. Hard copy or alternative systems, such as duplicates, tapes, or microfilm, designed to assure that backup data are exact and complete and that it is secure from alteration, inadvertent erasures, or loss shall be maintained.

Sec. 211.84 Testing and approval or rejection of components, drug product containers and closures

(d) (2) Each component shall be tested for conformity with all appropriate written specifications for purity, strength, and quality. In lieu of such testing by the manufacturer, a report of analysis may be accepted from the supplier of a component, provided that at least one specific identity test is conducted on such component by the manufacturer, and provided that the manufacturer establishes the reliability of the supplier's analysis through appropriate *validation* of the supplier's test results at appropriate intervals.

Sec. 211.110 Sampling and testing of in-process materials and drug products

(a) To ensure batch uniformity and integrity of drug products, written procedures shall be established and followed that describe the in-process controls, and tests, or examinations to be conducted on appropriate samples of in-process of each batch. Such control procedures shall be established to monitor the output and to *validate* the performance of those manufacturing processes that may be responsible for causing variability in the characteristics of in-process material and the drug product. Such control procedures shall include, but are not limited to, the following …

Sec. 211.113 Control of microbiological contamination

(b) Appropriate written procedures, designed to prevent microbiological contamination of drug products purporting to be sterile, shall be established and followed. Such procedures shall include *validation* of any sterilization process.

In 1987, the US FDA published its Guideline on General Principles of Process Validation.[1] A slightly shortened copy of this Guideline is given in Annex 1 to this chapter. In it, validation is defined as:

VALIDATION — Establishing documented evidence which provides a high degree of assurance that a specific process will consistently produce a product meeting its pre-determined specifications and quality attributes

The EC GMP Guide offers the following definition:

VALIDATION — Action of proving, in accordance with the principles of Good Manufacturing Practice, that any procedure, process, equipment, material, activity or system actually leads to the expected results (see also qualification).

Qualification is defined in the EC GMP Guide as follows:

QUALIFICATION - Action of proving that any equipment works correctly and actually leads to the expected results. The word validation is sometimes widened to incorporate the concept of qualification.

There will be those, including this writer, who consider that these two EC definitions, when considered together, are not entirely clear or definitive. It will also be noted that, whereas in the FDA Guideline *validation* is a term applicable to a (specific) *process*, in the minds of those who prepared the EC GMP Guide, *validation* is applicable to "any procedure, equipment, material, activity or system" as well as to a *process*.

The European Commission Directive 91/356/EEC, Medicinal Products for Human Use, and its companion directive 91/412/EEC, Veterinary Use, "laying down the principles and guidelines of good manufacturing practice," state as a requirement of European Law that:

EUROPEAN DIRECTIVES 91/356/EEC AND 91/412/EEC

Article 10 — Production

Any new manufacture or important modification of a manufacturing process shall be validated. Critical phases of manufacturing processes shall be regularly revalidated.

There are a number of references to validation in the EC GMP Guide. These include:

EC GMP GUIDE

Chapter 5 Production

Validation

5.21 Validation studies should reinforce Good Manufacturing Practice and be conducted in accordance with defined procedures. Results and conclusions should be recorded.

5.22 When any new manufacturing formula or method of preparation is adopted, steps should be taken to demonstrate its suitability for routine processing. The defined process, using the materials and equipment specified, should be shown to yield a product consistently of the required quality.

5.23 Significant amendments to the manufacturing process, including any change in equipment or materials, which may affect product quality and/or the reproducibility of the process should be validated.

5.24 Processes and procedures should undergo periodic critical re-validation to ensure that they remain capable of achieving the intended results.

Chapter 6 Quality Control

6.15 Analytical methods should be validated. All testing operations described in the marketing authorisation should be carried out according to the approved methods.

Annex 1 — Manufacture of Sterile Medicinal Products

55. All sterilisation processes should be validated ...
73. (re. Sterilisation by radiation) Validation procedures should ensure that the effects of variations in density of the packages are considered.
76. (re. Sterilisation with ethylene oxide) ... During process validation it should be shown that there is no damaging effect on the product and that the conditions and time allowed for degassing are such as to reduce any residual gas and reaction products to defined acceptable limits ...

Annex 11 — Computerised Systems

2. Validation: the extent of validation necessary will depend on a number of factors, including the use to which the system is to be put, whether the validation is to be prospective or retrospective and whether or not novel elements are incorporated. Validation should be considered as part of the complete life cycle of a computer system...

In July 2001, the European Commission issued a new Annex (no. 15) to the EC GMP Guide on Qualification and Validation. This is reproduced as Annex 2 to this chapter.

One of the simplest and most easily understood definitions comes from Fry[2] (at the time a senior official of the FDA), who in 1982 at a Conference of the Pharmaceutical Inspection Convention in Dublin said:

> To prove that a process works is, in a nutshell, what we mean by the verb to validate.

Viewed in this fashion, it all seems relatively simple, without a hint of the complexities and confusion that have arisen since the concept of validation arrived in the industrial pharmaceutical arena. Indeed, not the least of the many surprising things about validation is the way in which such large and arcane semantic structures, such mountains of paper, and (as some would argue) such unnecessary effort, consternation, confusion, and expense have arisen from such slight original regulatory bases.

One way of looking at the ongoing increasing emphasis on validation is as a stage in the general trend, over the last 50 years or so, *away* from regarding end-product testing as a sole determinant of product quality, and *toward* a more comprehensive view of what it takes to *assure* the quality of pharmaceutical products. This is expressed particularly well in the following passage from the FDA Guideline on General Principles of Process Validation (1987):

> Assurance of product quality is derived from careful attention to a number of factors including selection of quality parts and materials, adequate product and process design, control of the process, and in-process and end-product testing.

Due to the complexity of today's medical products, routine end-product testing alone often is not sufficient to assure product quality for several reasons. Some end-product tests have limited sensitivity. (For example, USP XXI states: "No sampling plan for applying sterility tests to a specified proportion of discrete units selected from a sterilization load is capable of demonstrating with complete assurance that all of the untested units are in fact sterile.") In some cases, destructive testing would be required to show that the manufacturing process was adequate, and in other situations end-product testing does not reveal all variations that may occur in the product that may impact on safety and effectiveness.

The basic principles of quality assurance have as their goal the production of articles that are fit for their intended use. These principles may be stated as follows: (1) quality. safety. and effectiveness must be designed and built into the product; (2) quality cannot be inspected or tested into the finished product; and (3) each step of the manufacturing process must be controlled to maximize the probability that the finished product meets all quality and design specifications. Process validation is a key element in assuring that these quality assurance goals are met.

It is through careful design and validation of both the process and process controls that a manufacturer can establish a high degree of confidence that all manufactured units from successive lots will be acceptable. Successfully validating a process may reduce the dependence upon intensive in-process and finished product testing. It should be noted that in most all cases, end-product testing plays a major role in assuring that quality assurance goals are met; i.e., validation and end-product testing are not mutually exclusive.

In addition to its Validation Guideline, the FDA has issued a Guideline on Sterile Drug Products Produced by Aseptic Processing (June 1987, currently under revision — draft circulated for comment August 2003), which contains useful guidance on the validation of this type of process. There is also an FDA Guide to Inspection of Computerized Systems in Drug Processing, issued as reference material for investigators, which contains relatively brief sections on hardware and software validation. The FDA has also published a Guide to Inspections of Validation of Cleaning Process (July 1993). The FDA Guide to Inspections of Pharmaceutical Quality Control Laboratories (July 1993) contains three short paragraphs on Methods Validation, and their Guide to Inspections of Microbiological Pharmaceutical Quality Control Laboratories (July 1993) also contains a short section on Methodology and Validation of Test Procedures. There are other FDA guidelines directed at the validation of a range of nonsterile manufacturing processes, and the general regulatory trend worldwide (either by explicit statement or by implication) is to expect at least some level of validation of many types of processes, nonsterile as well as sterile.

A few questioning voices have been heard. At the 1982 Dublin seminar, to which we have already referred above, where he "presented the only real definition that 'to validate' has ever needed" (Akers[3]), Fry (at the time, still a senior official of the FDA), in a paper generally in favor of validation but with a rational approach, commented *inter alia* that "… a magnificent selling job has been done on the concept of validation …"; that it " … is beginning to rank high on our list of sacred concepts, right up there with Motherhood, the Flag and apple pie"; and that " ... the validation requirements have resulted in virtually a new industry to be created, of consulting firms who perform validation studies for pharmaceutical manufacturing clients" (Fry[2]).

At the same seminar, Witschi[4] considered the extent to which validation is needed, and suggested a differential approach to the amount of validation effort required, based on such factors as potential hazard in use, mode of administration, criticalness of dose-level and value/reliability of in-process and end-product testing.

Sharp[5,6] attacked a number of the more bizarre manifestations of the validation cult, and suggested that "... the need for validation (is) perhaps inversely proportional to the adequacy of product design, raw material control, in-process control, and end product testing to provide assurance of routine product quality."

Anisfeld[7] questioned the cost/benefit advantages of validation. Anisfeld noted that 'the annual cost (in US understood) is running at just below a billion dollars," and asked "... are we seeing less product failures? ... less testing? ... no one seems sure." He stated that there had been no significant decline in product recalls in the 1980s as compared with the 1970s, and wondered "How come" ... (if as some validation pundits had suggested validation should ultimately give rise to cost-savings) "... drug prices keep climbing at three times the rate of inflation?" He declared that we need to "relate our validation work to the needs of patient safety and product quality."

In 1994, Fry[8] (by then wearing a different professional hat as president of PDA) argued that "every regulatory document ... such as the (FDA's) 'Guideline on Documentation of Sterilization Validation' (and) the 'Cleaning Validation Inspection Guide' ... adds to the cost of producing drugs. Many regulatory initiatives are quality-related and make good sense — but when these additional cost burdens do nothing discernable to improve quality, the patient ends up paying more for no good reason."

Akers[3] also joined the "challenge (to) some of the elaborate belief structures upon which validation is based," and expressed the views that 'validation is becoming in many ways a negative force within our industry," that 'outside of sterilization applications it has become mainly a verbal rather than a technical exercise," and that it is 'evolving into a bureaucratic exercise instead of a scientific one." He considered that 'validation is conceptually quite simple' but that 'the emphasis, terminology and definition has enabled a few individuals to build empires within their firms or create lucrative side businesses." He noted an "increasing tendency towards validation by the pound (i.e., pound weight) where the number of protocols written, and the amount of paper contained therein is equated with success' and urged a "back to basics' approach of simply demonstrating that "a process works."

This review of opposing views on validation has been necessary to show that alternative views are possible. Indeed, there are almost certainly more dissenters than is immediately apparent, but a number of these fail to speak out for fear of bringing regulatory wrath upon their companies.

Despite the alternative views, the current regulatory pressure to validate extensively is unlikely to "go away," and there is no probability (or even much likelihood) that the more moderate voices will be heard by the regulators. Furthermore, despite the manifold absurdities that have developed around it (see examples given in Sharp[5]), it cannot be doubted that *in basic essence* the concept of validation is a sound and valuable one.

HOT TOPICS

Current validation "hot topics," particularly insofar as regulatory bodies are concerned, are:

- Validation of sterilization processes
- Validation of cleaning procedures
- Computer systems validation
- Analytical methods validation

There can be no doubt that the field in which validation is crucially necessary, and is universally acknowledged to be so, is in the manufacture of sterile products. Because of the limitations of the standard sterility test as a means of giving assurance of the sterility of batch as a whole and the very serious potential hazards of these products if they are *not* sterile, there is an unquestioned need to validate the processes used to obtain and maintain that condition.

Thus discussion of *process* validation has indeed tended largely to focus on sterile products. On the validation of processes used for the manufacture of other dosage forms, for example, tablets, capsules, oral liquids, and nonsterile topical creams and ointments distinctly less has been said and written. This could, perhaps, lead us to reflect that, possibly, the need for validation is inversely proportional to the adequacy of product design, raw-material control, in-process control, and end-product testing to provide assurance of the quality of routine batch production. None of these things, alone or in combination, can provide the necessary assurance of the success of a sterilization process, hence, the crucial need, in that context, for process validation.

In contrast, consider a simple, nonmodified release tablet product. Given a well-designed process, and given that in routine production this process is followed:

- Using materials of confirmed quality
- Applying appropriate in-process tests and controls
- Assaying granules to determine even distribution of actives
- Using modern tabletting equipment (with, for example, automatic weight control)
- Applying regular checks on tablet weight, hardness, thickness, friability, disintegration, etc.
- Performing end-product tests, which provide independent checks on these latter parameters, and including, for example, individual tablet assays

Given that all these things are done, might it not be concluded that each production run is self-validating (or self-invalidating, as the case may be)?

Furthermore, may not the same sort of thinking be applied to the manufacture of other nonsterile products?

In sum, perhaps the extent of the need for, and the application of, validation should be determined by a consideration of an interaction between two hierarchies: (a) the critical nature and hazard of product; (b) the ability of factors other than validation to provide assurance of quality. On this basis, products intended for injection would come top of the list for extensive process validation, on grounds of critical use, potential high hazard, and the absence of any realistic test to confirm a crucial quality characteristic (sterility). On the other hand, a simple topical liquid preparation intended to soothe hot feet — simple to make, easy to assay, noncritical, and unlikely to represent a critical patient hazard — would come well down the list.

To hold such views as these, does not, of course, indicate or even imply a disregard for the need for validation, but merely a realization of the need rationally to evaluate the extent of that need in different contexts. *However*, it is not possible to be sure that regulatory officials will take a similar view.

TERMINOLOGY OF VALIDATION

In addition to the definitions of "validation" as given in the FDA Guideline and the EC Annex 15, and as quoted above, there are a number of other subsidiary terms, which are not always well understood. The definitions given in the two regulatory documents may be compared as follows:

Change Control

A formal system by which qualified representatives of appropriate disciplines review proposed or actual changes that might affect the validated status of facilities, systems, equipment, or processes. The intent is to determine the need for action that would ensure and document that the system is maintained in a validated state. **(EC — Not specifically defined in FDA)**

Cleaning Validation

Cleaning validation is documented evidence that an approved cleaning procedure will provide equipment that is suitable for processing medicinal products. **(EC — Not specifically defined in FDA)**

Concurrent Validation

Validation carried out during routine production of products intended for sale. **(EC — Not specifically defined in FDA)**

Design Qualification (DQ)

The documented verification that the proposed design of the facilities, systems, and equipment is suitable for the intended purpose. **(EC — Not specifically defined in FDA)**

Installation Qualification (IQ)

Establishing confidence that process equipment and ancillary systems are capable of consistently operating within established limits and tolerances. **(FDA)**

The documented verification that the facilities, systems, and equipment, as installed or modified, comply with the approved design and the manufacturer's recommendations. **(EC)**

Operational Qualification (OQ)

The documented verification that the facilities, systems, and equipment, as installed or modified, perform as intended throughout the anticipated operating ranges. **(EC — Not specifically defined in FDA)**

Process Performance Qualification

Establishing confidence that the process is effective and reproducible. **(FDA)**

Performance Qualification (PQ)

The documented verification that the facilities, systems, and equipment, as connected together, can perform effectively and reproducibly, based on the approved process method and product specification. **(EC)**

Product Performance Qualification

Establishing confidence through appropriate testing that the finished product produced by a specified process meets all release requirements for functionality and safety. **(FDA — Not specifically defined in EC)**

Prospective Validation

Validation conducted prior to the distribution of either a new product, or product made under a revised manufacturing process, where the revisions may affect the product's characteristics. **(FDA)**

Validation carried out before routine production of products intended for sale. **(EC)**

Retrospective Validation

Validation of a process for a product already in distribution based upon accumulated production, testing, and control data. **(FDA)**

Validation of a process for a product which has been marketed based upon accumulated manufacturing, testing and control batch data. **(EC)**

Re-Validation

A repeat of the process validation to provide an assurance that changes in the process/equipment introduced in accordance with change control procedures do not adversely affect process characteristics and product quality. **(EC — Not specifically defined in FDA)**

Risk Analysis

Method to assess and characterise the critical parameters in the functionality of an equipment or process. **(EC — Not specifically defined in FDA)**

Simulated Product

A material that closely approximates the physical and, where practical, the chemical characteristics (e.g., viscosity, particle size, pH, etc.) of the product under validation. In many cases, these characteristics may be satisfied by a placebo product batch. **(EC — Not specifically defined in FDA)**

System

A group of equipment with a common purpose. **(EC — Not specifically defined in FDA)**

Validation

Establishing documented evidence that provides a high degree of assurance that a specific process will consistently produce a product meeting its predetermined specifications and quality attributes. **(FDA)**

(Process) Validation

The documented evidence that the process, operated within established parameters, can perform effectively and reproducibly to produce a medicinal product meeting its predetermined specifications and quality attributes. **(EC)**

Note that whereas the EC definition of "validation" embraces equipment, materials, systems, etc. (see above), "process validation" is closer to the more limited meaning of the FDA's "validation."

Validation Protocol

A written plan stating how validation will be conducted, including test parameters, product characteristics, production equipment, and decision points on what constitutes acceptable test results. **(FDA — Not specifically defined in EC)**

Worst Case

A set of conditions encompassing upper and lower processing limits and circumstances, including those within standard operating procedures, which pose the greatest chance of process or product failure when compared to ideal conditions. Such conditions do not necessarily induce product or process failure. **(FDA)**

A condition or set of conditions encompassing upper and lower processing limits and circumstances, within standard operating procedures, which pose the greatest chance of product or process failure when compared to ideal conditions. Such conditions do not necessarily induce product or process failure. **(EC)**

It is not difficult to be confused by these terms and to wonder at the necessity for the various different "Qs." What now follows is an attempt to provide a simplified picture of the overall concept of validation:

Before process validation can commence, there must be what may be termed a prevalidation phase (or phases). This phase, in addition to such considerations as equipment specification, equipment design, and equipment purchase, requires attention to qualification of the processing equipment, sometimes called equipment qualification (EQ). This is a (possibly) redundant term, which means approximately the same thing as what has long been referred to in engineering circles as "commissioning."

EQ is often considered, in turn, to have three (at least) main phases:

- **Design qualification (DQ),** which, as we have seen, is defined in the EC GMP Guide (but not in the FDA Validation Guideline) as "the documented verification that the proposed design of the facilities, systems and equipment is suitable for the intended purpose."
- **Installation qualification (IQ),** which, for our purposes, may be defined as: "the action of demonstrating and certifying that a piece of equipment is properly installed, provided with all necessary (and *functioning*) services, subsidiary equipment, and instruments, and is capable of performing in accordance with its basic design parameters."
- **Operational qualification (OQ),** which, for our purposes, may be defined as "demonstrating that the equipment, as specified and installed, will perform consistently (within predefined limits) throughout its operating range (including, some have suggested, operating under 'worst-case' conditions)."

There have been those who wonder if this distinction between IQ and OQ is necessary or relevant and who find it difficult to understand where one ends and the other begins. If the reader feels that way, then he or she has this author's sympathy. It might help to know that it has been suggested that IQ is about specific static aspects of a piece of equipment or system, and OQ is about specific dynamic aspects. On the other hand, this distinction might not help at all. What can be said with certainty is that lengthy discussion over where, conceptually and practically, IQ ends and OQ begins is a total waste of time and effort. To add to the confusion in what is increasingly appearing to be a distinctly *in*exact science, it has also been recently suggested that IQ equals "inspection qualification." How that differs from "installation qualification" is not entirely clear.

On a purely common-sense view, none of these various phases need to be considered as entirely "water-tight" compartments. The divisions should exist solely as matters of convenience in discussion. In practice, there is likely to

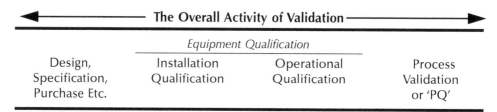

Figure 15.1   A view of the overall concept of validation

be some overlap, or merging, between the various components of validation and qualification. However, as Akers[3] has noted, there has been "impassioned and generally useless argument" over where a given activity should be classified in the IQ/OQ/PQ spectrum.

Some (including this writer) also find it difficult to understand what the difference is between "process validation" and "process performance qualification" or "performance qualification" (PQ) as defined, respectively in the FDA Guideline and the EC Annex. A simplified illustration of the relationships between these various phases is suggested in Figure 15.1.

Validation has also been considered by some to have a number of possible approaches or strategies. The three most commonly encountered (others have been suggested) are "prospective validation," "concurrent validation," and "retrospective validation." It is to be noted that, despite their definitions or mentions in various regulatory documents the necessity for, or the relevance of, these subsidiary terms is not universally accepted. They have been used to apply either to (a) just the process validation phase, or to (b) the complete qualification *plus* process validation cycle. The following are attempts to clarify the meanings of these, possibly unnecessary, terms:

"Prospective Validation" refers to new processes and new equipment, where studies are conducted and evaluated, and the overall process/equipment system confirmed as validated *before* the commencement of routine production.

"Concurrent Validation" applies to existing processes and equipment. It consists of studies conducted during normal routine production and can only be considered acceptable for processes that have a manufacturing and test history indicating consistent quality production. Although lack of suitable records relating to the qualification phases may not necessarily compromise concurrent validation of some processes, evidence of proper machine installation is important in some contexts.

"Retrospective Validation" applies to existing processes and equipment and is based on historical information. Unless sufficiently detailed past processing and control records are available, retrospective validation studies are unlikely to be either possible or acceptable. For example, it would be necessary to establish that the process had not been modified and that the equipment was still operating under the same conditions of construction and performance as documented in the historical records. Maintenance records and process change control documentation would be necessary to support any such claim. Furthermore, the incidence of process failures, and records of rejects or reworking would need to be carefully evaluated for evidence of inconsistency in the process. Manufacturing, maintenance, testing, and calibration data would all need to unequivocally demonstrate process uniformity, consistency, and continuity.

The meaning of "revalidation" would seem to be self-apparent, but it can, in fact, refer to the act of repeating all, or part, of a validation study in response to a modification to process or equipment (some would argue that this is validation, not revalidation, since this is now a new ball game) or to the regular planned repetition of validation, as, for example, in the context of regular revalidation of sterilization processes.

It is worth repeating that there is, worldwide, considerable variation in the understanding and use of the various terms discussed. It would probably be best if both manufacturers and regulators did not confuse or overconcern themselves (or each other) with the semantic niceties. The important issues are that:

- The overall process is understood
- Equipment is appropriately specified and designed
- Equipment is properly installed and maintained and is demonstrably operating as specified and designed
- The process is studied and monitored to ensure that it does achieve the desired and intended result

To quote Akers[3] again, perhaps "the time has come to clear away this terminological debris, and instead focus on what we want our process control programs to achieve in terms of product quality."

THE VALIDITY OF VALIDATION

For any validation study, itself, to be valid it is essential that:

- The measuring devices and instruments used in the study are properly calibrated.
- Personnel performing the study are competent and trained to undertake the work.

Following the study, for the results to remain valid, it is necessary to:

- Maintain all equipment to the same standards as used in the study
- Ensure that all measuring devices and process controlling instruments remain in calibration
- Ensure that operating staff remain properly competent to perform the process in routine production

■ Implement a system of change control, which will guard against unplanned or inadvertent process changes and will highlight any change that will require consideration of the need for further validation

VALIDATION DOCUMENTATION

Of the many forms of documentation that may be required, three types may be distinguished as of primary importance:

1. The validation *protocol*, which defines the procedure
2. The *report* on the validation study
3. The *supporting documentation* presented with the report

There may be, in addition, a need or requirement for a Validation Master Plan.

Validation Protocol

Each major stage in an overall validation process should be conducted in accordance with a detailed, written, preestablished, and formally approved validation protocol.

Written change control procedures should also be established, which will prevent unauthorized changes to either the process itself, or to the study protocol, and restrict change during any stage of the study until all relevant data have been formally evaluated for the effect(s) that any change may have on the study, and any conclusions derived from it.

Validation protocols should bear a title, date, and unique identification or reference number. They should be formally authorized and approved by person(s) with the competence and authority to do so.

Validation protocols should give in detail:

■ The objectives and scope of the study, that is, there should be a clear *definition of purpose*
■ A clear and precise definition of the process, equipment, system, or subsystem that is to be the subject of the study, with details of their performance characteristics
■ Installation and qualification requirements for new equipment
■ Any upgrading requirements for existing equipment, with justification for the change(s) and a statement of qualification requirements
■ Detailed, step-wise statement of actions to be taken in performing the study (or studies)
■ Assignment of responsibility for performing the study
■ Statements on all test methodology to be employed, with a precise statement of the test equipment or materials to be used
■ Test equipment calibration requirements
■ Reference-listing of relevant standard operating procedures (SOPs)
■ Statement of requirements for the content and format of the report on the study

- A statement of acceptance criteria, that is, the criteria against which the results obtained and documented in the course of the study are to be evaluated, in order to determine the success (or otherwise) of the study; and to decide whether or not the process is to be considered as validated
- A statement of the personnel responsible for evaluating, and certifying as acceptable, each stage in the study, and for the final evaluation and certification of the process as a whole, all as measured against the predefined acceptance criteria

The following is intended as a quick check-list of the basic requirements of a *validation protocol*:

- Title
- Date
- Unique identification (reference number)
- Formal authorization/approval (author, others involved)
- Objective(s)/scope (clear definition of purpose)
- Test equipment/materials to be used.
- Calibration requirements
- Description of procedure to be followed
- References to any relevant SOPs
- Report requirements/format
- Assignment of responsibility
- Acceptance criteria (i.e., the criteria against which the success, or otherwise, of the study is to be judged)

In developing validation protocols, consideration may also be given to the use of risk, or hazard, analysis (HACCP and the like) to determine the more critical operations within an overall manufacturing process, in terms of the potential risk to quality (and, hence, to patients) that a failure in any given process-step, or substep would represent, and thus to establish validation priorities.

Validation Report

A validation report should bear a title, date, and unique identification or reference number. It should include a copy of, or an unequivocal reference to, the validation protocol that was followed in order to generate the report. It should be possible to readily retrieve the protocol and relate it to the report.

The report should contain:

- Outline of procedure followed
- Copies of, or specific references to, test procedures followed
- Details of the calibration of test equipment and measuring devices
- Details of the results obtained in the study
- Formal assessment of those results against the acceptance criteria
- Decision, with signature(s)

Validation Report — Supporting Data

Documentation, which should be available, as relevant, to support the results and conclusions of the validation report, includes applicable SOPs, "raw" test results, recorder charts or printouts, calibration reports, equipment and work-flow diagrams, full analytical and test reports, and environmental monitoring reports

Some regulatory agents have considered that a company's overall validation policies, plans, and programs should be summarized in a validation master plan (VPM).

VALIDATION DATA REVIEW AND EVALUATION

All information and data generated during the course of a study, as run in accordance with a validation protocol, should be formally evaluated against the acceptance criteria (as defined in the protocol) and judged as meeting or failing to meet those criteria. All decisions thus made should be documented, with written evidence supporting the evaluations and the conclusions drawn.

If any such evaluation shows that protocol criteria have not been met, the study should be considered as having failed to validate the process, and the reasons should be investigated and documented. All necessary corrective action should be taken before the validation study is resumed or recommenced. Any failure to follow the procedure as laid down in the protocol should be considered as potentially compromising the validity of the study, and requires critical evaluation of the impact on the study.

Final certification, on completion of a validation study, should be made only and specifically in relation the predetermined acceptance criteria.

PERSONNEL

All personnel taking part (in any capacity) in validation work should be specifically trained in the tasks assigned to them in the validation protocol. Documented evidence of the relevant experience and training of all personnel involved in validation studies should be maintained.

Appropriately qualified and experienced supervisory personnel should ensure that the protocol and the testing methodology involved are based on sound scientific and engineering principles and that all studies are properly evaluated and certified. All personnel conducting tests should be trained and experienced in the use of the equipment and measuring devices used. All manufacturing and control operations that are the subject of the study must be conducted in accordance with Good Manufacturing Practice.

TESTING FACILITIES AND PROCEDURES

All tests and measurements (physical, chemical, and microbiological) conducted during validation studies should be performed by personnel, properly trained and competent to carry out the test procedures assigned to them, using appropriately calibrated equipment, instruments, and devices, in suitably equipped laboratories (or other testing facilities).

Instrument calibration should be performed in accordance with preestablished and approved programs and procedures, and calibration records maintained. The calibration status (when calibrated, next due date, etc.) of any measuring instrument or device should be readily apparent.

Detailed, authorized, written procedures setting out the relevant, validated test-methodology should be available for all tests that are to be carried out during the course of a validation study. These written procedures should be referenced in the validation protocol.

If external contract laboratory facilities are used in the course of a validation study, the competence of these laboratories to carry out the test(s) required should be determined in advance. This requirement should be stated in the validation protocol. The names and addresses of any contract laboratories used should be documented in the validation report.

REFERENCES

1. FDA, Guideline on General Principles of Process Validation, US Food and Drug Administration, Rockville, MD, 1987.
2. Fry, E., Validation — Theory and Concepts. Collected papers of PIC Seminar on Validation, Dublin, June 1982, EFTA Secretariat, Geneva, 1982.
3. Akers, J., Simplifying and improving process validation, *J. Parent. Sci. and Tech,* 47(6), 281–284, 1993.
4. Witschi, T., Validation from the inspector's standpoint, Collected papers of PIC Seminar on Validation, Dublin, June 1982, EFTA Secretariat, Geneva, 1982.
5. Sharp, J., Problems of Process Validation, *PJ,* 1986, Revised version of a paper presented at a colloquium on industrial pharmacy, Ghent, 1985, and later republished in Sharp, J., *Good Manufacturing Practice — Philosophy and Applications*, Interpharm Press, US, 1991.
6. Sharp, J., Validation — How much is required? *PDA J Pharm Sci and Tech,* 49(3), 111–118, 1995.
7. Anisfeld, M., Validation — How much can the world afford?, Proc. PDA International Congress, Basel, 191–197, 1993. (See also Anisfeld Validation — How much can the world afford? Are we getting value for money? *J. Parent. Sci. and Tech.,* 48(1), 45–48.)
8. Fry, E., PDA Letter, June 1994.

ANNEX 1 TO CHAPTER 15

Slightly abridged copy of the US FDA Guideline (May 1987) on:

General Principles of Process Validation

I. Purpose

This guideline outlines general principles that FDA considers to be acceptable elements of process validation for the preparation of human and animal drug products and medical devices.

II. Scope

This guideline is issued under Section 10.90 (21 CFR 10.90) and is applicable to the manufacture of pharmaceuticals and medical devices. It states principles and practices of general applicability that are not legal requirements but are acceptable to the FDA. A person may rely upon this guideline with the assurance of its acceptability to FDA, or may follow different procedures. When different procedures are used, a person may, but is not required to, discuss the matter in advance with FDA to prevent the expenditure of money and effort on activities that may later be determined to be unacceptable. In short, this guideline lists principles and practices which are acceptable to the FDA for the process validation of drug products and medical devices; it does not list the principles and practices that must, in all instances, be used to comply with law.

III. Introduction

Process validation is a requirement of the Current Good Manufacturing Practices Regulations for Finished Pharmaceuticals, 21 CFR Parts 210 and 211, and of the Good Manufacturing Practice Regulations for Medical Devices, 21 CFR Part 820, and therefore, is applicable to the manufacture of pharmaceuticals and medical devices.

Several firms have asked FDA for specific guidance on what FDA expects firms to do to assure compliance with the requirements for process validation. This guideline discusses process validation elements and concepts that are considered by FDA as acceptable parts of a validation program. The constituents of validation presented in this document are not intended to be all-inclusive. FDA recognizes that, because of the great variety of medical products (drug products and medical devices), processes and manufacturing facilities, it is not possible to state in one document all of the specific validation elements that are applicable. Several broad concepts, however, have general applicability which manufacturers can use successfully as a guide in validating a manufacturing process. Although the particular requirements of process validation will vary according to such factors as the nature of the medical product (e.g., sterile vs. non-sterile) and the complexity of the process, the broad concepts stated in this document have general applicability and provide an acceptable framework for building a comprehensive approach to process validation.

Definitions

Installation qualification — Establishing confidence that process equipment and ancillary systems are capable of consistently operating within established limits and tolerances.

Process performance qualification — Establishing confidence that the process is effective and reproducible.

Product performance qualification — Establishing confidence through appropriate testing that the finished product produced by a specified process meets all release requirements for functionality and safety.

Prospective validation — Validation conducted prior to the distribution of either a new product, or product made under a revised manufacturing process, where the revisions may affect the product's characteristics.

Retrospective validation — Validation of a process for a product already in distribution based upon accumulated production, testing and control data.

Validation — Establishing documented evidence which provides a high degree of assurance that a specific process will consistently produce a product meeting its pre-determined specifications and quality attributes.

Validation protocol — A written plan stating how validation will be conducted, including test parameters, product characteristics, production equipment, and decision points on what constitutes acceptable test results.

Worst case — A set of conditions encompassing upper and lower processing limits and circumstances. including those within standard operating procedures, which pose the greatest chance of process or product failure when compared to ideal conditions. Such conditions do not necessarily induce product or process failure.

IV. General Concepts

Assurance of product quality is derived from careful attention to a number of factors including selection of quality parts and materials, adequate product and process design, control of the process, and in-process and end-product testing. Due to the complexity of today's medical products, routine end-product testing alone often is not sufficient to assure product quality for several reasons. Some end-product tests have limited sensitivity. (For example, USP XXI states: "No sampling plan for applying sterility tests to a specified proportion of discrete units selected from a sterilization load is capable of demonstrating with complete assurance that all of the untested units are in fact sterile.") In some cases, destructive testing would be required to show that the manufacturing process was adequate, and in other situations end-product testing does not reveal all variations that may occur in the product that may impact on safety and effectiveness.

The basic principles of quality assurance have as their goal the production of articles that are fit for their intended use.

These principles may be stated as follows: (1) quality. safety. and effectiveness must be designed and built into the product; (2) quality cannot be inspected or tested into the finished product; and (3) each step of the manufacturing process must be controlled to maximize the probability that the finished product meets all quality and design specifications. Process validation is a key element in assuring that these quality assurance goals are met.

It is through careful design and validation of both the process and process controls that a manufacturer can establish a high degree of confidence that all manufactured units from successive lots will be acceptable. Successfully validating a process may reduce the dependence upon intensive in-process and finished product testing. It should be noted that in most all cases, end-product testing plays a major role in assuring that quality assurance goals are met; i.e., validation and end-product testing are not mutually exclusive.

The FDA defines **process validation** as follows:

Process validation is establishing documented evidence which provides a high degree of assurance that a specific process will consistently produce a product meeting its predetermined specifications and quality characteristics.

It is important that the manufacturer prepare a written validation protocol which specifies the procedures (and tests) to be conducted and the data to be collected. The purpose for which data are collected must be clear, the data must reflect facts and be collected carefully and accurately. The protocol should specify a sufficient number of replicate process runs to demonstrate reproducibility and provide an accurate measure of variability among successive runs. The test conditions for these runs should encompass upper and lower processing limits and circumstances including those within standard operating procedures, which pose the greatest chance of process or product failure compared to ideal conditions; such conditions have become widely known as "worst case" conditions. (They are sometimes called "most appropriate challenge" conditions.) Validation documentation should include evidence of the suitability of materials and the performance and reliability of equipment and systems.

Key process variables should be monitored and documented. Analysis of the data collected from monitoring will establish the variability of process parameters for individual runs and will establish whether or not the equipment and process controls are adequate to assure that product specifications are met.

Finished product and in-process test data can be of value in process validation, particularly in those situations where quality attributes and variabilities can be readily measured. Where finished (or in-process) testing cannot adequately measure certain attributes, process validation should be derived primarily from qualification of each system used in production and from consideration of the interaction of the various systems.

V. CGMP Regulations for Finished Pharmaceuticals

Process validation is required, in both general and specific terms, by the Current Good Manufacturing Practice Regulations for Finished Pharmaceuticals, 21 CFR

Parts 210 and 211. Examples of such requirements are listed below for informational purposes, and are not all-inclusive:

A requirement for process validation is set forth in general terms in section 211.100 — Written procedures; deviations — which states, in part:

"There shall be written procedures for production and process control designed to assure that the drug products have the identity, strength, quality, and purity they purport or are represented to possess."

Several sections of the CGMP regulations state validation requirements in more specific terms. Excerpts from some of these sections are:

Section 211.110, Sampling and testing of in-process materials and drug products.

(a) "...control procedures shall be established to monitor the output and VALIDATE the performance of those manufacturing processes that may be responsible for causing variability in the characteristics of in-process material and the drug product." (emphasis added)

Section 211.113, Control of Microbiological Contamination.

(b) "Appropriate written procedures, designed to prevent microbiological contamination of drug products purporting to be sterile, shall be established and followed. Such procedures shall include VALIDATION of any sterilization process." (emphasis added)

VI. GMP Regulation for Medial Devices

Process validation is required by the medical device GMP Regulations, 21 CFR Part 820. Section 820.5 requires every finished device manufacturer to: "... prepare and implement a quality assurance program that is appropriate to the specific device manufactured..."

Section 820.3(n) defines quality assurance as: "...all activities necessary to verify confidence in the quality of the process used to manufacture a finished device.'

When applicable to a specific process, process validation is an essential element in establishing confidence that a process will consistently produce a product meeting the designed quality characteristics.

A generally stated requirement for process validation is contained in section 820.100: "Written manufacturing specifications and processing procedures shall be established, implemented, and controlled to assure that the device conforms to its original design or any approved changes in that design."

Validation is an essential element in the establishment and implementation of a process procedure, as well as in determining what process controls are required in order to assure conformance to specifications.

Section 820.100(a)(1) states: "... control measures shall be established to assure that the design basis for the device, components and packaging is correctly translated into approved specifications."

Validation is an essential control for assuring that the specifications for the device and manufacturing process are adequate to produce a device that will conform to the approved design characteristics.

VII. Preliminary Considerations

A manufacturer should evaluate all factors that affect product quality when designing and undertaking a process validation study. These factors may vary considerably among different products and manufacturing technologies and could include, for example, component specifications, air and water handling systems, environmental controls, equipment functions, and process control operations. No single approach to process validation will be appropriate and complete in all cases; however, the following quality activities should be undertaken in most situations.

During the research and development (R&D) phase, the desired product should be carefully defined in terms of its characteristics, such as physical, chemical, electrical and performance characteristics. (For example, in the case of a compressed tablet, physical characteristics would include size, weight, hardness, and freedom from defects, such as capping and splitting. Chemical characteristics would include quantitative formulation/potency; performance characteristics may include bioavailability (reflected by disintegration and dissolution). In the case of blood tubing, physical attributes would include internal and external diameters, length and color. Chemical characteristics would include raw material formulation. Mechanical properties would include hardness and tensile strength; performance characteristics would include biocompatibility and durability.)

It is important to translate the product characteristics into specifications as a basis for description and control of the product. Documentation of changes made during development provide traceability which can later be used to pinpoint solutions to future problems. The product's end use should be a determining factor in the development of product (and component) characteristics and specifications. All pertinent aspects of the product which impact on safety and effectiveness should be considered. These aspects include performance, reliability and stability. Acceptable ranges or limits should be established for each characteristic to set up allowable variations. (For example, in order to assure that an oral, ophthalmic, or parenteral solution has an acceptable pH, a specification may be established by which a lot is released only if it has been shown to have a pH within a narrow established range. For a device, a specification for the electrical resistance of a pacemaker lead would be established so that the lead would be acceptable only if the resistance was within a specified range.) These ranges should be expressed in readily measurable terms.

The validity of acceptance specifications should be verified through testing and challenge of the product on a sound scientific basis during the initial development and production phase.

Once a specification is demonstrated as acceptable it is important that any changes to the specification be made in accordance with documented change control procedures.

VIII. Elements of Process Validation

A. Prospective Validation

Prospective validation includes those considerations that should be made before an entirely new product is introduced by a firm or when there is a change in the manufacturing process which may affect the product's characteristics, such as uniformity and identity. The following are considered as key elements of prospective validation.

1. **Equipment and Process** The equipment and process(es) should be designed and/or selected so that product specifications are consistently achieved. This should be done with the participation of all appropriate groups that are concerned with assuring a quality product, e.g., engineering design, production operations, and quality assurance personnel.

    a. **Equipment: Installation Qualification** Installation qualification studies establish confidence that the process equipment and ancillary systems are capable of consistently operating within established limits and tolerances. After process equipment is designed or selected, it should be evaluated and tested to verify that it is capable of operating satisfactorily within the operating limits required by the process. (Examples of equipment performance characteristics which way be measured include temperature and pressure of injection molding machines, uniformity of speed for mixers, temperature, speed and pressure for packaging machines, and temperature and pressure of sterilization chambers.) This phase of validation includes examination of equipment design; determination of calibration, maintenance, and adjustment requirements; and identifying critical equipment features that could affect the process and product. Information obtained from these studies should be used to establish written procedures covering equipment calibration, maintenance, monitoring, and control.

    In assessing the suitability of a given piece of equipment, it is usually insufficient to rely solely upon the representations of the equipment supplier, or upon experience in producing some other product. Sound theoretical and practical engineering principles and considerations are a first step in the assessment.

    It is important that equipment qualification simulate actual production conditions, including those which are "worst case" situations.

    Tests and challenges should be repeated a sufficient number of times to assure reliable and meaningful results. All acceptance criteria must be met during the test or challenge. If any test or challenge shows that the equipment does not perform within its specifications, an evaluation should be performed to identify the cause of the failure. Corrections should be made and additional test runs performed, as needed, to verify that the equipment performs within specifications. The observed variability of the equipment between and within runs can be used as a basis for determining the total number of trials selected for the subsequent performance qualification studies of the process. (For example, the AAMI Guideline for Industrial Ethylene Oxide Sterilization of Medical Devices approved 2 December 1981, states: "The performance qualification should include a minimum of

3 successful, planned qualification runs, in which all of the acceptance criteria are met (5.3.1.2.).")

Once the equipment configuration and performance characteristics are established and qualified, they should be documented. The installation qualification should include a review of pertinent maintenance procedures, repair parts lists, and calibration methods for each piece of equipment. The objective is to assure that all repairs can be performed in such a way that will not affect the characteristics of material processed after the repair. In addition, special post-repair cleaning and calibration requirements should be developed to prevent inadvertent manufacture a of non-conforming product. Planning during the qualification phase can prevent confusion during emergency repairs which could lead to use of the wrong replacement part.

b. **Process: Performance Qualification** The purpose of performance qualification is to provide rigorous testing to demonstrate the effectiveness and reproducibility of the process. In entering the performance qualification phase of validation, it is understood that the process specifications have been established and essentially proven acceptable through laboratory or other trial methods and that the equipment has been judged acceptable on the basis of suitable installation studies.

Each process should be defined and described with sufficient specificity so that employees understand what is required.

Parts of the process which may vary so as to affect important product quality should be challenged. (For example, in electroplating the metal case of an implantable pacemaker, the significant process steps to define, describe, and challenge include establishment and control of current density and temperature values for assuring adequate composition of electrolyte and for assuring cleanliness of the metal to be plated. In the production of parenteral solutions by aseptic filling, the significant aseptic filling process steps to define and challenge should include the sterilization and depyrogenation of containers/closures, sterilization of solutions, filling equipment and product contact surfaces, and the filling and closing of containers.)

In challenging a process to assess its adequacy, it is important that challenge conditions simulate those that will be encountered during actual production, including "worst case" conditions. The challenges should be repeated enough times to assure that the results are meaningful and consistent.

Each specific manufacturing process should be appropriately qualified and validated. There is an inherent danger in relying on what are perceived to be similarities between products, processes, and equipment without appropriate challenge. (For example, in the production of a compressed tablet, a firm may switch from one type of granulation blender to another with the erroneous assumption that both types have similar performance characteristics, and, therefore, granulation mixing times and procedures need not be altered. However, if the blenders are substantially different, use of the new blender with procedures used for the previous blender may result in a granulation with poor content uniformity. This, in turn, may lead to tablets having significantly differing potencies. This situation may be averted if the quality assurance system detects the equipment change in the first place, challenges the blender performance, precipitates a revalidation of the process, and initiates appropriate changes. In this example, revalidation comprises installation qualification of the new equipment and performance qualification of the process intended for use in the new blender.)

c. **Product: Performance Qualification** For purposes of this guideline, product performance qualification activities apply only to medical devices ...

2. System to Assure Timely Revalidation

There should be a quality assurance system in place which requires revalidation whenever there are changes in packaging, formulation, equipment, or processes which could impact on product effectiveness or product characteristics, and whenever there are changes in product characteristics. Furthermore, when a change is made in raw material supplier, the manufacturer should consider subtle, potentially adverse differences in the raw material characteristics. A determination of adverse differences in raw material indicates a need to revalidate the process.

One way of detecting the kind of changes that should initiate revalidation is the use of tests and methods of analysis which are capable of measuring characteristics which may vary. Such tests and methods usually yield specific results which go beyond the mere pass/fail basis, thereby detecting variations within product and process specifications and allowing determination of whether a process is slipping out of control.

The quality assurance procedures should establish the circumstances under which revalidation is required. These may be based upon equipment, process, and product performance observed during the initial validation challenge studies. It is desirable to designate individuals who have the responsibility to review product, process, equipment and personnel changes to determine if and when revalidation is warranted.

The extent of revalidation will depend upon the nature of the changes and how they impact upon different aspects of production that had previously been validated. It may not be necessary to revalidate a process from scratch merely because a given circumstance has changed. However, it is important to carefully assess the nature of the change to determine potential ripple effects and what needs to be considered as part of revalidation.

3. Documentation

It is essential that the validation program is documented and that the documentation is properly maintained. Approval and release of the process for use in routine manufacturing should be based upon a review of all the validation documentation, including data from the equipment qualification, process performance qualification, and product/package testing to ensure compatibility with the process.

For routine production, it is important to adequately record process details (e.g., time, temperature, equipment used) and to record any changes which have occurred. A maintenance log can be useful in performing failure investigations concerning a specific manufacturing lot. Validation data (along with specific test data) may also determine expected variance in product or equipment characteristics.

B. Retrospective Process Validation

In some cases a product may have been on the market without sufficient premarket process validation. In these cases, it may be possible to validate, in some measure, the adequacy of the process by examination of accumulated test data on the product and records of the manufacturing procedures used.

Retrospective validation can also be useful to augment initial premarket prospective validation for new products or changed processes. In such cases, preliminary prospective validation should have been sufficient to warrant product marketing. As additional data is gathered on production lots, such data can be used to build confidence in the adequacy of the process. Conversely, such data may indicate a declining confidence in the process and a commensurate need for corrective changes.

Test data may be useful only if the methods and results are adequately specific. As with prospective validation, it may be insufficient to assess the process solely on the basis of lot by lot conformance to specifications if test results are merely expressed in terms of pass/fail. Specific results, on the other hand, can be statistically analyzed and a determination can be made of what variance in data can be expected. It is important to maintain records which describe the operating characteristics of the process, e.g., time, temperature, humidity, and equipment settings. (For example, sterilizer time and temperature data collected on recording equipment found to be accurate and precise could establish that process parameters had been reliably delivered to previously processed loads. A retrospective qualification of the equipment could be performed to demonstrate that the recorded data represented conditions that were uniform throughout the chamber and that product load configurations, personnel practices, initial temperature, and other variables had been adequately controlled during the earlier runs).

Whenever test data are used to demonstrate conformance to specifications, it is important that the test methodology be qualified to assure that test results are objective and accurate.

IX. Acceptability of Product Testing

In some cases, a drug product or medical device may be manufactured individually or on a one-time basis. The concept of prospective or retrospective validation as it relates to those situations may have limited applicability, and data obtained during the manufacturing and assembly process may be used in conjunction with product testing to demonstrate that the instant run yielded a finished product meeting all of its specifications and quality characteristics. Such evaluation of data and product testing would be expected to be much more extensive than the usual situation where more reliance would be placed on prospective validation.

ANNEX 2 TO CHAPTER 15

Annex 15 to EC GMP Guide on:

Qualification and Validation

Principle

1.   This Annex describes the principles of qualification and validation which are applicable to the manufacture of medicinal products. It is a requirement of GMP that manufacturers identify what validation work is needed to prove control of the critical aspects of their particular operations. Significant changes to the facilities, the equipment and the processes, which may affect the quality of the product, should be validated. A risk assessment approach should be used to determine the scope and extent of validation.

Planning for Validation

2.   All validation activities should be planned. The key elements of a validation programme should be clearly defined and documented in a validation master plan (VMP) or equivalent documents.

3.   The VMP should be a summary document which is brief, concise and clear.

4.   The VMP should contain data on at least the following: (a) validation policy; (b) organizational structure of validation activities; (c) summary of facilities, systems, equipment and processes to be validated; (d) documentation format: the format to be used for protocols and reports; (e) planning and scheduling; (f) change control; (g) reference to existing documents.

5.   In case of large projects, it may be necessary to create separate validation master plans

Documentation

6.   A written protocol should be established that specifies how qualification and validation will be conducted. The protocol should be reviewed and approved. The protocol should specify critical steps and acceptance criteria.

7.   A report that cross-references the qualification and/or validation protocol should be prepared, summarising the results obtained, commenting on any deviations observed, and drawing the necessary conclusions, including recommending changes necessary to correct deficiencies. Any changes to the plan as defined in the protocol should be documented with appropriate justification.

8.   After completion of a satisfactory qualification, a formal release for the next step in qualification and validation should be made as a written authorisation.

Qualification

Design qualification

9.   The first element of the validation of new facilities, systems or equipment could *(sic)* be design qualification (DQ).

10. The compliance of the design with GMP should be demonstrated and documented.

Installation qualification

11. Installation qualification (IQ) should be performed on new or modified facilities, systems and equipment.
12. IQ should include, but not be limited to the following:
    (a) installation of equipment, piping, services and instrumentation checked to current engineering drawings and specifications;
    (b) collection and collation of supplier operating and working instructions and maintenance requirements;
    (c) calibration requirements;
    (d) verification of materials of construction

Operational qualification

13. Operational qualification (OQ) should follow Installation qualification.
14. OQ should include, but not be limited to the following:
    (a) tests that have been developed from knowledge of processes, systems and equipment;
    (b) tests to include a condition or a set of conditions encompassing upper and lower operating limits, sometimes referred to as "worst case" conditions.
15. The completion of a successful Operational qualification should allow the finalisation of calibration, operating and cleaning procedures, operator training and preventative maintenance requirements. It should permit a formal "release" of the facilities, systems and equipment.

Performance qualification

16. Performance qualification (PQ) should follow successful completion of Installation qualification and Operational qualification.
17. PQ should include, but not be limited to the following:
    (a) tests, using production materials, qualified substitutes or simulated product, that have been developed from knowledge of the process and the facilities, systems or equipment;
    (b) tests to include a condition or set of conditions encompassing upper and lower operating limits.
18. Although PQ is described as a separate activity, it may in some cases be appropriate to perform it in conjunction with OQ.

Qualification of established (in-use) facilities, systems and equipment

19. Evidence should be available to support and verify the operating parameters and limits for the critical variables of the operating equipment. Additionally, the calibration, cleaning, preventative maintenance, operating procedures and operator training procedures and records should be documented.

Process Validation

General

20. The requirements and principles outlined in this chapter are applicable to the manufacture of pharmaceutical dosage forms. They cover the initial

validation of new processes, subsequent validation of modified processes and revalidation.

21. Process validation should normally be completed prior to the distribution and sale of the medicinal product (prospective validation). In exceptional circumstances, where this is not possible, it may be necessary to validate processes during routine production (concurrent validation). Processes in use for some time should also be validated (retrospective validation).

22. Facilities, systems and equipment to be used should have been qualified and analytical testing methods should be validated. Staff taking part in the validation work should have been appropriately trained.

23. Facilities, systems, equipment and processes should be periodically evaluated to verify that they are still operating in a valid manner.

Prospective validation

24. Prospective validation should include, but not be limited to the following: (a) short description of the process; (b) summary of the critical processing steps to be investigated; (c) list of the equipment/facilities to be used (including measuring/monitoring/recording equipment) together with its calibration status (d) finished product specifications for release; (e) list of analytical methods, as appropriate; (f) proposed in-process controls with acceptance criteria; (g) additional testing to be carried out, with acceptance criteria and analytical validation, as appropriate; (h) sampling plan; (i) methods for recording and evaluating results (j) functions and responsibilities; (k) proposed timetable.

25. Using this defined process (including specified components) a series of batches of the final product may be produced under routine conditions. In theory the number of process runs carried out and observations made should be sufficient to allow the normal extent of variation and trends to be established and to provide sufficient data for evaluation. It is generally considered acceptable that three consecutive batches/runs within the finally agreed parameters, would constitute a validation of the process.

26. Batches made for process validation should be the same size as the intended industrial scale batches.

27. If it is intended that validation batches be sold or supplied, the conditions under which they are produced should comply fully with the requirements of Good Manufacturing Practice, including the satisfactory outcome of the validation exercise, and with the marketing authorisation.

Concurrent validation

28. In exceptional circumstances it may be acceptable not to complete a validation programme before routine production starts.

29. The decision to carry out concurrent validation must be justified, documented and approved by authorised personnel.

30. Documentation requirements for concurrent validation are the same as specified for prospective validation.

Retrospective validation

31. Retrospective validation is only acceptable for well-established processes and will be inappropriate where there have been recent changes in the composition of the product, operating procedures or equipment.

32. Validation of such processes should be based on historical data. The steps involved require the preparation of a specific protocol and the reporting of the results of the data review, leading to a conclusion and a recommendation.

33. The source of data for this validation should include, but not be limited to batch processing and packaging records, process control charts, maintenance log books, records of personnel changes, process capability studies, finished product data, including trend cards and storage stability results.

34. Batches selected for retrospective validation should be representative of all batches made during the review period, including any batches that failed to meet specifications, and should be sufficient in number to demonstrate process consistency. Additional testing of retained samples may be needed to obtain the necessary amount or type of data to retrospectively validate the process.

35. For retrospective validation, generally data from ten to thirty consecutive batches should be examined to assess process consistency, but fewer batches may be examined if justified.

Cleaning Validation

36. Cleaning validation should be performed in order to confirm the effectiveness of a cleaning procedure. The rationale for selecting limits of carry over of product residues, cleaning agents and microbial contamination should be logically based on the materials involved. The limits should be achievable and verifiable.

37. Validated analytical methods having sensitivity to detect residues or contaminants should be used. The detection limit for each analytical method should be sufficiently sensitive to detect the established acceptable level of the residue or contaminant.

38. Normally only cleaning procedures for product contact surfaces of the equipment need to be validated. Consideration should be given to noncontact parts. The intervals between use and cleaning as well as cleaning and reuse should be validated. Cleaning intervals and methods should be determined.

39. For cleaning procedures for products and processes which are similar, it is considered acceptable to select a representative range of similar products and processes. A single validation study utilising a "worst case" approach can be carried out which takes account of the critical issues.

40. Typically three consecutive applications of the cleaning procedure should be performed and shown to be successful in order to prove that the method is validated.

41. "Test until clean." is not considered an appropriate alternative to cleaning validation.

42. Products which simulate the physicochemical properties of the substances to be removed may exceptionally be used instead of the substances themselves, where such substances are either toxic or hazardous.

Change Control

43. Written procedures should be in place to describe the actions to be taken if a change is proposed to a starting material, product component, process equipment, process environment (or site), method of production or testing or any other change that may affect product quality or reproducibility of the process. Change control procedures should ensure that sufficient supporting data are generated to demonstrate that the revised process will result in a product of the desired quality, consistent with the approved specifications.

44. All changes that may affect product quality or reproducibility of the process should be formally requested, documented and accepted. The likely impact of the change of facilities, systems and equipment on the product should be evaluated, including risk analysis. The need for, and the extent of, re-qualification and re-validation should be determined.

Revalidation

45. Facilities, systems, equipment and processes, including cleaning, should be periodically evaluated to confirm that they remain valid. Where no significant changes have been made to the validated status, a review with evidence that facilities, systems, equipment and processes meet the prescribed requirements fulfils the need for revalidation.

GLOSSARY

Definitions of terms relating to qualification and validation which are not given in the glossary of the current EC Guide to GMP, but which are used in this Annex, are given below.

Change Control

A formal system by which qualified representatives of appropriate disciplines review proposed or actual changes that might affect the validated status of facilities, systems, equipment or processes. The intent is to determine the need for action that would ensure and document that the system is maintained in a validated state.

Cleaning Validation

Cleaning validation is documented evidence that an approved cleaning procedure will provide equipment which is suitable for processing medicinal products.

Concurrent Validation

Validation carried out during routine production of products intended for sale.

Design Qualification (DQ)

The documented verification that the proposed design of the facilities, systems and equipment is suitable for the intended purpose.

Installation Qualification (IQ)

The documented verification that the facilities, systems and equipment, as installed or modified, comply with the approved design and the manufacturer's recommendations.

Operational Qualification (OQ)

The documented verification that the facilities, systems and equipment, as installed or modified, perform as intended throughout the anticipated operating ranges.

Performance Qualification (PQ)

The documented verification that the facilities, systems and equipment, as connected together, can perform effectively and reproducibly, based on the approved process method and product specification.

Process Validation

The documented evidence that the process, operated within established parameters, can perform effectively and reproducibly to produce a medicinal product meeting its predetermined specifications and quality attributes.

Prospective Validation

Validation carried out before routine production of products intended for sale.

Retrospective Validation

Validation of a process for a product which has been marketed based upon accumulated manufacturing, testing and control batch data.

Re-Validation

A repeat of the process validation to provide an assurance that changes in the process/equipment introduced in accordance with change control procedures do not adversely affect process characteristics and product quality.

Risk Analysis

Method to assess and characterise the critical parameters in the functionality of an equipment or process.

Simulated Product

A material that closely approximates the physical and, where practical, the chemical characteristics (e.g., viscosity, particle size, pH etc.) of the product under validation. In many cases, these characteristics may be satisfied by a placebo product batch.

System

A group of equipment with a common purpose.

Worst Case

A condition or set of conditions encompassing upper and lower processing limits and circumstances, within standard operating procedures, which pose the greatest chance of product or process failure when compared to ideal conditions. Such conditions do not necessarily induce product or process failure.

16

VALIDATION — APPLICATIONS

The last chapter was concerned with more general aspects of validation. This one will consider some (but only some) of the more important specific applications of the concept. These are:

Validation of sterilization processes
Validation of cleaning procedures
Computer systems validation

In addition to being most important targets for validation, *per se,* they are also, at the time of writing, regulatory "hot topics," and are likely to remain so for the foreseeable future."

A fourth important validation application, analytical methods validation, stands somewhat apart from the others in having a long-standing acceptance and methodology, and it is, therefore, by no means a new issue. It has already been considered in Chapter 10.

VALIDATION OF STERILIZATION PROCESSES

It is crucially important that all processes used to sterilize pharmaceuticals, medical devices, and the like should be validated, for reasons already discussed. Not surprisingly, different approaches and techniques are employed for each of the major types of sterilization, *viz.*:

Filtration (or other bulk sterilization) with aseptic filling
Heat — steam and dry
Gaseous (e.g., ethylene oxide)
Radiation

Validation of a Filtration/Bulk Sterilization and Aseptic Filling Process

Discussion of this type of process will also serve to exemplify some of the general principles considered in the previous chapter. The treatment of validation of the other types of sterilization process will be covered in somewhat less detail.

It is worth reminding ourselves that sterile products may be broadly classified into two main categories, according to the manner in which they are produced: those that are sterilized after the product has been filled and sealed in the final container(s) (terminally sterilized products); and those where the sterilization stage (or stages) takes place before the bulk product is filled. In the latter instance, all subsequent processing (typically, the filling and sealing operations) must be conducted aseptically in order to prevent recontamination of the sterilized product. Given a properly sealed container, the integrity of which remains unbreached, terminal sterilization eliminates the possibility of recontamination. Thus, any product intended, required, or purported to be sterile, should be terminally sterilized, unless there are good reasons that dictate otherwise, for example, where terminal sterilization will adversely affect the product. Manufacturers who decide that terminal sterilization is inappropriate for any given product should be prepared to justify this decision.

The two most common pharmaceutical applications of aseptic processing methods are (a) the filling of liquid products following sterilization by filtration and (b) the filling of previously sterilized bulk powder products.

The main steps in the validation of any such process may be summarized as follows:

- As a prerequisite, all studies should be conducted in accordance with a detailed, pre-established, *protocol*, or series of protocols, which in turn is subject to formal change control procedures.
- Both the personnel conducting the studies, and those running the process(es) being studied should be appropriately trained and qualified and (in all respects) be suitable and competent to perform the tasks assigned to them.
- All data generated during the course of the studies should be formally reviewed and certified, as evaluated against predetermined criteria.
- Suitable testing facilities, equipment, instruments, and methodology must be available.
- Suitable clean room facilities should be available, in terms both of the "local" and "background" environments.
- Assurance that the clean room environment conforms to, and is maintained at, the standard specified should be secured through initial commissioning (qualification) and subsequently through the implementation of a program of retesting, in-process control and monitoring.
- All processing equipment should be properly installed and maintained.
- When appropriate attention has been paid to the above, the aseptic process may be validated by means of process simulation (or "media fill") studies.
- The process should be revalidated at defined intervals.

■ Comprehensive documentation should be available to define, support, and record the overall validation process.

Note: Although this discussion is concerned only with the validation of aseptic processes, it is crucial to the success of any such process that the product, materials, components, etc. that are being handled and processed aseptically (e.g., bulk solution or powder; containers, and closures) plus any equipment, vessels or surfaces (e.g., holding tanks, pipework, filling machines) that will or can come into contact with sterilized products or materials have themselves been previously sterilized by appropriate, *validated* sterilization processes. In any aseptic filling process, assurance of container and closure integrity is also vital. Evidence that *all* this is so should be maintained as part of the overall validation documentation.

Protocol Development and Control

Each stage in the validation of the overall process should proceed in accordance with a preestablished and formally approved, detailed, written protocol, or a series of related protocols. Prior to the commencement of the studies, written change control procedures should also be established, which will prevent unauthorized changes to either the process itself, or to the study protocol, and restrict change during any stage of the study until all relevant data are evaluated.

The protocols should have a title, date, and a unique identification or reference number. They should be formally authorized and approved by person(s) with the competence and authority to do so. Protocols should give in detail:

1. The objectives and scope of the study, that is, there should be a clear definition of purpose
2. A clear and precise definition of the process, equipment, system, or subsystem that is to be the subject of the study, with details of performance characteristics
3. Installation and qualification requirements for new equipment
4. Any upgrading requirements for existing equipment, with justification for the change(s) and a statement of qualification requirements
5. Detailed, step-wise statement of actions to be taken in performing the study (or studies)
6. Assignment of responsibility for performing the study
7. Statements on all test methodology to be employed, with a precise statement of the test equipment and materials to be used
8. Test equipment calibration requirements
9. References to any relevant standard operating procedures (SOPs)
10. Requirements for the content and format of the report on the study
11. Acceptance criteria against which the success (or otherwise) of the study is to be evaluated
12. The personnel responsible for evaluating and certifying as acceptable each stage in the study, and for the final evaluation and certification of the process as a whole, all as measured against the predefined acceptance criteria

Personnel

As with all process validation studies, documented evidence of the relevant experience and training of the personnel involved in conducting the studies should be maintained. Furthermore, the personnel actually performing the aseptic processing (both during the course of any validation studies, and in routine operation) can, and do, inevitably have a crucial effect on the quality of the end product. It is necessary, therefore, to consider not only the experience and training of the personnel involved in the performance of the validation studies, but also of the personnel performing the aseptic processing itself, both during the course of the validation work, *and* in routine processing. To use a handpicked "elite" team of operators to run the process when it is being validated is to defeat the whole objective.

Thus, appropriately qualified personnel should ensure that the protocol and the testing methodology are based on sound scientific principles and that all studies are properly evaluated and certified. All personnel conducting any test procedure should be trained and experienced in the use of the instruments, measuring devices, and materials used. Engineering and maintenance personnel also should be fully trained and competent in the operation and maintenance of the machines, equipment, and air control systems involved.

Although automated and barrier techniques may appreciably reduce the contamination risk, the significance of the human factor in all aseptic processing operations cannot be overstressed. For the results of any validation studies themselves to be valid, it is essential that the risk represented by so potentially random a variable as a human operator is kept as much under control as is possible — that is, steps must be taken to *reduce the risk* and to *minimize the variability*. This, in turn, means that operators performing the aseptic processing operation(s) that are the subject of a validation study should adopt the same techniques, disciplines, and standards of hygiene, clothing, and behavior as they would, and do, in normal routine manufacture. Everything should be done to simulate normal routine processing as closely as possible. Process operators should conduct themselves as they do in routine manufacture, neither better, nor worse. Furthermore, if the process operators conduct themselves, during routine production, in a manner that is different *in any way* from their behavior, etc. during the validation studies, then conclusions drawn from the validation will, themselves, be invalid.

It is therefore vital that all personnel involved in aseptic processing operations are trained in, and fully understand, the concepts and principles of GMP, and the relevant elements of microbiology. They must understand the importance of personal hygiene and cleanliness, and be made fully aware of the possible hazardous consequences of product contamination. They should be provided with suitable clean room clothing and trained in the appropriate gowning technique(s). The type of clothing to be worn, and the "scrub-up" and gowning process should be defined in written procedures, available to the operators, and preferably displayed in the changing room(s). The same clothing and gowning standards should be observed during validation studies as in routine production, and *vice versa*.

The maximum number of personnel permitted in the clean room during normal routine production should also be present in the clean room during any validation test runs. At all times, operators should be encouraged to report any infections, open lesions, or any other conditions that could result in the shedding of abnormal numbers of particles or microorganisms. As with routine manufacture, no person thus affected should be present in the clean room during validation test runs.

As in routine production, clean room operators involved in validation studies should be microbiologically monitored by taking test samples from gloves, gowns, and facemasks.

Normal routine process documentation should specify and record the numbers and types of operator interventions that are permitted during processing, and in what circumstances. A similar series of interventions should occur during any validation test runs. Details should be provided as part of the overall validation documentation.

Laboratory and Instruments

All laboratory tests (including physical, chemical, and microbiological determinations) should be performed by a competent laboratory, suitably equipped, and staffed with personnel properly trained and qualified to carry out the test procedures assigned to them. Detailed, authorized, written procedures defining the relevant, validated methodology should be available for all laboratory tests and determinations that are to be carried out during the course of the study. These procedures should be referenced in the study protocol.

If any external laboratory facilities are used, systems should be in place for determining the competence of these laboratories to carry out the tests required. This requirement should be referenced in the study protocol.

All measuring, recording, and indicating instruments employed in the studies should be adequate for the purpose, in terms of range, accuracy, reproducibility, etc. They must be calibrated in accordance with predefined written procedures before any validation studies are commenced. Records of each calibration should be maintained and should form part of the overall validation documentation.

It must be clearly understood that for the conclusions drawn from any qualification or validation studies themselves to remain valid during routine production, all controlling and recording instruments must be subjected to a written maintenance and calibration program.

Clean Room Standards Monitoring

For the results to have valid relevance to routine production, validation studies must be conducted under precisely the same environmental conditions as employed, or intended to be employed, during normal routine production. Confirmation and Certification that the room and the work station(s) used do, in fact, conform to the Environmental Standard specified may be considered as forming part of the "Installation Qualification" phase. To this end, the following basic work should be carried-out on the initial commissioning (or "Qualification") of a new Clean Room installation:

- Room air filter integrity tests
- Determination of air velocity at the face of each air inlet filter
- Room air change rate
- Room air particle counts
- Room air pressure differentials and air flow patterns
- Lighting, heating, humidity
- Work station(s) air filter efficiency tests
- Determination of air velocity at face of work station air filters
- Particle counts within work station areas

Following the initial commissioning, a regular retest program should be adopted, e.g.:

1. Room and Work Station Air Filter Tests: Repeat at least annually, unless results of normal in-process monitoring indicates a need for more frequent, or additional testing.
2. Air Velocity and Room Air Changes: Repeat at least twice a year.
3. Air Particle Counts: Determine as part of regular in-process monitoring, with formal certification by a competent specialist agency three times per year.

In addition, *room air-pressure differentials should be monitored on a continuous, ongoing, basis.*

Walls, floors, work stations, and surfaces generally should be subject to a predetermined program of cleaning and disinfection. In order to ensure that, during routine manufacture, products remain within the quality parameters established during the overall validation process, it is necessary to design and implement a program of in-process control and monitoring. Similarly, as part of the overall assurance that process validation studies are conducted under comparably normal processing conditions, *a similar in-process control and monitoring program should be operated during the process validation runs.*

In-process monitoring and control may be considered under two headings:

- Microbiological
- Environmental particulate

In addition, where sterile filtration of a liquid product is involved, filter integrity testing of the filter(s) used to sterilize that product must be performed. These filter integrity tests should be conducted *after* each use of the filters, in order to detect any leaks or perforations that may have occurred during the filtration process itself. Often, filter integrity testing is also done before the filtration of the product commences, and this is a generally sound practice. It is, however, the filter integrity test performed *after* the batch, or lot, has been filtered that is critical.

As appropriate to the type of manufacturing process, consideration needs to be given to the following microbiological monitoring and control procedures:

- Bioburden check on bulk solution, prior to sterile filtration
- Exposure of settle plates at defined critical positions within the general clean room environment and at the controlled work station(s)

- Use of air sampling devices to determine the number of viable organisms per cubic meter (or cubic foot) of air in the room and within the work station(s)
- Use of contact plates, or swabs, to check the microbiological quality of surfaces

Environmental particulate monitoring should be carried out using appropriate air particle counting devices to check that the general environmental and work station air remains in conformity with specification. All in-process monitoring and control should be conducted in accordance with a written, predetermined program, which includes specified test limits and standards, with all results formally reported and evaluated against those limits. This requirement applies as much to validation studies as it does to routine manufacture.

Equipment Qualification and Maintenance

Various items of mechanized equipment may be used in aseptic processing, for example, ampoule filling and sealing machines; vial, bottle, cartridge, tube or syringe filling, sealing and capping machines; powder fillers, freeze driers (lyophilizers), and so on. Before any process validation studies may be commenced, it is necessary that all such equipment should be properly installed and operationally qualified.

The essential requirements are that the equipment is:

a. Confirmed as having been constructed as specified
b. Properly installed and provided with all necessary *functioning* services, ancillary equipment and instruments
c. Confirmed as capable of operating consistently, within predetermined limits, over its defined operating range

Processing equipment must be confirmed as complying with a, b, and c above before any subsequent studies can be considered valid. For the results of any validation studies themselves to remain valid in routine manufacture, a comprehensive routine maintenance program must be developed, setting out each activity in detail along with the frequency in terms of real time, machine time, or other time base. The time base should be clearly defined for each procedure.

Unless such a program is developed and implemented and the manufacturing equipment and attendant instruments remain in the same state of maintenance and calibration as during the validation studies, then any assurance derived from those studies is to be considered as negated.

Media Fill Studies (Solution Products)

The "media-fill," or "broth-fill," technique is one in which a liquid microbial nutrient growth medium ("broth") is prepared and filled in a simulation of a normal manufacturing operation. The nutrient medium is processed and handled in a manner that simulates the normal manufacturing process as closely as possible with the same exposure to contamination risk (from operators, environment, equipment, and surfaces) as would occur during routine manufacture. The sealed containers of medium thus produced are incubated under prescribed conditions and then examined for evidence of microbial growth, and thus of an indication of the level of contaminated units produced.

It is important to recognize that, in many instances, media fills are, among other things, a test of the human operators' aseptic techniques. In this test situation, the operators can hardly remain unaware that nutrient medium is being filled, and that they themselves are, to an extent, "under test." There is, therefore, the possibility that they will take more than their usual care, and thus the normal process will not be precisely simulated. Every effort should be made to ensure that the operators *do* behave normally during the media fills, and conversely (and perhaps more importantly) that during routine production they do not deviate in any way from the high standards adopted during the simulation studies.

A further difficulty that needs to be noted is the possibility of contamination of the facility and equipment by the nutrient medium. Some writers have seen this possibility as a strong argument against the media-fill technique. However, if the process is well controlled and the media fill is promptly followed by cleaning and disinfection, and (as necessary) sterilization of equipment, this should not be a problem. Nevertheless, it *is* important to recognize the potential hazard, and to respond accordingly.

It must also be reemphasized that the filling of a nutrient medium solution *alone* does not constitute an acceptable aseptic process validation. The whole manufacturing cycle must be simulated, from the dispensing and reconstitution of the powdered medium under normal manufacturing conditions to the filling and sealing process itself. Operators (and numbers of operators), numbers and types of filtrations, etc. should all be as normal, as should holding times in any mixing vessels, or interim holding tanks. General activity should be at a normal level, and no attempt should be made to take any special precautions to ensure that the test run is successful. If any deviation from the normal is permitted, it should only be in the direction of presenting a *greater*, rather than a lesser, microbiological challenge to the process.

The liquid nutrient medium used should meet the following criteria:

Selectivity: The medium should have *low* selectivity, that is, it should be capable of supporting growth of the widest possible range of microorganisms that might reasonably be expected to be encountered. As a minimum requirement, it should support the growth of cultures of organisms normally found in the manufacturing environment, as well as such organisms as:

> *Escherichia coli*
> *Pseudomonas aeruginosa*
> *Staphylococcus aureus*
> *Candida albicans*
> *Aspergillus niger*

Clarity: As "made up," the medium should be clear, to allow for the observation of any turbidity (that is, growth) following incubation.

Filterability: Where the process being simulated includes a filtration stage, the liquid medium should be capable of being filtered through the same grade and type of microbial retentive filter as that through which the actual product is, or will be, filtered.

Liquid soybean casein digest (SCD), also termed "tryptic soy broth" (TSB) is perhaps the liquid medium most frequently employed. However, other formulations (for example, liquid tryptone glucose yeast extract, brain heart infusion, etc.) may be used, provided they meet the criteria set out above.

The liquid medium should be sterilized by filtration (if such a stage is part of the normal operation being simulated), in the same way, using the same grade and type of filter and housing, and in the same sequence as normal. If it is presterilized by heat, or subjected to any form of heat treatment, it must be *cooled to ambient temperature before proceeding*.

The *number of units* to be filled per run should be sufficient to provide a high probability of detecting a low incidence of microbial contamination. For example, in order to give 95% confidence of detecting a contamination rate of 1 in 1000 units filled with medium, 3000 units need to be filled. (In fact, based on an assumption of a Poisson distribution, or of a binomial distribution, of contaminated units, the precise figures are, respectively, 2986 and 2995. Traditionally this is rounded up to 3000. However, see later in this chapter)

For the initial validation of a new process or facility, or after any major change in the process or the equipment, sufficient consecutive media-fill runs should be performed to provide assurance that the results obtained are consistent, meaningful, and provide an acceptable level of sterility assurance. At least three separate, consecutive, successful runs per operator team, or shift, should be performed to provide acceptable initial validation of a given processing line. There is no statistical basis for this "rule of three." It just seems like a good idea, and it probably is until somebody thinks of something that is both better and practicable.

The *volume to be filled* per unit should be the normal production fill volume, where possible. In the case of high-volume containers, a lesser quantity may be used, provided steps are taken to ensure wetting of *all* the inner surface of the container, and any closure, by the medium, e.g., by shaking or inversion, or by inverting the containers part way through the incubation period. It is a good practice also to take similar steps to ensure complete inner-surface wetting, even when normal full volumes are filled. Immediately following filling, all units filled should be examined for leaks and damage. In this context, any leak-test method in which heat is employed must obviously not be used. Any leaking or damaged units should be rejected. The *incubation* of the filled units should follow immediately after filling and leak testing, and should be for a minimum period of 14 days.

Opinions tend to vary regarding the *incubation temperature* (PDA,[1] Prout[2]) to be used, but 25 to 35°C is a reasonable compromise. Whatever temperature range is chosen, it should be carefully controlled, monitored, and maintained throughout the incubation period. For strict comparability between results obtained in different test runs, in practice the incubation temperature should be controlled, on each and every occasion, between tighter limits than the range suggested.

Test Controls: To demonstrate the nutritive properties of the medium used, a few filled units from each run should be inoculated with low levels of challenge organisms, and then incubated, suitably labeled, along with the test units. A number of texts appear to suggest that this test-control operation should be conducted almost as something separate and divorced from main media-full run. It clearly makes better sense, and is better science, to randomly select a number of filled units from the run (that means a few extra will need to be filled), inoculate them, mark them carefully and conspicuously to avoid confusion, and then incubate them, in the same incubator with the other media-filled units. Thus, a suggested control procedure is as follows:

Take 6 medium-filled units from each run and inoculate in 3 sets of 2 each with the following organisms, at a level of 100 organisms per unit:

2 x *Staphylococcus aureus*
2 x *Bacillus subtilis*
2 x *Candida albicans*

Label and incubate these 6 inoculated control units along with the test units.

Organisms other than those suggested may be used, provided they represent a similar range of microbial type. An alternative is to use cultures of organisms found in the manufacturing environment. A combination of these environmental isolates and "standard" organisms such as those listed is much to be recommended.

Reading of Results: All units filled and incubated should be visually examined for microbial growth after at least 14 days incubation. Any contaminated units will be identifiable by the turbidity of the medium. Any contaminated units that are found should be examined in the laboratory, and the contaminating organisms identified, at least to genus level, so that appropriate preventative action may be taken. For the results of the media-fill run to be considered valid, all the inoculated control units should display growth.

The flow diagram shown at Figure 16.1 illustrates the type of process simulation described above, in relation to liquid-filled vial product.

The percentage *contamination level* found in a media fill run is calculated as follows:

$$\% \text{ Contamination} = \frac{\text{No. of contaminated units}}{\text{No. of units incubated}} \times 100$$

Media Fill Acceptance Criteria

The process simulation (or media-fill) test, as outlined above, may well appear to be a sound and useful technique. It has long been accepted, is widely practiced and has regulatory and compendial recognition. Unfortunately, as a truly scientific method, it tends somewhat to totter, if not fall, at the last hurdle — that is, when the problem of deciding the criteria by which the success, or failure, of a media-fill run is confronted (or not, as the case may be), in the face of statistical, and some very considerable practical, difficulties.

The most widely quoted acceptance limit remains not more than 1 in 1,000 (0.1%) contaminated media-filled units. (This limit was considered acceptable, for example, in the FDA. Guideline on Sterile Drug Products Produced by Aseptic

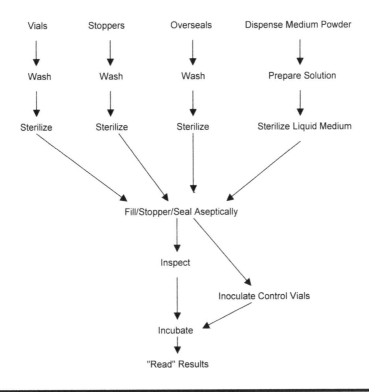

Vials   Stoppers   Overseals   Dispense Medium Powder

Wash   Wash   Wash   Prepare Solution

Sterilize   Sterilize   Sterilize   Sterilize Liquid Medium

Fill/Stopper/Seal Aseptically

Inspect

Inoculate Control Vials

Incubate

"Read" Results

Figure 16.1   Process Flow Diagram of Liquid Media Filling of Vials
Notes: **(1) Different types of containers will require different methods of sterilization. For example, glass vials are likely to be dry-heat sterilized; plastic vials may be sterilized by irradiation or ethylene oxide. (2) Any other components, e.g., teats/droppers, will also need to be presterilized by some suitable validated method.   (3) The process flow for liquid media filling of ampoules will be analogous to the above, without the operations involving stoppers, overseals, etc.**

Processing,[3] where it is also stated that acceptance of this level in a test does not mean that an aseptically processed lot of product purporting to be sterile may contain 1 nonsterile unit in every 1,000 filled and that it is merely a recognition of the scientific and technical limitations of the test procedure itself.)

It has to be recognized, however, that a media-fill run of 3,000 units only gives 95% confidence of detecting 0.1% microbial contamination *in those 3,000 units.* It does not serve as a direct prediction of a contamination of 0.1% or less in normal production runs, which are normally much larger. The fairly common 3,000-unit media fill thus represents only a simulated sample of a normal production run, and any contamination rate calculated from a media fill will therefore be subject to sampling error. Thus, 3 contaminated units found in a media fill of 3,000 may well be indicative of a potential contamination rate *in actual production* that is significantly greater than 0.1%. Although it has been suggested by various regulatory and other authorities, that a limit of 1 in 10,000 (0.01%), or better, should be the aim, it must be understood that to demonstrate a likelihood of compliance (on a normal production scale) with limits of that order, impractically large numbers of units may need to be filled with medium and incubated. In addition to the time, space, and cost implications, there is the not inconsiderable problem of the disposal of the large numbers of broth-filled units after the test.

Table 16.1 shows the number of units that would need to be media filled, with the respective numbers of contaminated units that could be permitted in order to provide 95% confidence of achieving 0.1% contamination, or less, in normal production runs of more than 100,000 units. (Values are rounded off.)

Thus, for example, to provide confidence (95%) of complying with the 0.1% limit in a production batch in excess of 100,000 units, 4,750 media-filled units would be required with no more than one unit found contaminated, or 6,300 units with no more than 2, and so on.

Figures relevant to other production batch sizes (100,000 or less) and a 0.1% acceptance limit (95% confidence, values rounded) are shown in Table 16.2. Thus, for a process in which the normal batch size is 20,000 units, a media fill of 4,340 would provide 95% confidence of a limit of 0.1% contamination if no more than 1 contaminated media-filled unit was found.

If the acceptance criterion is tightened, it is then that the real problems of scale become apparent. To demonstrate compliance with a contamination limit of one in 10,000 (0.01%), notably larger numbers of units would need to be filled with broth. For example, in relation to a normal production run of 50,000 units, more than 46,000 units would need to be filled with medium, with no more than 1 unit found contaminated on incubation.

These statistical considerations reveal a real practical problem with regard to the number of units that may need to be filled with medium and incubated, particularly in any attempt to demonstrate a probability of a low (for example, less than 0.1%) level of contamination in "standard" production batch sizes. Although there has been the odd report of media fills (by automated means) of the order of one million units per run, such numbers are hardly practical for routine validation use. And this is where most, if not all, apparently authoritative texts fail to provide any useful guidance to anyone wishing to demonstrate contamination rates of, say, no more than 0.001%. The British Pharmacopoeia, for example, while commending the media fill in general terms, makes absolutely no comment on the number of units to be filled, nor on the acceptance criteria to be adopted. In the circumstance, all one can suggest is that, purely on the basis of the practical limitations of the test procedure,

Table 16.1 Maximum permitted number of contaminated units, per various media-fill run-sizes to indicate 0.1% contamination limit in a production run in excess of 100,000 at 95% confidence level.

Media Fill Units	Contaminated Units Permitted
3,000	0
4,750	1
6,300	2
7,750	3
9,150	4

Table 16.2 Maximum permitted number of contaminated test unit to indicate 0.1% contamination (95% confidence) in production runs of not more than 100,000 units.

Production Batch Units	Media Fill Units	Permitted Contaminated Units
5,000	2,470	0
5,000	3,680	1
5,000	4,680	2
10,000	2,670	0
10,000	4,050	1
10,000	5,210	2
20,000	2,810	0
20,000	4,340	1
20,000	5,670	2
50,000	2,910	0
50,000	4,580	1
50,000	6,040	2
100,000	2,950	0
100,000	4,670	1
100,000	6,210	2

a contamination level of 0.1%, detected infrequently in media fills, may be considered to be acceptable. Regular, or common, contamination levels (in media fills) of 0.1% or above should be regarded as unsatisfactory. While it may be statistically unsound to summarize in a simple fashion data from a series of discrete events, and then treat these data as if they had been derived from a single event, a series of "good" media fill results over a period of time (assuming reasonable comparability of conditions, etc.) may be regarded as strengthening confidence, if not in any precisely quantifiable fashion. That this is a common sense, gut feeling viewpoint will be apparent. This weakness inherent in the media-fill approach to aseptic process validation must, until someone suggests something better, be considered as further strengthening the argument that, whenever possible, products intended or purported to be sterile should be terminally sterilized.

Media Fills Applied to Nonsolution Products

The same general principles, conditions, and statistical considerations as set out above apply, but the various types of nonsolution sterile products require various adaptations to the approach already described.

Sterile Powders: The use of the media-fill technique in the validation of the filling of sterile powder products presents certain special problems, arising from the probable necessity to employ additional equipment, techniques, or manipulations that are different (or additional) to those used in routine production. In such circumstances, the media fill cannot unequivocally be said to be a *precise* process simulation. This inevitable shortcoming may, however, have to be accepted. A number of different approaches have been proposed and used, as follows:

a. The normal process is simulated as closely as possible, but instead of filling a powder, a sterile liquid medium is filled. This approach is virtually the same as that described above for a solution product and fails to simulate the actual *powder* fill.

b. The normal process is simulated as closely as possible, with a sterile, dry, inert powder filled in place of the normal product or material. Lactose, mannitol, and polyethylene glycol 8000 are examples of simulation powders that have been used. There are two possible variations on this approach:

   i. Fill the chosen inert powder into the containers (e.g., ampoules or vials), which are already filled with sterile liquid medium.

   ii. Fill the inert powder first, and then add the sterile liquid medium. In both these variations, a powder fill *is* simulated, but an additional, nonroutine step (i.e., the filling of the liquid growth medium) is involved.

c. Fill sterile, dry, powdered medium into the containers, in simulation of the normal powder filling operation, aseptically adding sterile aqueous diluent on-line, to form liquid medium solution. Here, a powder fill is simulated, but an additional operation is involved.

Whichever approach is adopted, it is important to ensure that any powder/medium/diluent combination used does not cause growth inhibition through hyperosmolar or other antimicrobial effects.

Suspension Products: Simulate the entire normal process as closely as possible, including any micronization (if this is part of the normal process), using a sterile inert powder in place of the normal powder ingredient. Form the suspension, using sterile liquid growth medium in place of the normal liquid phase of the suspension product. Fill as normal and incubate.

Freeze-Dried Product: Simulate the entire normal process (i.e., preparation of bulk solution, filling of solution, loading of freeze dryer, running of freeze-drying cycle, sealing and closing of containers, and inspection) but using a liquid growth medium (dispensed as a powder, dissolved, and sterilized) in place of normal product. Actual freeze drying of the medium solution is not usually practicable, but exposure and holding times in the freeze dryer should be as normal.

Semi-Solid Products (e.g., sterile ointments and creams): Simulate the normal process cycle as closely as possible, filling a sterile liquid growth medium made to a similar consistency as the normal product by the addition, for example, of agar (ca. 4 g per liter) or carboxymethylcellulose.

For more information on validation, by process simulation, see Prout.[2,4]

Note: All the approaches and techniques outlined in this section must be performed in conformity with the general principles of GMP. In all procedures involving the use of growth media, it is vital to control any contamination of equipment, surfaces, etc. by the medium used. All media-fill studies should be promptly followed by the application of thorough cleaning, disinfecting, and sterilization procedures.

Revalidation

Following initial aseptic process validation, media-fills and process simulations should be repeated to an extent, and at a frequency, that will depend on the occurrence of events or changes that may bear upon the potential microbial hazard to the process or product. Significant modifications to equipment or facilities, changes in personnel, undesirable trends in environmental monitoring results, and sterility test failures may all indicate an immediate need to implement a full process validation protocol (i.e., a minimum of three consecutive successful media-fill runs), with the facility in question taken out of service until any problems have been resolved, and the results of the three media fills evaluated and found acceptable.

In the absence of any significant changes, or of any other events giving cause for concern, a minimum retest frequency should be twice per year per operator shift or team, for each process line. For single shift operations, the minimum frequency should be three times for each process line per year.

Data Review

All information or data generated as a result of implementing the study protocol should be evaluated by authorized persons against the protocol criteria and formally judged as meeting or failing the requirements. Written evidence supporting the evaluation and the conclusions drawn should be available.

The evaluation should be made as the information becomes available, and if it shows that protocol criteria have not been met, the study should be considered as having failed to demonstrate acceptability. The reasons should be investigated and documented. Any failure to follow the procedure as laid down in the protocol must be considered as potentially compromising the validity of the study itself, and requires critical evaluation of the impact on the study.

Final certification of the validation study should specify the predetermined acceptance criteria, against which success or failure was evaluated.

Summary of Documentation Requirements (Aseptic Process Validation)

Documents that should be available to define, support, and record the overall validation process include:

a. Protocol(s) covering the overall process, with all relevant written change control procedures and records
b. Documented evidence that the product, materials, components, etc. that are being handled or processed aseptically (e.g., bulk solution or powder; containers, and closures) plus any equipment, vessels, or surfaces (e.g., holding tanks, pipework, filling machines) that will, or can, come into contact with sterilized products or materials have themselves been previously sterilized by appropriate and validated sterilization processes
c. Documented evidence of the competence and training of *all* personnel involved in the studies
d. SOPs defining clothing requirements and gowning procedures

    e. Copies of all other relevant SOPs, e.g.:
- Dispensing ingredients.
- Water quality and supply
- Cleaning, disinfection, and sterilization (as appropriate) of all equipment, surfaces, and services
- Sterilization of equipment, vessels, and pipelines
- Filter integrity testing
- Machine setup, startup, and adjustment

    f. Written procedures for all laboratory tests

    g. Formally recorded results of all laboratory tests, with a recorded evaluation of those results against criteria established in the study protocol(s)

    h. Written calibration program and procedures covering all controlling, measuring, and recording instruments, with the results obtained during those calibrations

    i. Design specifications for major items of mechanized equipment

    j. Written installation qualification procedures, with report(s) confirming successful installation in accordance with those procedures

    k. Written operational qualification procedures, with reports certifying that equipment, as installed, will perform consistently within defined limits

    l. Statement of the environmental standards, as designated for each stage of the manufacturing process

    m. Certification of conformity of any controlled environment with the designated standard(s)

    n. Environmental retest program with evidence that this program is routinely implemented, with a record of the results obtained

    o. Written routine planned machine maintenance program, with documented evidence of the regular implementation of that program

    p. Written in-process monitoring and control procedures, with records of results obtained, both during process validation, and in routine manufacture

    q. Policy or records relating to permitted operator interventions

    r. Full process validation report, including:
- Medium used
- Volume filled
- Number of units filled
- Number of leakers rejected
- Number of units incubated
- Incubation temperature
- Incubation time
- Control organisms used
- Filter integrity test results
- Record of all in-process monitoring and control results
- Results of examination of incubated units
- Confirmation of growth in inoculated control units

    s. Final, formal evaluation of results against established criteria, with pass/fail decision

Validation of Steam Sterilization Processes

It is commonly considered (e.g., PDA,[5] Soper[6]) that there are too main approaches to the design, operation, and validation of steam (i.e., autoclave) sterilization process:

1. The probability of survival approach
2. The overkill approach

each with two aspects, in the context of process validation:

■ Physical validation
■ Biological (or microbiological) validation

The relative effort demanded in terms of physical, as compared with biological, validation by the two approaches will vary.

There could also be said to be a possible third approach, which we will term the "simplified approach," which eliminates the mental effort of juggling with the kinetics of microbial thermal death rates (which some, but not all, find difficult and confusing), and reduces the time and effort to be spent on detailed microbiological laboratory studies.

The simplified approach is to "take as read" the standard compendial cycle, or cycles, and to assume (as seems entirely reasonable) that such a cycle, if operated properly on a relatively low bioburden load, will indeed provide the level of assurance of sterility required. Thus, the British Pharmacopoeia states "...the reference conditions (for steam sterilization) of aqueous preparations are heating at a minimum of 121°(C) for 15 minutes." Assuming the adoption of this time/temperature cycle for routine use, it then simply becomes necessary to ensure in batch production, and to confirm by process validation studies, that the coldest part of the coldest item in the coldest position in the autoclave load attains a temperature of 121°C and is held at that temperature for at least 15 minutes. It is, of course, necessary to ensure in this, or any other, approach that the product or material is able to withstand this level of heat input without being degraded.

It is also necessary to be aware of variations in temperature and heat penetration throughout the load, so as to ensure the attainment of 121°C at the "coldest part," etc. does not mean that other parts of the load are exposed to heat inputs that would cause degradation. Thus, in the simplified approach, as in the others, it is not only necessary to pay attention to autoclave design and operation, and to presterilization bioburden, but also to determine a) heat distribution throughout the chamber load and b) heat penetration into the load.

This suggested simplified approach could well be regarded as neither more nor less than an overkill approach, *always provided that an overkill cycle is employed*. Soper[6] declared that "compendial sterilization cycles have been devised using overkill methods." This is a statement that is only accidentally correct.

The most commonly cited compendial cycles (with the over-pressure that is required to achieve the corresponding steam temperature) over the years have been:

Temperature (°C)	Pressure (PSI)	Time (min)
115–118	10	30
121–124	15	15
126–129	20	10

The neat progression (10, 15, 20 PSI) is no coincidence. These cycles were originally derived at a time when it was common, among the unenlightened, to speak of operating an autoclave at so many pounds *pressure*, rather than at a stated *temperature*. Thus, traditional compendial cycles were based more on "seems-like-a-good-idea-at-nice-round-number-pressures," than on overkill concepts. This attitude was common in the days before the current understanding of thermal death rates of bacteria. Nevertheless, it is indeed true that 15 minutes at 121°C, in an autoclave, does represent an overkill. To illustrate this point, it is necessary to understand something of the F_o concept, which will be covered in a little more detail later. For the present, F_o may be simply regarded as an index of the heat lethality delivered by the sterilization process. The F_o values, calculated for the various cycles listed above, are shown in Table 16.3.

This table illustrates two things. It shows the lack of comparability, in terms of lethality, of these historical compendial steam sterilization cycles. It also shows that, in comparison with a common view that a minimum acceptable F_o value is 8 (e.g., Akers and Anderson,[7] and various regulatory edicts), that 121°C for 15 minutes, in delivering an F_o of 15, does indeed represent an overkill.

The Probability of Survival Approach

This approach originated in the food canning industry (Stumbo[8]), many years before it became manifest in the pharmaceutical industry, and it is probably worth reflecting on the essential differences between the sterilization of cans of food, and the sterilization of pharmaceuticals.

Cans of food are relatively high-volume (in terms both of numbers of units and of space occupied), low-value items. Pharmaceuticals and the like are, by comparison, low volume and high value. All foodstuffs run the risk of "overcooking" in the sterilization process. By no means are all pharmaceuticals degraded by normal heat sterilization cycles. It is of particular interest to food processors to determine and apply a *minimum* acceptable cycle. A powerful motive, in addition to avoiding degrading the food and spoiling its flavor, is to reduce the very

Table 16.3 F_o values, calculated for the various sterilization cycles.

Temperature (°C)	Holding time (min)	F_o value
115–118	30	7.5–15
121–124	15	15–30
126–129	10	32–63
134–138	3	60–150

considerable time and energy costs of the sterilization of large numbers of food cans. The food producer is concerned with what is the minimum requirement. Although contaminated food is dangerous, it is probably not as dangerous as a contaminated parenteral, although the point is perhaps debatable. As a function of turnover, the cost of sterilization of pharmaceuticals is by no means as great as it is in relation to food. The manufacturer of parenterals will be rather more concerned with the best possible assurance of sterility. This is not to say that there is anything gravely wrong with the "probability of survival" approach. It is merely useful to know that, in origin, it is coming from a different direction.

Various academics and other pundits have held that when a population of microorganisms is exposed to a lethal agent (specifically, lethal heat treatment) the number of surviving organisms decreases exponentially with the extent (or time) of exposure. Thus, it is postulated that the process of microbial inactivation is analogous to a first-order chemical reaction, and may be represented thus:

$$N_t = N_0 e^{-kt}$$

where N_t is the number of surviving organisms after time t, N_0 is the number of organisms at time zero (that is, it is the pretreatment bioburden) and k is the microbial inactivation rate constant. If the logarithm of the fraction of survivors (N_t/N_0) is plotted against exposure time, the result is a curve (the "survivor curve") that is linear with a negative slope (Figure 16.2). The slope of the curve is k/2.303, from which k, the microbial inactivation rate constant, can be calculated.

D, IF, Z, and F Values

On the basis of the log-linear survivor curve, a number of values have been defined. Instead of k (see above) as a measure of microbial inactivation rate, the D value is more often cited and used. This, in the context of a heat-sterilization process, is the time (in minutes), at a given temperature, required to reduce the number of microorganisms by 90%, that is to 10% of the original bioburden, or a one-log cycle decrease in the survivor curve. For an assumed log-linear curve, D value is equal to 2.303/k. Both are measures of the resistance of an organism to a stated temperature (or other sterilizing agent). For heat inactivation, D value is the time in minutes required to achieve the one log cycle reduction of a population of an organism at a specified temperature, which is usually shown as a subscript, e.g., "D_{121}." A D value refers to the resistance (in minutes), at a given temperature, of a specific organism. It is meaningless if the temperature is not stated, or understood. (That applies in the context of heat treatment. D values can be, and have been, used in relation to both radiation and gaseous sterilization, where they are expressed in terms of absorbed dose and time of exposure, respectively.)

The inactivation factor, or IF, is a measure of the total microbial inactivation achieved by a given process. It is defined as the reduction in the number of viable organisms brought about by the process. The relationship between inactivation factor and D value is expressed as:

$$IF = 10^{t/D}$$

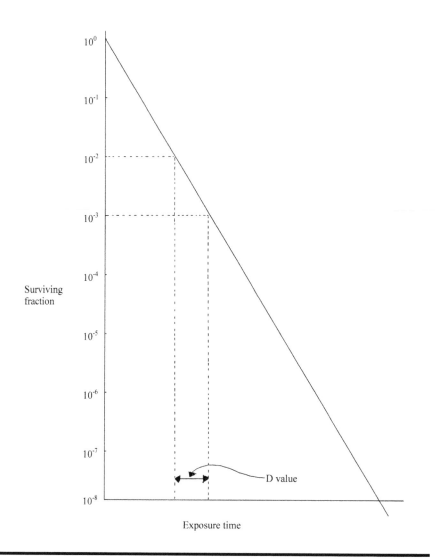

Figure 16.2   Survivor Curve

where t is the exposure time and D is the D value of the organism at the specified temperature.

If the logarithms of the D values of an organism are plotted against the temperatures at which those D values were determined, the result is a linear curve of negative slope. The Z value of the organism is the negative reciprocal of the slope of this line, and represents the increase in temperature required to reduce the D value of the organism by 90%, that is, to produce a one-log-cycle reduction in the thermal resistance curve. Thus, for a specified organism, a Z value defines the relationship between the time required to achieve a given thermal reduction at one temperature and the time required to achieve the same effect at another temperature. For all practical purposes it may be considered to be constant, for a given organism, over the relatively small temperature ranges normally used for heat sterilization (115° to 135°C for steam and 170 to 190°C for dry heat). Z values can be determined thus:

$$Z = (T_2 - T_1)/(\log D_1 - \log D_2)$$

where D_1 is the D value of the microorganism at temperature T_1, and D_2 is the D value of the microorganism at temperature T_2.

The Z value is a specific characteristic of a species, or type, of microorganism. Z values vary quite widely from one organism to another. The more heat-resistant organisms (e.g., *Bacillus stearothermophilus, B. subtilis,* and *Clostridium sporogenes*) have Z values of around 10. Thus, often when a Z value is not known, or has not been determined empirically, a value of 10 is assumed.

An F value is a measure of heat sterilization efficiency. It may also be defined as a measure of the overall lethality of a process. It equates a heat treatment at any temperature (with regard to its ability to destroy microorganisms) with the time in minutes required, at a defined reference temperature, to destroy a reference organism of a stated Z value. For steam heat at a reference temperature of 121°C, and in relation to an organism with a Z value of 10°C, the F value is termed the F_0 value, (usually articulated as either "Eff-oh" or "Eff-sub-zero"). Thus, to say that a steam sterilization process has an F_0 of 8 means that the sum of all the effects of the process is equivalent, in terms of lethality in relation to an organism with a Z value of 10, to 8 minutes at 121°C. F_0 may be expressed mathematically as:

$$F_0 = D_{121}(\log N_0 - \log N) = D_{121} \log IF$$

where
D_{121} is the D value of the reference organism at 121°,
N_0 is the initial number of the reference organism,
N is the final number of organisms, and
IF is the inactivation factor.

In most steam sterilization processes the reference organism is usually taken to be *B. sterothermophilus*, with an assumed Z value of 10 and a D_{121} of 1.5 minutes in aqueous systems. However, components of a formulation can affect thermal resistance of organisms, and if sterilization cycles are designed and validated based upon F_0 concepts, it becomes necessary to show that the formulation does not increase thermal resistance above the assumed value(s), or to determine the D_{121} value of the reference organism *in the product to be sterilized.*

Usually, heat sterilization protocols (compendial and otherwise) define a temperature and a holding time at that temperature, and take no account of the heating-up and cooling-down phases of the overall cycle. But the heat (say above 100°C) imparted to the load, during heating up and cooling down can contribute significantly to the overall lethality of the process (and, potentially, to product degradation). Application of the F_0 concept permits integration of the total lethality of a process, including the heating-up and cooling-down phases, and forms the basis for the microprocessor control of autoclaves, and for the commercially available multipoint thermocouple, with automatic printout integrator, sets.

Application of F_0 to the Validation of a Steam Sterilization Process

As an important preliminary, it needs to be emphasized that it is meaningless to think, or talk, of the validation of an autoclave cycle, or of an autoclave load, in isolation. It is the validation of a defined time/temperature cycle, when applied to a defined (in terms both of content and loading pattern) load, in a specified autoclave, that is crucially important. It is not possible to extrapolate results obtained with one cycle, applied to one load, in one autoclave to any other cycle, load, or autoclave. It is, however, not unreasonable to argue that a cycle validated for a maximum load in a given autoclave will be more than adequate for a smaller load of the same product or material in the same autoclave.

Although the theoretical basis for the design and validation of steam sterilization processes, based upon the probability of survival and the F_0 concept, is derived from studies on, and hypotheses about, the thermokinetics of microbial death rates, the main emphasis in practice is usually on *physical* (as distinct from *biological*) validation. Physical validation is aimed at ensuring a defined, reproducible F_0 value throughout a load, and specifically at the coolest part of that load. However, the use of biological indicators (BIs) in addition to temperature-sensing devices is often recommended for validation studies. The use of BIs as monitoring devices in routine manufacture is generally not to be recommended, on the common-sense ground that, if it can be avoided, it is a far-from-good idea to knowingly introduce heat-resistant spores into a sterile products manufacturing area.

There have been suggestions that not only the design and validation, but also the routine operation of steam sterilization cycles, should be based upon the determination of the thermal resistance of organisms found a) in the manufacturing environment or b) in or on the product, immediately pre-sterilization. Such a position presents both theoretical and practical difficulties. True, it is important to carry out microbiological monitoring of the manufacturing environment, and to perform batch-wise checks on presterilization bioburden. However, a knowledge of "typical" environmental microflora will not necessarily predict the bioburden of any one specific product batch. Furthermore, while a knowledge of presterilization bioburden is an essential element in making a final product-release decision, it is hardly a practical or sensible proposition to determine a presterilization bioburden, and then delay sterilization (with the bioburden multiplying all the while) until laboratory studies on the thermal resistance of that bioburden have been completed, and a sterilization cycle tailored to suit. In practice, test organisms of known heat resistance (e.g., *B. sterothermophilus*) are used, with a Z value of 10 assumed (Haberer and Wallhaeusser[9]).

For any attempt at the validation of a steam sterilization process to yield meaningful and reliable information, it is an essential prerequisite that the sterilizer (autoclave) has been properly designed and built, has been properly installed, has been supplied with all necessary services (including steam of the required quality), is fitted with instruments (temperature and pressure gauges) of known accuracy and precision, and is functioning as desired and intended. That is, it must be qualified, in terms both of installation and operation. It is also essential that it is subject to a program of planned preventive maintenance (PPM), aimed at ensuring that it remains at the same operational standard as it was at the time of process validation. Data acquired during process validation is invalid in relation to a sterilizer that has subsequently been modified, or that has deteriorated operationally.

Heat Distribution and Heat Penetration Studies

These are performed by using heat sensing probes, normally thermocouples, connected to an electronic recording instrument. The most commonly employed thermocouples are Teflon™-coated copper/constan (type T). Resistance temperature detectors (RTDs) have been used. They are sensitive, but are not sufficiently corrosion resistant for regular use in autoclaves. Before and after each use, it is essential that the thermocouples are calibrated. This can be done by immersing the probe tips in a water and/or heating bath along with a certified reference standard thermometer immersed to the same depth. The use of two reference temperatures is recommended; 0°C (ice/water bath) and around 120 to 125°C (oil or glycerol bath). Calibration equipment, designed for just this purpose, is also commercially available. The recorder readings at the reference temperature(s) are compared with those of the reference thermometer, and the recorder is adjusted to bring the temperature sensed by each probe in line with the reference thermometer. The accuracy of the thermocouples should be ±0.5°C. Thermocouples that do not display that level of accuracy, as compared to the reference thermometer reading, should be checked for bad connections, short circuits, and the like. If the problem cannot be resolved, they should be discarded.

Initial heat distribution studies are often performed on an empty chamber to investigate if and where any cold spot, or spots, are to be found. Between 10 and 20 thermocouples should be used per cycle, distributed throughout the chamber in a predetermined (and recorded) pattern. The probes can be temporarily fixed to the chamber walls by means of adhesive tape capable of withstanding the sterilizing conditions. (Rightly or wrongly, autoclave tape is often used.) Great care is necessary to ensure that the thermocouple tips make no contact with the chamber wall, or with any other metallic object. It has been suggested (Akers and Anderson[7]) that two probes from the set should remain, for reference, outside the chamber, each immersed in one of the two temperature baths.

Following the empty-chamber studies, further cycles are run with full-, half-, and minimum loads in order to study the effects of chamber loads on the location of the cold spot(s). The difference between the temperature at the coldest spot and the mean chamber temperature should not exceed ±2.5°C.

Following the heat-distribution studies, heat-penetration studies should be performed. The successful validation of a steam sterilization cycle depends upon being able to demonstrate the delivery of the desired F_0 to the coldest part of the coldest article located at the cold spot(s) determined in the heat-distribution studies. To this end, thermocouple probes are inserted into containers of liquid products, or into packages of devices or dressings, or deep into items of equipment. Further probes located in the chamber, close to the articles with probes inserted, should also be used. The crucial issue is whether or not the F_0 achieved in the coldest part of the coldest item in the coldest location is equal at least to the value required to ensure the desired level of probability of microorganism survival (or its inverse "sterility assurance level"). If it is not, then it is necessary to redesign the time/temperature cycle.

The use of biological indicators (BIs), in addition to physical methods, is fairly common, and some regulatory authorities have required it as a component of the process validation of steam sterilization. The most commonly used are preparations of the heat-resistant spores of *B. sterothermophilus*. It is necessary that, before use, the D_{121} and Z values, and the number of organisms present, is accurately known. Because of potential changes in storage, it is necessary to experimentally determine D values, whether the BIs have been purchased, or prepared in-house.

Validation of Dry Heat Sterilization Cycles

Dry heat sterilization takes two main forms: sterilization using the conventional hot air oven, and sterilization by means of a sterilizing tunnel. A major difference from a validation aspect is the additional variable introduced by the conveyor system in the tunnel, which may operate at different speeds, either by intention or inadvertence. The basic validation objectives are the same as for steam sterilization. That is, they ensure that, repeatably under all normal conditions of use, all items being treated reach the required temperature (which as we have seen will be higher than for steam sterilization) for the required time, and thus render them sterile. Often there is the additional objective of rendering them pyrogen-free.

As ever, an essential prerequisite is that the oven or tunnel has been properly designed and built, has been properly installed, has been supplied with all necessary services, fitted with instruments of known accuracy and precision, and is functioning as desired and intended. It is also essential that the equipment is subject to a program of planned preventive maintenance (PPM), aimed at ensuring that it remains at the same operational standard as it was at the time of the process validation. Data acquired during process validation is invalid in relation to equipment that has subsequently been modified, or that has deteriorated operationally. Examples of engineering aspects that need to be investigated, specified, and maintained under control before, during, and after the process validation are:

Hot Air Oven
- Air intake system and filtration
- Air exhaust system and filtration
- Fan speeds and internal air circulation
- Stability of current to fan(s) and heaters
- Functioning of heaters
- Temperature sensing, measuring, and indicating devices
- Integrity of door seals

Sterilizing Tunnel
- Relative air pressures at entrance and exit
- Filtration of cooling and exhaust air
- Current to heaters and fans
- Functioning of heaters
- Control of conveyor speed
- Particulate control (in tunnel and at tunnel entrance to clean room)

In a hot air oven, heat-distribution and heat-penetration studies can be performed in a manner similar to that employed in relation to steam sterilization. Temperature variation throughout a load should not exceed +/– 5°C. If biological validation is performed, in addition to physical methods, the indicator organisms commonly used, and recommended by the pharmacopoeias, are spores of *B.subtilis* var. *niger.*

Validation of a sterilization tunnel, although based upon the same principles, presents some additional challenges. There is the question of the air flow within the tunnel, which must be carefully balanced to ensure that validation and routine use conditions remain constant. It must be possible closely to control, and maintain under control, conveyor speed, and, hence, time of exposure. Heat-distribution and heat-penetration studies are performed using calibrated thermocouples, with leads sufficiently long to allow transportation (and subsequent recovery) along the entire length of the tunnel. To avoid a "birds nest" of wires, some form of harness will probably be necessary to keep things neat and tidy.

Validation of Ethylene Oxide Sterilization

The success, or failure, of an ethylene oxide ("EtO" — scientifically dubious, but common usage) sterilization process is dependent on the interaction of five physical variables — EtO concentration, humidity, temperature, time, and pressure/vacuum — in addition to such biological variables as bioburden and EtO resistance of bioburden. That is why considerable emphasis needs to be placed upon equipment qualification, maintenance, and control, and also why microbiological monitoring by use of BIs is usually considered necessary both in process validation, and in routine manufacture.

An EtO cycle must be designed taking into account the chemical and physical nature of the product or material to be sterilized and the nature and EtO-perme-ability of its packaging. Careful attention needs to be given to the calibration of the instruments used to monitor the conditions within the sterilizing chamber — heat-sensing thermocouples and recorders, humidity sensors and recorders, pressure gauges and recorders, gas chromatography instrumentation for determination of EtO concentration within the chamber, and timer controls.

Preliminary heat-distribution studies should be performed on the empty chamber, in order to determine cold spot(s), and thus the crucial locations for subsequent loaded chamber heat penetration runs. The final step is to perform a series of cycles (a minimum of three) on a fully loaded chamber ("load" defined as *for routine manufacture*) using both thermocouples and BIs. In the validation of EtO sterilization, the use of BIs is usually considered mandatory, and not an optional extra. (Note that the pharmacopoeias require, for EtO sterilization, the use of BIs in each routine manufacturing load.) The recommended BI is spores of *B. subtilis* var. *niger.*

These should be placed within the load (with emphasis on the previously determined cold spots), at a rate of around 10 BIs per 100 cubic foot of chamber space, at the same locations as the thermocouples. Throughout the test runs the temperature(s), humidity, gas concentration, and pressure should all be carefully monitored and recorded. The process may be considered validated if it can be shown that, in all the test runs, the desired conditions were achieved throughout the load, and that all the indicator organisms were destroyed.

Validation of Radiation Sterilization Processes

The dose of sterilizing radiation delivered to a load depends upon the strength of the source, the distance of the load from the source, the density (in radiation–penetration terms) of any material between the load and the source, and the total exposure time. Physical process validation consists of performing test runs with calibrated dosimeters inserted throughout the load so as to ensure that the required lethal dose is delivered to the entire load. Biological indicator organisms of suitable D value may also be used. In the context of radiation sterilization, D value is defined as the dose of radiation required to produce a 90% reduction in the number of organisms. The pharmacopoeias recommend spores of *Bacillus pumilus*, each indicator preparation to carry at least 1 x 10^7 viable spores.

VALIDATION OF CLEANING PROCEDURES

Unless otherwise stated, the term "cleaning procedures" is generally taken to mean, in the context of validation, procedures for the cleaning of equipment. The FDA Guide to Inspections of Validation of Cleaning Processes,[10] quite clearly refers only to equipment cleaning. The prime motive is to ensure that procedures used for cleaning are adequate to ensure the very minimum possibility of contamination of one product by another. Along with mix-ups with printed packaging materials, cross-contamination is one of the major reasons for product recall, and the major cause of cross-contamination is product residues inadequately cleaned from manufacturing equipment.

The essential preliminary is that there must be authorized written procedures. It is totally impossible to validate procedures that are conceptual, all in the mind, ad hoc, or rule of thumb. Written procedures are needed for each piece, or type, of equipment. If different approaches are adopted for cleaning between batches of the *same* product, as compared with cleaning between batches of *different* products, then these both should be clearly covered in the written procedure(s), as should the measures taken to remove traces of any agents (detergents, solvents) used in the cleaning process. The validation of the cleaning processes, implemented in accordance with these written cleaning procedures, should be subject to the same documentation requirements as outlined in Chapter 15. It is generally considered that equipment cleaning between batches of the same product may be considered adequate if the equipment is visibly clean, with no further validation required, a view endorsed by the FDA.[10]

A "traditional" method for measuring the effectiveness of a cleaning process has been to sample the final rinse liquid and examine the sample in the laboratory for traces of the previous product or material. While still used, this method has, of late, encountered some disfavor. The FDA, for example, with a rather touching fondness for the homely analogy, consider that it is like looking at the washing-up water to see if a cooking pot is clean, rather than looking at the pot itself.

Currently, the more favored method is direct surface sampling, using swab samples taken from defined areas (say, 100 cm²) at defined locations. The advantage of swab sampling (which can also be used to evaluate surface microbial contamination) is that it enables the targeting of the more obviously difficult-to-clean surfaces. The disadvantages are the more difficult-to-clean surfaces may well be the most inaccessible, and may require a level of disassembly that would not occur in routine manufacture (thus introducing variables not encountered routinely). The very use of swabs may, in itself, also introduce chemical or microbial contamination.

A third method, which some manufacturers have employed, is to manufacture a placebo batch in the cleaned equipment under the same conditions as normal and then examine samples of the placebo for contamination. This is hardly a very good idea, not least from the aspect of the cost in terms of time, effort, and money.

Not for the first time, a problem is encountered when it comes to the determination of *acceptance criteria* for validation studies. (It seems strange that, in the context of all the emphasis that has been placed upon validation, in at least two major validation areas, aseptic process validation and cleaning process validation, there is a notable degree of indeterminacy over the criteria to be adopted for deciding whether or not a process has been validated.) The FDA[10] evades the issue by overtly declaring that it "... does not intend to set acceptance specifications ...." Suggestions that have been offered — and they are no more than that (e.g., Cook[11] and McCormick and Cullen[12]) — include levels of contamination that would consistently ensure no more than 10 ppm of the contaminant in subsequent batches of product, or to not more than 1/1000th of the minimum daily dose of the contaminant in the maximum daily dose of subsequently manufactured product.

COMPUTER SYSTEMS VALIDATION

Although the expression "computer systems validation" is commonly used and heard, it needs to be realized that it is not only mainframe computer hardware and software that needs to validated, but all computer and microprocessor control systems. Thus, better expressions would be "validation of automated systems" or "validation of computer-related systems."

The validation process should establish documentary evidence that provides a high degree of assurance that an automated system will consistently function as specified and designed, and that any manufacturing process involving the automated system will consistently yield a product of the required and intended quality.

User specifications for both the hard- and software that comprise the overall system should be subject to design review and qualification to ensure that the system will be, and remain, fit for the purpose intended. This design review and qualification should include a careful consideration of potential system failures and of the possibility (and the consequences) of any undetected system failure that could adversely affect product quality.

Hardware must be:

    a. Suitable, and of sufficient capacity, for the tasks required of it
    b. Capable of operating, not merely under test conditions, but also under worst-case production conditions (e.g., at top machine speeds, high data input, high or continuous usage)

Hardware should be tested to confirm the above, with the tests repeated enough times to ensure an acceptable level of consistency and reproducibility. Hardware validation and revalidation studies should be documented, in accordance with the basic documentation requirements outlined in chapter 15.

Software should be validated to ensure that it consistently performs as intended. Test conditions should simulate worst-case production conditions, e.g., of process speed, data volume, and frequency. Tests should be repeated a sufficient number of times to ensure consistent and reliable performance. Software validation and revalidation studies must be documented as for hardware validation.

Much of the necessary microprocessor and computer hardware and software validation may well be performed by the machine, hardware, or software supplier. However, it must be stressed that the final responsibility for the suitability and reliability of any automated system used in pharmaceutical manufacture must rest with the pharmaceutical manufacturer.

Manufacturers should obtain (and retain) from the relevant third party sufficient data (specifications, programs, protocols, test data, conclusions, etc.) to satisfy themselves, and any enquiring regulatory body, that adequate validation work has been carried out to assure system suitability.

For all involved, or interested in automated systems in pharmaceutical manufacturing, the following is strongly recommended: The *GAMP (Good Automated Manufacturing Practice) Supplier Guide for Validation of Automated Systems in Pharmaceutical Manufacture Version 3.0, pub. GAMP Forum,* 1998. The PDA *Validation of Computer-Related Systems* — Technical Report No. 18, PDA Journal of Pharmaceutical Science and Technology, Supplement to Vol. 49/1, January/February 1995 is also useful.

REFERENCES

1. PDA Technical Report No. 22, Process Simulation Testing for Aseptically Filled Products, Supplement to *PDA J. Pharm. Sci. Technol.,* 50, S1, 1996.
2. Prout, G., Ed., The use of process simulation tests in the evaluation of processes for the manufacture of sterile products, Parenteral Society Technical Monograph no.4, 1993.
3. US FDA, *Guideline on Sterile Drug Products Produced by Aseptic Processing,* Rockville, MD, 1987.
4. Prout, G., Validation and routine operation of a sterile dry powder filling facility, *J. Parent. Sci. Tech.,* 36(5), 199, 1982.
5. PDA, Technical Monograph No. 1, Validation of Steam Sterilization Cycles, Parenteral Drug Association, Philadelphia, 1978.
6. Soper, C., Sterilization, in *The Pharmaceutical Codex*, 12th Ed., Pharmaceutical Press, London, 1994.
7. Akers, M. and Anderson, N., Sterilization validation of sterile products, in *Pharmaceutical Process Validation*, 2nd edition, Berry and Nash, Eds., Marcel Dekker , New York, 1993.
8. Stumbo, C., Thermobacteriology in Food Processing, 2nd ed., Academic Press, New York, 1973.
9. Haberer, K. and Wallhaeusser, K.H., Assurance of sterility by validation of the sterilization process, in *Guide to Microbial Control in Pharmaceuticals*, Denyer and Baird, Eds., Ellis Horwood Ltd., New York, 1990.
10. US FDA, *Guide to Inspections of Validation of Cleaning Processes,* Rockville, MD, 1993.

11. Cook, R., Validation, in *The Pharmaceutical Codex*, The Pharmaceutical Press, London, 1994.
12. McCormick, P. and Cullen, L., Cleaning validation, in *Pharmaceutical Process Validation*, 2nd edition, Berry and Nash, Eds., Marcel Dekker, New York, 1993.

17

SELF-INSPECTION AND QUALITY AUDIT

The International Standards Organisation (ISO) has defined a quality audit as follows:

> QUALITY AUDIT
>
> A systematic and independent examination to determine whether quality activities and related results comply with planned arrangements, and whether these arrangements are implemented effectively and are suitable to achieve objectives (ISO 8402: 1986, Quality Vocabulary)

In addition to its appearance in the ISO Quality Vocabulary, this definition is also repeated in a number of other ISO documents, e.g., the ISO 9000 series. It is all rather vague, not to say "woolly," and makes no mention of the crucial issue of what is to be done as follow-up to the audit.

By comparison, a definition offered by the European Organisation for Quality (EOQ) was relatively laconic:

> QUALITY AUDIT
>
> A systematic and independent examination of the Quality System or of its parts (EOQ)

The use of the definite article "the" is odd. What is *the* quality system one might ask; is there only one?

The second edition of the French national GMP Guide (Bonnes Pratique de Fabrication, or BPF, since superseded by third edition, which is "harmonized" with the EC GMP Guide) provided the following two definitions (author's translations):

> SELF INSPECTION consists of a periodic detailed examination of conditions and working procedures by a team from the production site, with the aim of verifying that good pharmaceutical manufacturing practices are being applied and to propose any necessary corrective measures to responsible management.

and

> A QUALITY AUDIT consists of an examination and an evaluation of all or part of a system of quality assurance. It must be carried out by a specialist or a team designated for this purpose. It may be extended, as necessary, to suppliers and sub-contractors.

These two, taken together, are useful definitions that do make the important point that one of the objectives is to propose corrective methods. It is not, however, explicitly stated that there is a need to ensure that the proposed corrective steps are, in fact, taken — and that if they are not, the whole point of the exercise is lost. A perhaps inappropriate distinction is also drawn between a "self-inspection" and a "quality audit." It is more logical to consider self inspection as a subclass of the more general class, quality audit, and then derive the the following classification of audit types:

QUALITY AUDITS: BASIC TYPES

1. Imposed *upon* manufacturer or supplier
   a. Regulatory
   b. Customer, or potential customer
   c. Third party (on behalf of customer)
2. Performed *by* manufacturer
   a. Internal (i.e., self-inspection)
      i. Overall
      ii. Departmental
      iii. Product-orientated
      iv. System-orientated
   b. External, e.g.:
      i. Of supplier
      ii. Of contract manufacturer
      iii. Of contract packager
      iv. Of contract warehouse/distributor

Some writers have variously drawn a distinction between an inspection and an audit, but it is difficult to see why any such distinction is necessary. The exercise, whatever it may be called, can vary in length, depth, and intensity as circumstances dictate, and the only distinction necessary is "internal" vs. "external," and even then the ultimate objectives are similar.

REGULATORY REQUIREMENT

The US cGMPs (*for Drug Products*) do not have an explicit requirement for internal company audits and self-inspections. (cf. the US Medical Device GMPs [21CFR 820] and the ICH Q7A guidelines on APIs, which do). However, the preamble to the US cGMPs (21 CFR 210/211) states *inter alia* that it "encourages quality assurance program audits that are candid and meaningful." Furthermore FDA Compliance Policy Guideline no. 7151.02 (last revised 01/03/ 1996) states that "during routine inspections and investigations conducted at any regulated entity that has a written quality assurance program, FDA will not review or copy reports or records that result from audits and inspections of the written quality assurance program ... FDA may seek written certification that such audits and inspections have been implemented, performed, and documented and that any required corrective action has been taken ...."

It would therefore seem that self-inspections are *expected* by the FDA, even if they are not explicitly or expressly *mandated*.

Inspection of suppliers of materials is, perhaps, implicit in such statements as:

US cGMPs

> 211.84 (d) (3) .... Provided that the manufacturer establishes the reliability of the supplier's test results through appropriate validation of the supplier's test results at appropriate intervals

The regulatory requirement for the performance of *internal* quality audits is clear and unequivocal in both the European GMP Directive (91/356/EEC, Article 14) and in the EC GMP Guide.

The European GMP Directive (91/356/EEC, Art 14) states:

> Self-Inspection: The manufacturer shall conduct repeated self-inspections as part of the quality assurance system in order to monitor the implementation and respect of Good Manufacturing practice, and to propose any necessary corrective measures. Records of such self-inspections and any subsequent corrective action shall be maintained.

It is not clear what is meant by "repeated" (every hour? once every 50 years?), but that is what the Directive states.

The EC GMP Guide states, in its first chapter, Quality Management, 1.2, that:

> The system of Quality Assurance ... should ensure that ...:

> ix There is a procedure for self-inspection and/or quality audit which regularly appraises the effectiveness and applicability of the quality assurance system.

The meaning of "regularly," like "repeated," is ambiguous, and the intended meaning of "applicability" is unclear. Again, there would also appear to be only the one ("the") quality assurance system.

In addition to the above statement, the EC GMP Guide devotes a whole short chapter to self-inspection, thus:

EC GMP GUIDE

Chapter 9 Self-Inspection

> **Principle:** Self-inspections should be conducted in order to monitor the implementation and the respect of Good Manufacturing Practice (GMP) principles and to propose necessary corrective measures.

> 9.1 Personnel matters, premises, equipment, documentation, production, distribution of the medicinal products, arrangements for dealing with complaints and recalls, and self-inspection, should be examined at intervals following a pre-arranged programme in order to verify their conformity with the principles of Quality Assurance.
>
> 9.2 Self-inspections should be conducted in an independent and detailed way by designated competent person(s) from the company. Independent audits by external experts may also be useful.
>
> 9.3 All self-inspections should be recorded. Reports should contain all the observations made during the inspections and, where applicable, proposals for corrective measures. Statements on the actions subsequently taken should also be recorded.

This short chapter from the EC GMP Guide, while it could hardly be considered a model of precise, lucid English, nevertheless does embrace a number of important, key, auditing issues, *viz*:

- Follow a "pre-arranged program" (i.e., inspection should be *planned* not impromptu)
- Inspection should be "independent" (Presumably this means *unbiased*)
- It should be conducted "by competent persons"
- Results and findings should be recorded
- Report should contain "all observations"
- Report should make "proposals for corrective measures"
- All actions subsequently taken to be recorded

Regulatory statements on *external* audits in the EC GMP Guide are relatively sparse, but there are passages where the need to perform external quality audits is at least implied e.g.,:

EC GMP GUIDE

> 5.25 The purchase of starting materials.... should involve staff who have a particular and thorough knowledge of the suppliers.
>
> 5.26 Starting materials should only be purchased from approved suppliers.
>
> 5.40 The purchase of.... packaging.... materials should be accorded similar attention.

Reasons for Quality Auditing

In addition to any regulatory requirement, these may be summarized as follows:

1. Internal — in order to:
   - Determine the level of compliance
   - Build confidence (hopefully) in GMP and the QA system
   - Build interdepartmental trust, understanding, and communication (if the audit is done properly and tactfully)
   - Determine measures necessary to improve, e.g.,:
     - Premises, equipment, environment
     - Operations, actions, procedures
     - Personnel/training
   - Provide a stimulus for improvement
   - Recommend corrective action
   - Monitor improvement

2. External — in order to:
   - Establish and monitor capability of supplier or contractor to deliver goods and services that are fit for purpose (and on time, and in the quantity required)
   - Build mutual confidence
   - Promote understanding and communication between the parties involved (both sides can learn!)
   - And in general, as listed for "internal"

STEPS TO PERFORMING A QUALITY AUDIT

A fundamental prerequisite for the successful performance of any quality audit is the availability of competent, trained auditors. Given their availability, the key steps are:

- Plan and prepare
- Arrange and announce (?)
- Arrive at site of audit, meet, explain purpose
- Perform audit
- Informal oral report of finding
- Formal report, with recommendations
- Follow-up

The only one of these steps against which there is a question mark is the issue of whether or not the auditor(s) should announce, in advance to the auditee, their intention to audit at a stated time. In general, the only good reason to spring an audit would be when a regulatory body wants to try to catch a manufacturer suspected of improper practices. For nonregulatory audits (internal or external) it can be said that there is more to be gained from announcing intentions in advance (availability of key staff, cooperative attitude, etc.) and, potentially much to be lost (possible nonavailability of staff, lack of cooperation, and resentment) for failing to do so.

THE TOOLS OF THE AUDITOR

In planning and preparing to audit, certain tools are available (or should be) to the auditor:

1. Documents
   - Quality and GMP regulations, standards, and guidelines (local, national, and international)
   - Previous audit and follow-up reports
   - Auditee's own documents and records
   - Audit checklists and *aides memoires*
2. The auditors own eyes, ears, brain, words, character, etc.
3. The auditing plan
4. And, of course, paper, pen or pencil, and so on

To consider some of these tools in a little more detail, examples of the Quality regulations, standards, and guidelines are:

1. LOCAL (or corporate), e.g.,:
   - Audited company's own internal regulations, codes of practice, and guidelines
   - Their quality manual (if one exists)
   - Their site master file (if one exists)
2. NATIONAL, e.g.,:
   - US cGMPs (and relevant FDA compliance guidelines)
   - UK Medicines Act and Regulations
   - UK Orange Guide (medicinal products)
3. INTERNATIONAL,e.g.,:
   - ISO Standards (including the 9000 Series)
   - WHO GMP Guide
   - EC (EU) GMP Directive (91/356)
   - EC (EU) GMP Guide
   - PIC GMP Guide

The important thing here is to determine and agree, on all sides, the guideline, regulation, or standard against which the audit is, or will be, performed. An audit that is performed against no more than a vaguely notional, general concept of quality is hardly likely to be successful or of any great use.

Checklists

Another possible auditing tool is the checklist. A number of preprepared examples have appeared in both commercial and official publications. Some include a points-scoring system, with scores being accumulated to make a pass or fail decision.

There are both advantages and disadvantages to the use of checklists. The advantages are that they keep the auditor's mind on the job at hand and force a structured approach. They provide a way of making notes, in short form and in a structured manner. They also can provide a prop for the rookie auditor who, under pressure, may become uneasy about what to ask, or look at, next. The disadvantages are that they tend to be inflexible, and the audit (because of a company's geographical, functional, and administrative structure) may not happen in the same order as the checklist. That is, the structure of the checklist may not reflect the structure "on the ground." The checklist may also become the auditor's master rather than his servant. There is a danger of spending more time on completing the checklist than on looking, seeing, and listening, and thus failing to see the quality wood for the checklist trees.

The awarding of marks for each item on the checklist, which are then added together in order to make a pass or fail decision, is a dubious activity unless appropriate weightings are given to the specific elements. For example, the presence of litter on the factory surrounds, although indicative, is clearly not as critical as an inadequately validated sterilization cycle. If a "prop" is felt necessary for the auditor, any checklist should be as brief and as simple as possible.

THE AUDITOR

In the achievement of the objectives of an audit, and to ensure that it is well and effectively conducted, the most important of the tools listed above is the auditor himself — and his eyes, ears, brain, character, and communication skills. It is thus worth asking the question, what sort of person makes a good quality auditor or inspector?

It is probably true to say some people have a natural talent for the job. Equally, there will be others who will never be any good at it. And, as ever, in the middle there is a fair-size group of people who, if they put their minds to it and have the relevant knowledge, *and* gain the relevant experience, will make a reasonable job of it. The essential qualities of a good auditor are:

- Appropriate range and depth of knowledge, both of QA/GMP and of the relevant technology
- Range and depth of relevant experience
- Good powers of observation
- An enquiring, yet open, mind
- Able to think on feet
- Articulate — good communication skills
- "Unflappability" — able to stay cool
- Able to take a constructive approach
- Able to make sound judgments on the matters observed
- Able to be persistent, yet patient and diplomatic, in pursuit of the objectives of the audit
- Able to *listen*
- In good health

Many of these are obvious. It is important for an auditor to understand that there may well be ways for a manufacturer to achieve the desired quality objectives, other than those with which he or she (the auditor) is familiar. The auditor should be prepared to listen patiently and sympathetically to explanations offered, and to make a sound, open-minded judgment on the validity of those explanations. Whatever happens, he or she must be able to stay calm, and not enter into contentious argument. Quality auditing can be both mentally and physically demanding. It is not a job for the unfit.

The personal skills and qualities required of an auditor are thus extensive, although many of them, if not innate, can be acquired or developed. It is important, however, that no person who is entirely unsuited to the job, or to whom it is an anathema, should be coerced into doing it. An unwilling auditor is a bad auditor. Possibly even worse are the types of persons (and they have existed, even in official regulatory bodies) who see the job as a means of satisfying their appetite for power.

AUDIT PLANNING AND PREPARATION

Given a competent auditor, or auditing team, the success (or failure) of an audit in achieving its objectives greatly depends on the quality of the advance planning and preparation. It is vital that all concerned are aware of the type and objectives, and the date and time, of the audit and of the areas and systems to be covered. If the audit is to be performed by a team, then the team leader or lead spokesperson should be decided in advance.

It is important to learn as much as possible (and to think as much as possible) about the site or area to be audited, in advance.

Taking these various points, the following Audit Preparation Checklist may be set down:

- Agree date and time of audit
- Clarify and communicate objectives and the standard, regulation, or guideline against which audit will be performed
- Decide type of audit and auditing strategy (see below)
- Define areas or systems to be covered
- Inform auditee
- Obtain and review details of site or area to be audited, e.g.:
    - Site or area plans or drawings
    - Personnel organization charts
    - Products manufactured
    - Quality Manual, or Site Master File (if available)
    - Any available records of complaints and recalls
    - Any reports of previous audits and follow-up records
- Prepare a structure for notetaking, or devise a checklist
- Hold final team briefing meeting

Auditing Strategies

A number of different possible approaches to conducting an audit, or audit strategies have been proposed. These may be summarized as follows:

- Forward trace — i.e., following production flow from receipt of components or materials through to dispatch of finished products
- Backward trace — i.e., the reverse of the above
- Product-orientated
- Documentation-based
- Problem-orientated (e.g., centred around complaints or recalls, and any corrective action taken)
- System-based
- Completely random (??)

(Note: Mixes of strategies are entirely possible.)

By far the most common, and most simple and logical, approach is the first (forward trace). It can be applied to an entire manufacturing site or to a single department. A normal logical work-flow is followed from where it starts to where it ends. Doing things in reverse (backward trace) has had its advocates, but it is difficult to see any advantages, and it surely makes the task more difficult for the auditor(s). There is absolutely no merit in a completely random approach, unless the main objective is all-around confusion.

A product-orientated approach is a perfectly viable option. Here, the focus is on just one product, or family of similar products. It leads to an intense, in-depth concentration on the manufacture of just that one product (or family), but it is vitally important to remember that it is also necessary to evaluate other, more general, departments or functions (e.g., stores, dispensary, laboratories, data processing) for their impact on the quality of that product.

An audit can be documentation-based, as tends to be favored by the FDA. Such an approach is easy on the feet, but of limited value in terms of getting to grips with the whole quality picture. Other workable audit types can be problem-orientated (a special case, when a known problem needs to be investigated to discover causes and to propose corrective action), or directed at a system (e.g., water supply and quality, air supply, engineering, computer systems).

ARRIVING AND STARTING

All necessary arrangements and preparations having been made, background information gathered and strategies determined, auditors arrive at the site or the department to be audited. They should arrive:

- On time
- Well organized
- Well prepared

They should aim to look smart in appearance, with a well-mannered, professional style and approach that is sensitive to the thoughts, feelings, attitudes, procedures, safety and security requirements, and indeed the "culture," of the people to be audited. There is nothing to be gained, and much to be lost, by presenting as scruffy, boorish, insensitive auditors.

THE OPENING MEETING

There will be, or should be, some form of opening meeting. This is likely to be more formal at an external than at an internal audit. A suggested agenda for this meeting is:

- Introductions
- Auditors explain purpose and objectives
- Meet those who will be escorting the auditors and responding to their questions
- Auditors inquire of company rules, work practices, and the like (if this is an unfamiliar site)
- Agree programming/timing

Here is also the chance to establish good working relationships between the two sides, and also to gather general information (e.g., site size and geography, manufacturing capability and capacity, product types and range, any contract work or services used or supplied), if this is not already known.

Style and Conduct of Audit

The manner and style in which the audit is conducted bear heavily on the success (or failure) of the audit. Auditors should be themselves and not act out a role, or "kid." It can be fatal to pretend to know more than you do, and then get caught. Although an essential air of seriousness should pervade, a little leavening of humor can help things along. But it should be remembered that what sounds funny to some people can be offensive to others, and there are few things more deflating than a joke that falls flat. Under no circumstances should an auditor mock or ridicule any person, system, or institution, no matter how ludicrous they, or it, may seem to be.

Within the bounds of normal politeness, and while keeping as far as possible to the agreed program, auditors should aim to lead, rather than be led. That is, they should persistently, but patiently, seek to see what they came to see, and not just those things the escort(s) want to show them.

Questions asked should be pointed, probing, specific, and precise and should require a specific answer. Questions in the form "no doubt you ..." invite, almost demand, the response "yes we do." Auditors should not necessarily avoid asking the obvious question. They should never think, "nobody could be that stupid or careless," and thus avoid asking the obvious question. It is entirely possible that someone, somewhere, could indeed be "that stupid or careless." Where appropriate, evidence should be required of the veracity of responses made to questions asked, or statements made. (For example: Question: "How often do you check the quality of the water in your ring-main supply?" Answer: "Twice a week" Question: "May I see your record of the results over the past year?")

The value of silence should be remembered. Auditors should wait patiently for the response to their questions. They may not get an adequate answer, but the urge to fill an oral vacuum is a powerful one, and the auditor may well hear something else of interest. A golden rule for auditors is to *listen* harder than they talk. As the audit proceeds, the auditor should make careful notes, and record all relevant data.

A few more points on the personal approach of auditors: They should aim to be constructive, and certainly avoid destructive or personal criticism. An auditor should stay cool, and no matter the provocation, avoid any heated argument. If an auditor makes a mistake, or misunderstands something, he or she should admit it, and seek to understand better — rather than cover up. Under the stress of an intense audit, it is entirely possible to misunderstand something, but this is no crime, or slur upon the auditors competence. It is better to clarify things at once, rather than jump to possibly false conclusions. However, it is worth following up on hunches, which may well prove not to be as irrational as at first they may seem. (Potentially hazardous malpractices have been unearthed by an auditor acting upon a hunch.)

A novice auditor will probably have preconceived ideas, based on his own practical experience, as to the way things should be done, and in what order. Auditors must be flexible, and not necessarily expect everybody to be doing things in their own (the auditor's) preconceived way, in their preferred order. (See the point made above regarding the rigid approach that can be imposed by checklists.) It is important to remember that it is the end result that matters, and be prepared to consider alternative means of achieving the desired quality result.

AUDIT OBSERVATIONS

So, what are auditors looking for on an audit? What do they need to see, to learn, to observe, to investigate and take note of? These may be classified under three main headings:

1. Basic general company information
2. Departmental conditions and practices
3. Systems

1. General company information

The basic general information that auditors need to have, or acquire, includes:

- Where is the plant location and what is the immediately surrounding environment?
- Are the factory surrounds kept neat, clean, and tidy? (Such things may not directly affect product quality, but scruffy grounds are indicative of a poor attitude.)
- Are there any undesirable activities (sewage works, rubbish tips, etc.) going on in the immediate vicinity?
- How is the company managed?
- Is the technical, manufacturing, and quality management on site, or is the site managed (?) from afar, and merely an outpost of a large industrial empire?

- Is there a clear and explicit chain of command, made clear in organization chart(s), with written job descriptions for supervisory and management positions?
- What is the company's training policy, and how is it implemented, programmed, and recorded, both in relation to induction and continuing training?
- From where does the company obtain its starting and packaging materials?
- Does it use contractors to manufacture or supply any of its products, or to provide any services (analytical, engineering or maintenance)?
- Does it provide, under contract, any goods or services to other organizations?

2. Departmental conditions and practices

Most medicinal and other healthcare products are manufactured in a series of different stages in different departments or sections. Some of the things that auditors need to note are common to all (or most) departments and sections, for example, security, cleanliness and good order, the physical measures taken to avoid cross-contamination and mix-up, the availability of written procedures, and whether or not these are being followed. Other departmental points to be noted will be specific to the work of a given department or section — stores, tablet or capsule manufacture, sterile products, packaging, testing laboratory, or whatever. Here, the auditor's knowledge and understanding of the relevant science, technology, and techniques is crucial.

Despite the departmentalized manner in which manufacture tends to proceed, it is important for auditors to avoid a compartmentalized attitude, and to think instead in terms of the overall picture. In all but the very simplest and smallest of operations, production does tend to progress by the movement of materials, products, and documents from one department to another. Each of the individual departments or sections may be fine in isolation, but how does it all fit together? Is one department supplying the next with what it really wants? And is this second department getting what it thinks it is getting? The auditor should probe the interfaces between departments and attempt to evaluate how the things that one department does, or produces, impact the others.

Short, relevant, story: Imagine a large, famous-name pharmaceutical manufacturing company. In the dispensing area, the auditor notices a dispenser, weighing-up materials for an oral liquid product, paying scant attention to accuracy of weighing. The auditor asks, "Why are you not weighing the exact amounts, as specified on the batch document?" The dispenser replies, "Oh, I don't have to do that. I just have to get it roughly right. They check it all again in manufacturing." The auditor proceeds immediately to the relevant manufacturing area and, on seeing operators using previously dispensed ingredients without any further check, asks "Don't you check the weights/volumes of ingredients before you add them to the batch?" The operator replies, with a pitying smile, "Oh no, we don't have to do that, it has all been carefully weighed-up in the dispensary, you see." The batch manufacturing instructions and SOPs had nothing to say on this point, one way or another. This story is true. The author of this book can vouch for it. He was the auditor.

3. Systems

The systems that should be examined include:

- Quality system (of course)
- Overall documentation and document control
- Change control systems
- HVAC (Heating, ventilation, and air-conditioning)
- Water supply system
- Engineering maintenance
- Plant services generally

DIVERSIONARY TACTICS

As all hardened auditors and inspectors are aware, it is not unknown for manufacturers to attempt to waste auditors' time so that they become rushed to complete the audit and thus less observant and perceptive, or even to indulge in diversionary tactics. Techniques that have been employed include:

- Delaying the start by prolonging the opening discussion, and requesting an early finish
- Prolonged tea and coffee breaks
- Variably timed staff breaks, making it time consuming to summon "the one who knows the answer"
- Conducting the auditor the long way around the site
- Confusing the auditor by surrounding him or her with many escorts or experts
- Giving lengthy and elaborate explanations that overexaggerate the complexity of the operation
- Deliberate planting of minor faults, on the theory that auditors feel that they must justify their existence by finding *some* faults, and that having done so, they will relax
- Strategically timed fire alarms or drills
- Taking auditors out for a lavish lunch ("It won't take long. Just as quick as the canteen") — by the scenic route

These, of course, largely apply to external audits, but they do serve to remind potential auditors that they need to beware of deliberate delays and diversions, which some auditees may see as in their short-term interests. The good auditor should stick to the matter at hand, and *not* be diverted from it.

The Concluding Summary Session

Most audits, of whatever type, conclude with a summary session, and it is greatly to the benefit of both parties that they should. Here is the opportunity for the auditors to outline their findings, and to make their recommendations, and for the auditees to comment, offer explanations (or to challenge as appropriate), and for both parties to agree on any necessary corrective action.

The time for the summary session should be agreed upon in advance, and the auditors should strive to keep to that timing, in order to ensure that key auditee personnel will be available, and hopefully, present. The auditors should leave themselves time to get their notes in some sort of order, so as to be in a position to make a structured presentation of the audit findings. This does not have to be a highly polished performance, but it should be in a logical sequence, with the right degree of emphasis placed upon the most important issues. If the audit has been a team effort, then the auditors should appoint a lead spokesman, in advance of the session. There is no reason why the other auditors should not join in, but this should be done in a controlled fashion, with just one speaker at a time. The presentation of the audit findings should be factual rather than conjectural. It should be as concise as possible, and trivialities avoided. Efforts should be made to reach agreement, on both sides, of the justice of the findings, and on any corrective action that needs to be taken. The auditors should be ready and willing to offer constructive advice.

The Audit Report

After the audit comes the business of writing the report. This is something that few people like doing, but it must be done, and it is better (and in the end, easier) that it is done as quickly as possible, while the experience of the audit is fresh in mind. The report is the auditors' formal record of what was seen, done, heard, and agreed upon. Whether or not a copy of the full report is sent to those audited or just a summary of findings and recommendations for corrective action will be a matter of judgment, depending on circumstances. Thus, the report should be written, and a copy (whole or in part) sent to the auditee without delay. It should include all significant, nontrivial findings. It should be well-structured, and be compatible with the oral summary given at the concluding summary session. That is, it should contain no surprises.

A suggested structure for an audit report is:

1. Site or area audited
2. Date and time of audit
3. Auditor(s)
4. Objectives and purpose of audit
5. Personnel encountered (names and positions)
6. Changes since any previous audit (organization, premises, equipment, procedures, products, etc.)
7. Observations — in logical order
8. Corrective measures requested
9. Oral responses of auditee
10. Final conclusions

Follow-Up

The final phase of the audit is the follow-up. It is at least as important as any of the others, since there is no point in quality auditing unless any necessary corrective action is taken *and confirmed*. In fact, the follow-up could be said to have commenced at the concluding summary session, with the oral requests made and the written assurances given. These should be confirmed by written requests, and hopefully, written assurances from the auditee that the requested corrective actions have been taken. For full and final confirmation, a further on-site meeting may be held, combined with a partial or total re-audit.

Corrective Action Report

For internal audits (and perhaps some external audits) the use of a simple corrective action report (see Figure 17.1) is a simple and easy way of requesting action, receiving information on action taken, and keeping track of any follow-up required.

CORRECTIVE ACTION REPORT   **Ref. No.** ....

To: ....   **Date** ....

During an audit of ....   On ....

The following was/were rated as unacceptable/immediate action required:

The following corrective action is required:

Signed ....   Date ....

Corrective action taken/comments:

Signed ....   Date ....

Auditor review of corrective action:

Problem resolved ....   Action unsatisfactory ....

Comments:

Signed ....   Date ....

Figure 17.1   Corrective Action Report

18

US CGMPS AND EC GMP GUIDE — CONCLUDING COMPARISON

The main body of the text of the EC Guide is presented in nine chapters, approximately (but not precisely) in line with the Principles of the corresponding EC Directive 91/356. The chapter headings of the EC GMP Guide are:

Chapter 1 Quality Management
Chapter 2 Personnel
Chapter 3 Premises and Equipment
Chapter 4 Documentation
Chapter 5 Production
Chapter 6 Quality Control
Chapter 7 Contract Manufacture and Analysis
Chapter 8 Complaints and Product Recall
Chapter 9 Self Inspection

These nine chapters are followed by 18 annexes as follows:

Annexes to EC GMP Guide

1. Manufacture of Sterile Medicinal Products
2. Manufacture of Biological Medicinal Products for Human Use
3. Manufacture of Radiopharmaceuticals
4. Manufacture of Veterinary Medicinal Products other than Immunologicals
5. Manufacture of Immunological Veterinary Medicinal Products
6. Manufacture of Medicinal Gases
7. Manufacture of Herbal Medicinal Products
8. Sampling of Starting and Packaging Materials
9. Manufacture of Liquids, Creams and Ointments

10. Manufacture of Pressurised Metered Dose Aerosol Preparations for Inhalation
11. Computerised Systems
12. Use of Ionising Radiation in the Manufacture of Medicinal Products
13. Manufacture of Investigational Medicinal Products
14. Manufacture of Products derived from Human Blood or Human Plasma
15. Qualification and Validation
16. Certification by a Qualified Person and Batch Release
17. Parametric Release
18. Good Manufacturing Practice for Active Pharmaceutical Ingredients

The US cGMPs for Finished Pharmaceutical Products are in two parts: 21 CFR Part 210 General, which deals with such matters as status, applicability, and definitions of terms; and 21 CFR Part 211, which contains the specific "meat" of the cGMPs. This Part 211 has 11 subparts:

A. General Provisions
B. Organization and Personnel
C. Buildings and Facilities
D. Equipment
E. Control of Components and Drug Product Containers and Closures
F. Production and Process Controls
G. Packaging and Labelling Control
H. Holding and Distribution
I. Laboratory Controls
J. Records and Reports
K. Returned and Salvaged Drug Products

The following is intended only as a summary of significant differences and similarities:

PERSONNEL

The EC Guide, reflecting the European mandatory requirement (Directive 91/356) for the independence of Quality Control states that "the heads of Production and Quality Control must independent from each other." The US cGMPs do not appear specifically to require this separation of responsibility and authority, although it is perhaps implicit in the statements on the responsibility and authority of the "Quality Control Unit" that appear in 211.22. In any event, this separation does seem to be standard practice in US pharmaceutical manufacturing companies.

Both documents stress that there should be an adequate number of appropriately qualified, trained, and experienced personnel. The EC GMP Guide requires that "the manufacturer must have an organisation chart" and that "people in responsible positions should have specific duties recorded in written job descriptions, and adequate authority to carry out their responsibilities."

Both documents indicate the need for initial and ongoing training, in GMP *and* in specific task-related skills and knowledge. They both require hygienic personal practices and the wearing of suitable protective clothing.

The US cGMPs require that "persons shown … to have an apparent illness or open lesions which may adversely affect … drug products shall be excluded from contact with components, … containers, closures, in-process materials, and drug products …" The EC GMP Guide has a similar requirement, and also requires that "all personnel should receive medical examination on recruitment," to be repeated thereafter "when necessary." Manufacturers are also required to see that there are "instructions ensuring that health conditions that can be of relevance to the quality of products come to the manufacturer's knowledge."

The EC GMP Guide states that "eating, drinking, chewing or smoking, or the storage of food, drink, smoking materials or personal medication in the production or storage areas should be prohibited." Such matters are not covered explicitly in the US cGMPs.

The US cGMPs accept the use of consultants "advising on the manufacture, processing, packing etc. …." The EC Guide states that "normally key posts should be occupied by full-time personnel."

BUILDINGS/PREMISES

In general, the requirements of both documents are quite similar.

While in the EC Guide the need for the segregation of various different types of operation or activity is indicated, the cGMPs spell-out 10 different types of "defined area."

One surprising aspect of the US cGMPs is that although there is a requirement for a separate area for aseptic processing (with special requirements for walls/floors/surfaces, air supply, environmental monitoring, disinfection, etc.), this does not appear to be required for other forms of sterile production. (This may be due to a different shade of meaning of the word "aseptic" in American English, as compared to British English.)

In general, the EC Guide covers all, or most of, the points in the cGMPs in more-or-less similar terms. The special requirements for sterile products manufacture are not detailed in Chapter 3, Premises and Equipment, but are covered (far more extensively than in the US cGMPs) in the special Annex 1.

In addition, points specifically mentioned in the EC GMP Guide include: prevention of contamination from external environment; segregation of animal houses, maintenance workshops, and toilet facilities; general requirements for walls floors and ceilings; appropriate working conditions in terms of lighting, temperature, humidity, and ventilation.

EQUIPMENT

A number of the points made are similar in both documents, at least in terms of general requirements (design, construction, location, cleaning, and maintenance).

The US cGMPs include, in Sub-part D, statements on "automatic mechanical and electronic equipment" (211.68) which embrace requirements for "computers and related systems," and a section (211.72) on "fiber-releasing filters." The former is covered, in somewhat more detail, in Annex 11 of the EC Guide. The EC Annex 1, on Sterile Products, simply states "Fibres (sic)-shedding filters should not be used." The EC GMP Guide further requires that "measuring, weighing, recording and control equipment should be calibrated and checked at defined intervals by appropriate methods" and records maintained.

INGREDIENTS AND PACKAGING MATERIALS

(NB: The US cGMPs refer to ingredients ["starting materials" in EC terms] as "components" and packaging materials as "drug product containers and closures.")

Overall, on these aspects the two documents are similar in content, but rather different in emphasis. The US cGMPs cover documentary aspects under the Subpart E on Control of Components and Drug Product Containers and Closures. In its detailed coverage in the separate chapter on Documentation, EC Guide is more specific on materials specifications and records. The US cGMPs are rather more detailed on sampling of these materials than the EC GMP Guide, but lack the particular emphasis of the latter on the need to guard against mislabeling by material suppliers (see EC GMP Guide, Annex 8).

PRODUCTION AND PROCESS CONTROLS

The relevant sections are Subpart F of the US cGMPs and Chapter 5 of the EC Guide. The latter includes Packaging, which is accorded a separate Subpart (G) in the US cGMPs. It also includes a four-paragraph section on Validation (in addition to the relatively new Annex 15 on Qualification and Validation) to be compared with the occasional scattered references in the US cGMPs. (This, perhaps, seems a little odd in the context of the FDA's considerable emphasis on validation.)

Overall, requirements are basically similar, although the US cGMPs give a number of examples of the types of in-process controls to be exercised, and is also different in requiring "time limits for the completion of each stage of manufacture." Section 211.115 of Subpart F consists of two paragraphs on Reprocessing. The EC Guide has, in Chapter 5, a six-paragraph section on "rejected, recovered and returned materials."

PACKAGING AND LABELING

The relevant portion of the US cGMPs is Subpart G, Packaging and Labelling Control. Sections 211.122 (Materials examination and usage criteria), 211.125 (Labelling issuance), and 211.130 of this Subpart are broadly similar in content and emphasis to the sections on Packaging materials (5.40 to 5.43) and on Packaging operations (5.44 to 5.57) in the EC GMP Guide.

A relatively lengthy section of Subpart G of the US cGMPs is concerned with "Tamper-resistant packaging requirements for over-the-counter (OTC) human drug products." This is all very sound and reasonable stuff, but some might question whether this is strictly a GMP issue.

THE CONTROL LABORATORY

The requirements for Laboratory Controls, as set out in Subpart I of the US cGMPs are broadly similar to the guidance given in Chapter 6, Quality Control, of the EC Guide, as amplified by the other sections and chapters to which it refers.

Subpart I, Section 211.166 of the US cGMPs goes into some considerable detail on stability testing; the EC GMP Guide does not, although the matter is covered elsewhere in other EC guidelines and regulations. (It might be argued that although a sound stability-testing program is of crucial importance, it is not part of the day-to-day process of manufacture and, therefore, comes not within the scope of GMP.)

DOCUMENTATION/RECORDS

These are both covered in considerable detail, but with differences of emphasis, in EC GMP Guide Chapter 4, Documentation, and US cGMPs Subpart J, Records and Reports, respectively.

The final Subpart of the US cGMPs is Subpart K, Returned and Salvaged Drug Products. These aspects are covered in the section on "rejected, recovered and returned materials" (5.61 to 5.65) of the EC GMP Guide.

OTHER TOPICS

The EC GMP Guide deals with a number of topics that are either not covered at all, or covered only relatively briefly in the US cGMPs. (See, for example, EC Chapter 7 on Contract Manufacture and Analysis and Chapter 9 on Self Inspection, and a number of the 18 annexes listed above.)

GENERAL

In general, there are a number of similarities in terms of the content of the US and EC documents. There are certain differences, and a striking contrast is the very detailed separate treatment, in the EC GMP Guide, of sterile products manufacture, as compared with odd few paragraphs, scattered through various subparts of the cGMPs. Throughout, here are a number of differences of emphasis.

Three important broad distinctions may be drawn:

1. The EC Guide is usually more detailed where it has common ground with the US cGMPs and, in addition, covers topics not considered in the latter.
2. The US cGMPs are patently and explicitly a set of legally enforceable regulations, written in a style, and with an emphasis, that reflects this. The EC GMP Guide, as its title suggests, is indeed intended (or is purported) to be a guide. *But*, as commented in the first chapter of this book:

> However, careful attention needs to be paid to the statement that appears in the second paragraph of the foreword to the EC GMP Guide (see above) — "… (this) Guide to Good Manufacturing Practice which will be used in assessing applications for manufacturing authorisations and as a basis for inspection of manufacturers of medicinal products." Since the result of an adverse assessment of an application, or of an inspection, can be the refusal by the regulatory (or competent) authority to grant a manufacturing authorization, or to suspend or revoke such an authorization, the EC GMP Guide can be said to have significantly powerful teeth. As such, and in practice, it hardly has less regulatory force in Europe than do the US cGMPs in the USA.

3. With a few exceptions, the expression in the US cGMPs is precisely and explicitly meaningful. In contrast, the EC GMP Guide is often muddled, ambiguous, and equivocal. The main reason for this difference doubtless lies in the differences in origin and status. The US cGMPs is a statutory document, written by Americans to be read by Americans. In contrast, the EC GMP Guide is presented as a guide, and thus the drafting rigor normally required of a statutory document is not imposed. Furthermore, the EC GMP Guide and its seemingly endless series of annexes is produced by multinational committees, with each member probably having his or her own agenda. And it shows.

It is perhaps in the differences in the implementation and enforcement of Good Manufacturing Practice, and in inspection practices, that the most striking contrasts lie.

The task of FDA investigators is, fundamentally, to look for violations of the cGMP regulations. "Traditionally" European (and notably, UK) inspectors are looking for compliance with the EC GMP Guide, around which they aim to build a constructive dialogue with the manufacturer inspected. Their remit is, however, rather broader than that. An interesting and constructive light is shed upon the auditing style and approach envisaged in a statement from the European Commission:

> The task of an inspector is not limited to the disclosure of faults, deficiencies and discrepancies. The inspector should connect an observation with assistance in making the necessary improvements. An inspection should normally include educational and motivating elements. (European Commission Working Party on Control of Medicines and Inspections, January 1995)

It does seem that the investigators of the US FDA behave in a rather different way, their task being largely to find, and collect evidence of, "violations" of the US regulations. Even if they were inclined to give constructive advice or assistance, many of them would not be in a position to do so, lacking, as so many of them do, the relevant technical background and practical experience.

However, it is not possible to say whether or not the Inspectorates of *all* the Member States of the European Union do or will fully adhere to the admirable precepts laid down by the EC Working Party (above). Indeed, the whole issue of the commonality of standards of manufacture, and of inspection and enforcement, across the entire European Union is a distinctly cloudy one. With (at the time of writing) the imminent expansion of the Union to embrace a further 10 member states, it is unlikely to become clearer.

It is fervently hoped that, on both sides of the Atlantic and indeed worldwide, there is a firm realization that in the manufacture of pharmaceutical products, the major concern is for the safety, protection, and well-being of patients.

INDEX

Milton Keynes UK
Ingram Content Group UK Ltd.
UKHW052025071024
449327UK00027B/2424